Designing Tomorrow's Cars

DESIGNING TOMORROW'S CARS

FROM CONCEPT - STEP-BY-STEP - TO DETAIL DESIGN

WALTER H. KORFF

1980
M-C Publications
449 N. Lamer St.
Burbank, Calif. 91506

Library of Congress Catalog Card No. 80-80522
ISBN 0-9603850-0-2

First published 1980

Manufactured in the United States of America

– ACKNOWLEDGEMENTS –

Few projects are undertaken singlehandedly. Our shop work was no exception. Special thanks go to the following:

– George Irving, who helped set up shop originally and helped build the first front-wheel drive with automatic V-belt drive.

– Rudy Makela, who built the fiberglass bodies and molds and indulged my many changes and modifications.

– Bob Summers, for his many hours on experimental parts of all kinds.

– Nick Nicholas, for his excellent assistance on the MINI-K and other projects.

– Curt Smith, for his help in cost estimating and experimental work that includes previous projects in aircraft as well as small cars.

– Chuck Slover, who continued throughout several projects over a period of about ten years on a part-time basis and for whom no task was impossible.

Grateful appreciation goes to Dr. Joseph J. Cornish III, the Aerophysics Department, Mississippi State University, and the students who participated in the wind tunnel work on the model of my small streamliner in 1963. Dr. Cornish is now Vice-President, Engineering and Advanced Programs, Lockheed-Georgia Co., a division of Lockheed Aircraft Corporation.

The book also contains material from the following:

– UNIQUE V-BELT DRIVE – DAF, Netherlands.

– GOLDENROD data from the Summers brothers.

– VIKING studies from Dr. M. R. Seal, Director, Vehicle Research Institute, Western Washington University.

– MID-ENGINE MUSTANG SPORTS CAR – Ford Motor Co.

Experimental urban cars of yesterday and tomorrow include:

– THREE WHEEL COMMUTER CAR and SPECIAL PURPOSE URBAN CARS – General Motors

– GE–100, "Centennial Electric" – General Electric Co.

– U.S. Department of Energy (DOE) electric test vehicle, ETV-1, developed by General Electric Research and Development Center and Chrysler Corporation.

Grateful acknowledgement is also given to the Society of Automotive Engineers (SAE) and those magazines and others listed in the references for use of their valuable material.

TABLE OF CONTENTS

TABLE OF CONTENTS (Cont'd)

TABLE OF CONTENTS (Cont'd)

TABLE OF CONTENTS (Cont'd)

TABLE OF CONTENTS (Cont'd)

Chapter I

OPPORTUNITIES UNLIMITED

THE MOTOR VEHICLE INDUSTRY IS LARGE

There are over 286 million cars in the world. About ten percent (10%) – 28 million are replaced every year. Of these, nearly half are in the United States. Further, these numbers are only of automobiles. As we shall see there are many types of motor vehicles other than automobiles, each of which come in many varieties and sizes. For example, there are about 10 million motorcycles in this country with annual sales of approximately one million, again in various sizes and types.

The after market in motor vehicles is also immense and will be discussed separately a bit later.

MOTOR VEHICLES COME IN GREAT VARIETY

Before world war II American families looked forward to owning an automobile – the one car family. Today the two car or three car family is not unusual and is growing. Further, the vehicles differ by types, purposes, and sizes. Some families have a small sports car for dad and a station wagon for mom and the children. There may also be an RV. The recreation vehicle may be a motor home, a trailer, a boat, and/or one or two motorcycles (often off-road types). Mopeds are becoming popular in the U.S.A. for students and others. Still others have a van or pickup. As our society becomes more affluent we want (and sometimes need) different and additional motor vehicles. Population growth is another factor to justify the increasing numbers and variety.

Motor vehicles also include construction, commercial, and utility vehicles for business – all of which must be designed and manufactured.

OPPORTUNITIES FOR THE DESIGNER AND ENGINEER

A person may have a strong interest in a particular field but not realize how large and varied the field is. The motor vehicle designer's field is very large and offers unlimited opportunities as we shall soon see. Like a doctor or attorney, he may, if he chooses, specialize in a portion of motor vehicle design. Further, he may choose to work for one of the largest automobile companies in Detroit or a small company that has off-road or other products, or he may eventually set as his goal to be in business for himself. In this book we will introduce you to these facets of the motor vehicle industry as a matter of orientation so that you may begin to think of your own long range goals.

YOUR LONG RANGE GOALS

Many young engineers feel they will be happy for the rest of their life if they can devote their working hours to the design of motor vehicles, specifically automobiles. Thousands are. If that is your choice your strong creative interest has a technical basis – you are most interested in the product and how it evolves from a clean sheet of paper. You may even specialize further – into body design, powerplants, chassis, aerodynamics, or some aspect of research and development.

If you decide to work for a large company you will be placed in a design group that deals with just one portion of the product. To obtain a broad background of experience you will have to indicate your desire to work in other areas thru periodic transfers. A smaller company may have fewer projects but you will have larger assignments with broad experience opportunities.

MANAGEMENT POSITIONS APPEAL TO SOME

Moving into management is neither good nor bad for an engineer. It is a matter of interest and opportunity. It almost always comes along slowly and in small changes. Would you like to be a leadman with a small group of four or five engineers – you'll be working along with them on the technical problems. But you also become spokesman for the group, you issue assignments and follow-up on each person's progress (as well as your own). The next step might be group supervisor, then department head with schedules, budgets, and meetings but always less actual design work. This requires a technical background for good decision making, but it also requires an increasingly greater knowledge of other operations within the company. It also requires an engineer to be, or become, people oriented as well as technically oriented.

If you do not feel comfortable in such an atmosphere request to return to what you prefer. You will be more successful and often the wages can be just as high. Technical competence, creative ability, and specialization can be very rewarding in accomplishment as well as in money.

However, if you like management jobs you will have to learn additional skills. Planning, budgeting, and scheduling have to be learned. That is just the beginning. Writing and other communication skills must be developed. The ability to prepare a proposal and then orally present it to higher management is required. Company policy and decisions must also be presented and discussed with those within your department. Inter-department activities require broad knowledge of the work done in other areas of the company.

As stated earlier these changes come slowly with improved personal ability in alignment with opportunity within the company or elsewhere. Increased responsibility goes with upward movement in management along with greater pressure and longer hours. Everyone has his own level of competence and job at which he finds greatest satisfaction. Ambition is fine but don't let it become overpowering.

BEING AN ENTREPRENEUR – THE ONLY WAY TO GO FOR OTHERS

Many people, including some engineers and designers, are happiest only if they are running their own show. They want to be in business for themselves. It may not pay as well, require more sacrifice, longer hours, and the risks are greater, but this is still their choice.

If you are among this group you will do well to prepare yourself first in the technical field, then devote some time in the management area, and learn the skills of finance, marketing, and other aspects of general business.

The giants of the motor vehicle industry are not asleep but being large they are slow to move into *SPECIAL SITUATIONS.* The small entrepreneurs ideally get in first and are well established when the rest of the industry starts to play catch-up. This is what free enterprise is all about. The entrepreneur sees the opportunity in a *SPECIAL SITUATION,* offers value, and finds his market. Growth follows if he doesn't make too many mistakes, and he soon becomes well established. Many start as inventors or designers with a hobby who, after a development period, take the big step into their own successful business.

THE DESIGNER/ENTREPRENEUR – A FEW EXAMPLES

It is one thing to discuss opportunities in this business in a general overall manner, but it is a bit more to the point to provide examples of well known people who have done it – the hard way.

HOW THE PORSCHE CAME INTO BEING

The Porsche sports car is of course, internationally known and has a solid background based on endless racing victories. But do you know about the first Porsche sports car? The complete story along with later developments is beautifully recorded in reference 1. Briefly, in July 1948 the remnants of the Porsche "works" had found refuge on the site of an old sawmill in Carinthia. Work was done in primitive huts – without access to a railway. Dr. Porsche and his son Ferry had been in custody for more than two years after WWII and the Professor's health had suffered badly. The staff, in order to live, was making wheelbarrows, potato layers, and mowing machines. They also repaired old Volkswagens left behind by the Wehrmacht after the war. Dr. Porsche and his associates in his design office had designed the VW and other cars previously. Ferry, then age 39, was still driving an old Volkswagen which gave him slight satisfaction. He decided to design and build a sports car for his own use. Due to the shortage of materials, Volkswagen parts such as engine and transaxle were used. Engine horsepower was raised from 25 to 40 by the usual porting and other modifications. A new chassis and lightweight body was built – looking very much like the Porsche "Speedster" of later years. But the company needed money. So the roadster was taken to Switzerland and sold to an enthusiast for 7,000 Swiss francs.

Then the design evolved into a coupe along with a few other changes, and the first of these cars was shown at the January 1949 Geneva motor show. Orders followed and the first 50 or so cars were built in Carinthia. Later, due to demand for increased production, the company moved to Stuttgart and this outstanding success story continues to this day.

MICKY THOMPSON AND HIS PRODUCTS

Micky is of course well known so only a few reminders will suffice. He was a pressman for a Los Angeles newspaper when he began "hot-rodding" and setting records as a part-time hobby. This avocation became his vocation as his experience and reputation grew on the dry lakes and Utah Salt Flats, then later in off-road racing. A real market for his products followed and his excellent business ability created the numerous enterprises that carry his name today.

THE SUMMERS BROTHERS

Bob and Bill Summers also started with little more than a dream, good ideas, and lots of hard work in lieu of hard cash. They too, have a background in hot-rodding on the dry lakes and the Utah Salt Flats. Bob worked as a machinist and Bill drove big over-the-highway tractor-trailers to survive and help support their avocation. They too, had definite plans about how to get their business started. They would go after the World Land Speed Record for wheel driven cars. Their *GOLDENROD* did just that – and holds the record·to this day. More about the record car later in the book. It was this writer's privilege and honor to have been chosen by Bob and Bill to do the aerodynamics on the record car. Working with them was and is always a great pleasure.

After that the record car was put on exhibit throughout the world and the Summers brothers launched their business in a very small way in a small building that formerly was used to sell fruit. Their reputation went into their first product, a special precision rear axle shaft for hot-rods – guaranteed unbreakable with a money back guarantee. Other products followed and they moved into larger facilities with more employees, to meet the demands of an ever growing business. Their business is now a corporation.

Needless to say, these examples don't tell the whole story – the heartaches – the problems – the endless faith, patience, and stamina required. But the fact remains – it can be done and is being done day after day. The list of examples if not limited to these three. Look at the SEMA (Specialty Equipment Mfgr's Assoc.) membership. Many of them, in fact most of them, began in a small way and built fine businesses in the lucrative after-market field.

YOUR NEW PRODUCT – CAN IT COMPETE WITH THE PRODUCTS OF THE GIANTS OF THE U.S. MOTOR VEHICLE INDUSTRY?

It can be argued, very convincingly, that it is impossible to successfully introduce a new vehicle or portion thereof in the automotive field. Of course, failures are numerous just as in other industries. Names come to mind quickly such as Tucker, Davis, Kaiser-

Frazer, and more recently Bricklin. Observers say no one can compete head-on with the entrenched mass-production car market held by the Detroit manufacturers. Three of the four named companies produced only a few cars and the other did mass-produce but went out of business after but a few years. Many feel that no new automotive product has a chance; its initial higher cost to manufacture prices it out of the market. This is the "learning curve" problem and is discussed later.

So much for the negative side. Now let's look at the positive side. Needless to say, the large automobile companies employ thousands of designers. Many times as many motor vehicle designers are working elsewhere. The 1978 SAE Congress and Exposition in Detroit had 121 exhibitors – suppliers of products to the big companies. They use hundreds of designers. The 1978 Design Show in Chicago had approximately 500 companies exhibiting – admittedly not all in transportation. But they too use hundreds of designers.

LOOK FOR THE SPECIAL SITUATION

Earlier Porsche, Thompson, and the Summers brothers were discussed. Each recognized the opportunity in a "special situation" and did something about it. SEMA was also mentioned.

THE IMMENSE AFTER-MARKET – SEMA

How large is SEMA? The 11th Annual SEMA Show for 1977 had 274 exhibitors listed in their reference book (reference 2). It continues to grow year after year. Each member company came into being because its founder recognized a "special situation" in the motor vehicle industry. What "special" item did they design and now manufacture? To name a few – light alloy wheels, superchargers, headers and many other parts for engines, heavy duty suspension items, sun roofs, and deluxe seats. These and many other parts are the AFTER-MARKET in stock cars, vans, pickups, racers, trucks and motorcycles.

Business remains good even after Detroit manufacturers include some of these items as standard equipment on certain new models. "Special situations", every one of them. This is big business that has many opportunities for good designers.

A NEW COMPLETE VEHICLE AND ITS CHANCES IN THE MARKETPLACE

Let's now move along to complete vehicles. More race cars are built by small entrepreneurs than by Detroit manufacturers. Again, a "special situation". The same is true for REPLICARS. Fiberglass bodies for VW "Beetles" and other engine/chassis combinations are another case in point. Admittedly, these are not high production items – so Detroit isn't interested. But the market is established.

Next let's move up to high production complete vehicles – competing for a portion of the giant U.S. auto market. Foreign sports cars made major in-roads in this special field starting in the fifties. Then the Corvette appeared, and although its volume is good, so also remains the market for foreign sports cars.

From sports cars to small foreign economy cars was the next logical step. The VW "Beetle" sold only two cars in the U.S. the first year (1949), but went on to become as popular as the old Model T Ford in its day, until it too became obsolete. On the West Coast nearly 50% of the cars sold there in 1977 were foreign. Now who is following whom?

Detroit manufacturers must down-size to meet government edicted fuel mileage and so the small foreign front-drives set the example. Lots of them from Europe and Japan are doing just that. Detroit manufacturers even import certain small cars for sale through their dealerships along with U.S. models – at least until U.S. designs can be produced here. The "Special Situation" is now somewhat reversed.

In the case of certain foreign cars prices are far higher than U.S. makes, but sales remain brisk.

INDUSTRIAL, UTILITY, AND SPECIAL PURPOSE VEHICLES HAVE A PLACE TOO

Let's now get back to competing with large companies with other U.S. products in other divisions of the motor vehicle industry. Golf cars for instance. These started out powered by gasoline engines but now are totally electrically powered. Again – new companies sprang up with a new idea – a "special situation" all over again. How many golf cars? Estimated production for 1977 was 22,000 according to ELECTRIC VEHICLE NEWS. Industrial forklift trucks are another example – 21,200 (same source).

Special bodies for ambulances, limousines, small fire trucks, and buses are another field. Factory maintenance and small utility vehicles are another. All sorts of recreational, industrial, and utility vehicles are designed and built by small companies. They all need good designers.

THE NEXT BIG "SPECIAL SITUATION" IS NOW

What new large special situation is now on the horizon? Electrics – passenger cars, small city cars, utility, and many other types. The Government is getting interested with financial aid favoring electrics. Batteries, motors, and controls are greatly improved. Chassis and body developments reduce weight, improve ride and handling. Aerodynamics further improve performance and/or reduce the need for power. Over 2,500 electric cars (in addition to golf cars) have been sold and another 2,000 home-builts are running or under construction. Consumer interest in a small urban vehicle for two people and parcels is evident. When the smooth, quiet, reliable-operation of electrics is augmented with improved performance and fuel economy that more than pays the increased cost, we have a *SPECIAL SITUATION.*

RESEARCH AND DEVELOPMENT – ANOTHER OPPORTUNITY

Many companies specialize in R & D work. They do it for other companies on a contract basis. This is a high-technology business that is growing rapidly. Government contracts should also be looked into. The Federal government, primarily Department of Transportation (DOT) awards contracts to individuals, universities, small businesses and large businesses in specific fields of interest. From time to time contracts are let for R & D in electric and hybrid vehicles and components such as batteries, motors, and controls. Successful bidding for a government or private company award requires a carefully prepared proposal as well as technical know-how. This is a skillful selling job in written form.

SELLING PRICE VS VALUE

Selling price is important but value is more important. The relationship between price and value is not constant. For example–basic transportation in the form of a small, simple vehicle must have a low selling price – that is the main reason for its simplicity. A sports car may be just as small but is more sophisticated in design, and can justify a higher price. In fact, many buyers of sports cars "want what they want" and are more motivated by emotions than by utility or price. The price of a sports car is not as critical as its appeal or "value" to a certain market segment. Of course, it can be priced too high – out of the range of its intended market. The point is, sports car buyers first look for "what turns them on" – then see if they can afford it.

Selling pricc and value are both sensitive factors that must be carefully considered by the designer. The most value for the lowest selling price obviously is what most customers seek.

BUYING HABITS OF THE GENERAL PUBLIC – THE BELL CURVE

Inventors and designers, particularly those in research and experimental work, usually expect their ideas will meet with immediate wide customer acceptance. It just doesn't happen that way. In fact, they often have difficulty convincing their own boss. People in general exhibit a certain inertia in their buying habits – they resist change. Not everybody of course – but the great mass of customers. This can be depicted somewhat by a "bell curve" as shown in figure 1. This is a curve of public buying habits of typical consomer products – and it applies to motor vehicles as well. Note that buyers may be divided into three basic types. We have friends of each type. The "avant garde" type wants everything as soon as available – with little thought for price. This group provides the

THESE VERTICAL DIVISIONS TYPICAL FOR PROVEN MARKETING METHODS, GOOD DISTRIBUTION, AND COMPETITIVE PRICING. MAY VARY ONLY SLIGHTLY DUE TO ECONOMIC CONDITIONS, OTHER FACTORS.

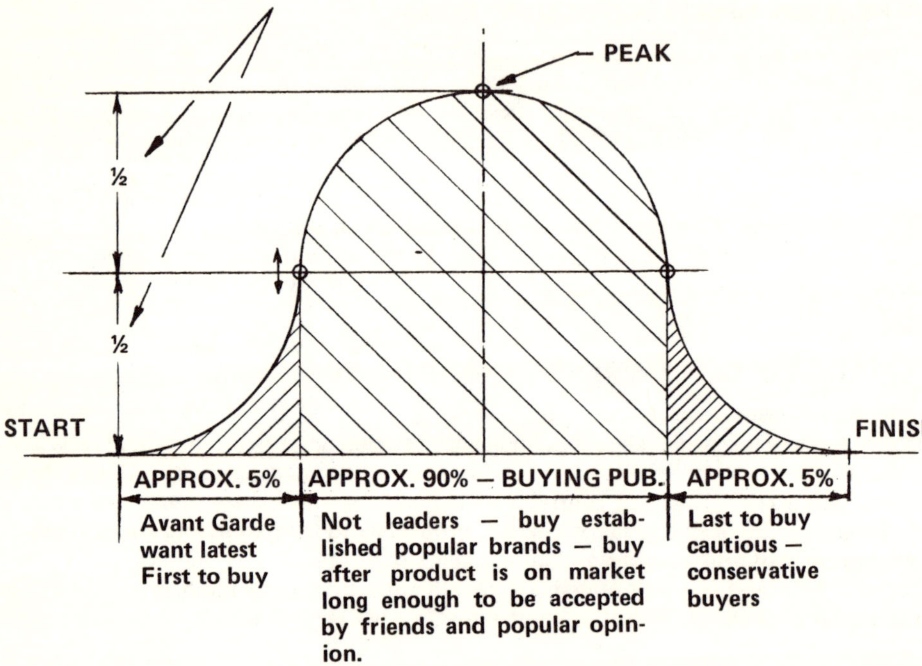

FIGURE 1

limited market for the small entrepreneur with a new product. These are the people who paid $12 each for the original Reynolds ball point pen when it first appeared on the market. Now, years later, ball point pens are sold everywhere for as low as 39 cents each. The businessman needs this "avant garde" group in order to get started – to pay for his immense start-up costs. Later, when fully established, he is ready with mass production to reach his share of 90% of the market in the middle of the bell curve.

Eventually all products become obsolete if they are not periodically up-dated. Even when practically obsolete there still remains a small market for them as shown on the opposite side of the curve. In the meantime, the progressive businessman has moved his company back to the start of the bell curve, again with new products ahead of the competition. Capable designers are needed in every facet of the motor vehicle industry to constantly design new products, then redesign, modify, and up-date these same products. Research and development is never-ending.

EXAMPLES OF THE BELL CURVE IN THE TRANSPORTATION FIELD

Henry Ford's Model T, affectionately referred to as the "tin lizzie", was basic transportation that certainly filled a need in its time. It began life in 1906 selling for from $750 to $850 depending on equipment. Automobile mass production became a fact with it. By 1916 half a million "Ts" were built and sold in a year at a price of $345 each. Profits reached nearly $60 million that year. In all, over 15 million Model T's were built and sold by the summer of 1927. Production stopped shortly thereafter because of its constantly declining market. A year was lost in coming out with the Model A in December 1928. The company also lost $30 million during that time. It is generally recognized that Mr. Ford stayed with the Model T too long and did not start the Model A soon enough.

The automobile has passed through many stages. Enclosed bodies replaced side curtains next. Then cars became more sophisticated with more and better features. Cheap fuel made it easy to build in more performance as well as larger cars after that. Later came "muscle cars" and better handling. Each began slowly, peaked, and became obsolete in time, although the features were improved. The VW "Beetle" is another case in point. Its life in the U.S. began with the sale of only two cars in 1949. Growth without interruption in the U.S. reached 1968 VW sales of 569,292 of which 70% were Beetles. VW had by then captured 57% of the total market for imports. Up-dating, changes and improvements continued but like the Model T, the age of the basic design was beginning to show. The last Beetle rolled off the assembly line in Germany early in 1978. It was number 16,255,500. However, more than 3 million additional Beetles have been made abroad and will continue for awhile longer in Mexico, Brazil, and South Africa. Its life span of about 30 years is something of a record, too. Its sales also follow the bell curve principle.

Airplane design, like everything else, also follows the bell curve. First, wood and wire biplanes, then all-metal monoplanes, later jets, now wide-body jets. Next, more fuel efficient aircraft must come along.

The "bell curve" principle is important to the designer and his company. The time span may be short or long depending on many factors such as, how outstanding a design is, competitive designs, customer acceptance of each, aggressive marketing, government regulations, and changing economic conditions. The bell curve will be discussed again in the chapter on forecasting and trends.

TRADING UP – HOW IT AFFECTS DESIGN

Traditionally, Americans like to trade up. There are good reasons why this happens. A young couple's first car often is an old "used" car. They seldom have the money for anything better and their limited money is needed for so many other things. With the first raise plus some savings they move up to a better car – later a "deluxe" model, then maybe two cars when the children start arriving. Eventually, if car buffs, a sports car is added, and so on.

If a youngster has a moped to go to school on, this often leads to a larger motorcycle later, and if he is really into "bikes" an "off-road" or other special type is added – and always the latest and best model he can afford. The same thing happens with boats, homes, everything the family grows into. Why? – pride, needs, desires, excitement of new acquisitions, higher incomes.

And how does this affect design? In working up a "family of designs" it is advantageous to provide for trading-up with added models that are: 1–more powerful; 2–more "deluxe" – added features; 3– special types such as sports car or personal cars.

People like to stay with the same company and the same dealer if they are happy with their first purchase and the dealer has models to trade-up to. Repeat customers are the best kind. If the manufacturer has only one model or size the dealer is handicapped. He must add to his showroom by going to other manufacturers for their products. Manufacturers have long recognized this and so endeavor to provide for trading up to keep their dealers as well as their customers.

As every new design starts out on the drawing board, every effort should be made to provide a "family of designs" for economical manufacture and customer continuity in trading up.

Chapter II

NEW CONSTRAINTS ON VEHICLES REQUIRE A NEW BREED OF HIGHER QUALIFIED DESIGNERS

The opportunities and special situations in today's motor vehicle industry can no longer be met by designers in the traditional manner. The approach to design problems in this business is now far more sophisticated than in the past. The designer's job admittedly was never simple. But the job must now be handled a bit differently. Many observers have noted that the motor vehicle designer must now deal with many more and conflicting problems of design and that the best way is in much the same manner as that used by the aerospace designer. This is why we say a new breed of higher qualified designer is required. This does not mean that existing or older designers cannot learn additional skills or new ways of doing things. Today no one designer can any longer do his job alone. He not only needs the help of specialists in many fields; he must understand what they are telling him; he must be knowledgeable enough to apply the specialized information to his problem; he must also evaluate and resolve conflicting solutions through scientific methods of "trade-offs" to arrive at the optimum design solutions. Let's now look at how it was formerly done and why that method is no longer good enough.

BASIC RESPONSIBILITY OF THE DESIGNER

The basic responsibility of a motor vehicle designer was and still is to create a product that functions in a predictable and reliable manner. For many years this was his sole responsibility. He could live in his private world at work and let the creative juices flow.

FORM FOLLOWS FUNCTION – WITH A LITTLE CLASS – THE "STYLIST" IS BORN

The old rule "Form follows function" remains but form took on a bit more importance in time. It was quickly discovered that appearance or eye appeal, later called sex appeal, helped immensely to sell the product. It still does. So the designer added this new facet of design to his responsibilities. In fact, many devoted all their time to appearance and became known as "stylists" rather than designers. Appearance was not enough – *changing appearance* or "style" became important. The design had to represent the latest – the "in thing". "Planned obsolescence" in motor vehicle design did not mean the vehicle no longer functioned or that it could not be repaired to restore reliability – rather that it was "out of style". The annual model change sold cars. The automobile became an art form.

For decades the motor vehicle designer's responsibilities were rather well-defined and limited to creating a product that:

1 – Functioned as intended with but little thought for efficiency. Competitive price – not efficiency – was more important.
2 – Is predictable as to performance and handling.
3 – Is reliable.
4 – Is "in style"

ADDED RESPONSIBILITIES FOR THE NEW BREED OF HIGHER QUALIFIED DESIGNER

Those responsibilities have not gone away for today's new breed of designer – but they are now just the beginning – the tip of the iceberg. Today's designer must add much more to his responsibilities, such as:

a) – Government regulations dealing with safety.
b) – Government regulations dealing with emissions (exhaust and noise)
c) – Efficient use of fuel because the cheap fuel of the past no longer exists as well as Government regulations dealing with mpg (Cafe rules)
d) – Alternate fuels due to depletion of fossil fuels.
e) – Economic factors of design due to inflation and tax law changes relative to manufacturing.
f) – Company inter-department interfaces relating to tooling, fabrication and assembly operations, procurement, marketing, and capital requirements.
g) – Since many of these influences are in opposition to each other the designer must become skilled at analyzing "trade-offs" to determine the optimum design.
h)) An ability to deal with far more details all at the same time – a real juggling act.

A reasonable knowledge of these additional factors and their effect on design and how to develop "trade-offs" plus the great mass of details make the designer's job far more complicated today. He simply must know and apply a great deal more engineering information. This requires more training – broader experience – a new breed.

It may be argued that the automobile designers, primarily in Detroit, have always had and used staff specialists. True – but now there is a difference. In the past the specialists dealt with fewer problems in traditional materials – usually steel. Today's material specialists deal with many more materials such as plastics, elastermers, aluminum, high strength body steels in addition to the traditional steels. Saving weight now has a high priority and is given far more attention than ever before. Specialists in safety and emissions control provide the top priority assistance needed in these fields along with extensive laboratory work. In the past aerodynamics was rarely considered except for racing. Now passenger cars are put in the wind tunnel and undergo extensive aerodynamic refinement to provide more efficient operation. All of this additional work plays havoc with costs – costs to design – direct cost to manufacture – capital investment for facilities. And today's designer must deal with these costs throughout his work. Greater efficiency at every step is a necessity. His job is immeasurably more complicated. Organization and teamwork, as well as knowledge, are required. This leads us to organization charts.

"CHAIN OF COMMAND" ORGANIZATION CHART

This is the typical organization chart of every company. In fact, there are many charts – at least one for each major department of the company. The engineering department starts with the Director of Engineering at the top, then branches out like a pyramid to show who reports to him, and then at the next lower level who reports to them. This continues down through each level of supervision, divisions, departments, groups, and is further divided laterally between projects (line supervision) and staff support and laboratories and other pertinent areas of responsibilities. These charts differ with companies, company size, and according to the nature of the products manufactured.

An engineer new to a company is exposed to these organization charts along with company policies and procedures to show him where he fits in the scheme of things.

These charts are unique to each company and are subject to change with each reorganization that occurs. Therefore, there is no point in attempting to depict one here.

FUNCTIONAL CHART

A functional chart can be depicted, although it too varies with companies and products, but not as much as the "chain of command" type. Automotive engineering departments, whether large, small, or even a one-man organization must deal with essentially the same functions in the development of a vehicle. Many staff and laboratory functions are then combined or contracted for outside the company on a part-time consulting basis. But they cannot be avoided or ignored without risk to the vehicle being developed. These functions are depicted in the blocks of the functional chart shown in figure 2.

Note that the chart is divided vertically between "Line Operations" and "Staff Support". Line operations deal solely with a particular model or specific project. Staff support is provided to all models and projects on a request basis. For example, a designer in a line operation on a specific project is considering the use of an alternative material or a new component but lacks sufficient information to proceed with his design. Rather than investigate it himself he requests assistance from the proper staff specialist. The investigation may even require laboratory testing and coordination with the supplier company. The designer is free to proceed with other aspects of his design while the staff specialists are handling the investigation. The staff specialists are expected to be just that – specialists in their particular field. They are responsible for providing a true, unbiased opinion that carefully shows the disadvantages as well as the advantages and to provide a recommendation for or against use. The "Line operations" designer has the final responsibility (in most companies) for his design and after discussions and review of all data provided by the staff specialist(s) makes his decision. It may or may not agree with the staff specialist(s) recommendations, possibly for other (or unrelated) reasons.

ALL DESIGNERS ARE NOT IN LINE OPERATIONS

It is interesting to note that all designers are not in "line operations". Many staff specialists are excellent designers. They create, invent, and come up with new ideas that ultimately are used on new vehicles. They often are closer to research and to what is going on throughout the industry. The same is true for many of the engineers in the laboratories.

RECOGNIZE YOUR OWN UNIQUE ABILITIES AND LIMITATIONS

Designers don't always remain designers, although the creative urge is not necessarily lost. An outstanding designer is recognized, honored, and often offered a "promotion" into management. This may or may not be a good thing for him or his company. It is a wise person who recognizes his or her own unique abilities and limitations. This can be frustrating. A happy designer then becomes an unhappy manager. Many companies recognize this and have upgraded top designers to "design specialists" or other titles on a par with management in recognition and pay.

It is also true that many designers take their creative nature with them into management to become outstanding managers. Some even branch out into new opportunities in their own businesses. It is not unusual for a young engineer to work up to designer classification in a large company, then after gaining much experience, move to a small company to broaden his background, and eventually leave to start his own company.

DO YOU PREFER A "LINE" OR A "STAFF" JOB?

Many engineers prefer a "line operations" job in which they work almost entirely on a drawing board on a single specific project until the project is completed. Then they are reassigned to another project along with others. It is quite gratifying to be a member of a team that designs a highly successful vehicle. But not all new vehicles become "winners". In fact, not all projects are completed or reach the production stage. Many, for various reasons, are cancelled or delayed or modified far beyond the original concept. Although it is always thrilling to start a new exciting project, there is no guarantee it will not become very routine near the end – after the major design problems are solved.

Many other engineers prefer a "staff support" job in which they specialize in a field, do research, work with the laboratories, contact suppliers, and provide their information

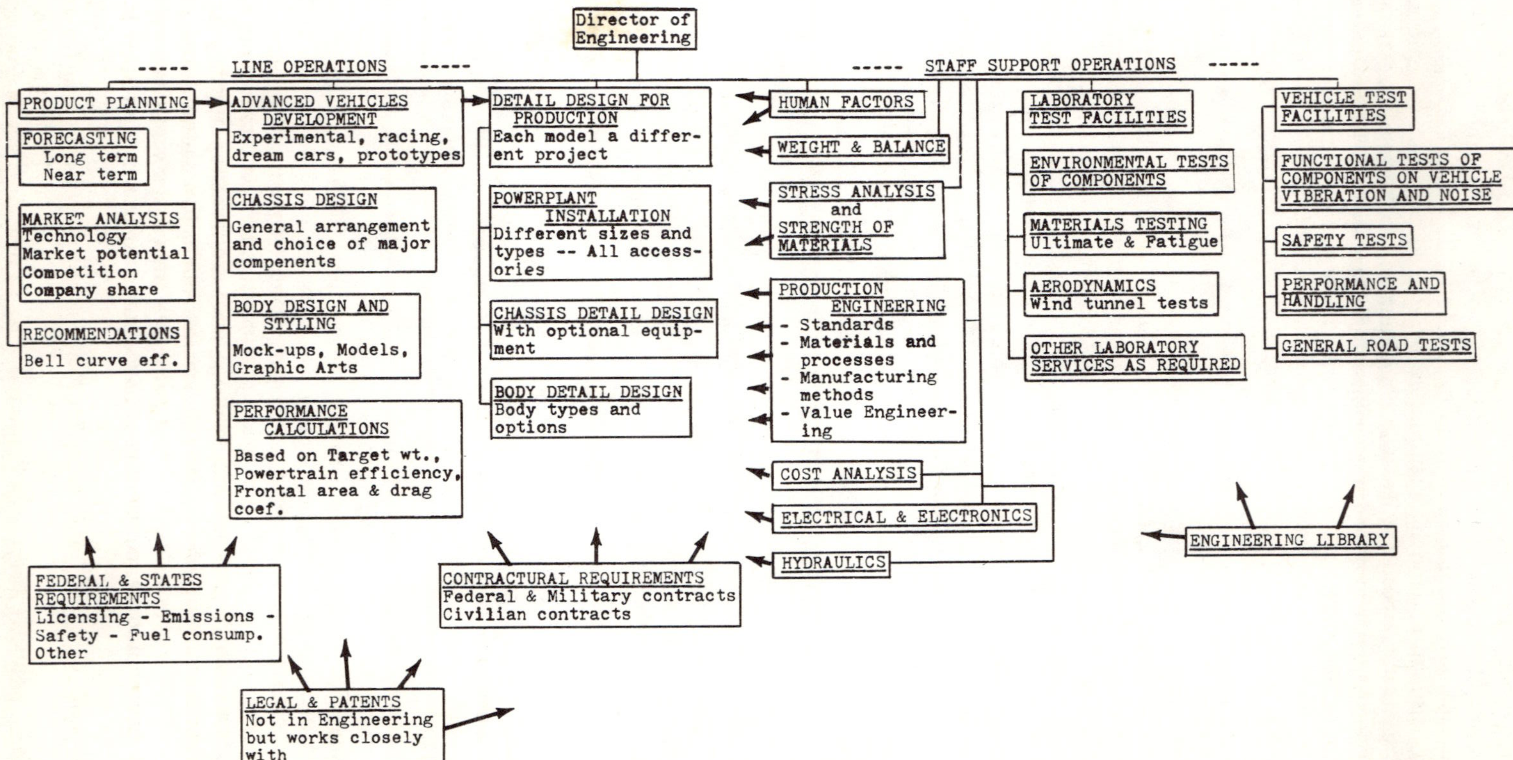

ENGINEERING FUNCTIONS

— FIGURE 2 —

to the designers on the various projects. Although a "staffer" is rarely assigned to a specific vehicle project his work can be very rewarding also. Many new and innovative ideas originate with staff people.

Some engineers have no difficulty in going from "line" to "staff" jobs or vice-versa, while others are "like fish out of water" if changed. If you know which you prefer and are best at early in your career – fine – but if not, ask for an opportunity to be transferred to learn and to work at what you excell at.

Don't worry about which may or may not be the best from the standpoint of advancement. Capable designers are found both in line and staff assignments. They go up the ladder in title and pay from both sides.

Chapter III

PRODUCT PLANNING

FORECASTING, MARKET RESEARCH, AND RECOMMENDATIONS

The upper left side of the functional chart (figure 2) depicts this function. We will begin here and work somewhat across the chart.

This group need not be located in Engineering but usually is, or at least works very closely, with engineering people. In a large company this group is well defined and works full time. In a smaller company the group may consist of assigned personnel from Engineering, Marketing, Finance, Manufacturing, and/or others who meet periodically on a part-time basis and each separately take on assignments or investigations. Very small companies usually lack this group. The owner handles the function himself, or may, at times, hire outside consultants to do portions of it such as market research. Often even large companies will hire outside consultants or pollsters to do independent surveys or other types of market research.

PURPOSE

The purpose of FORECASTING, MARKET RESEARCH, AND RECOMMENDATIONS is to provide direction and proper timing to company management in deciding on the need for new products (vehicles), what the trends will be in marketing, government regulations, engineering developments, consumer interest, economics, and any other pertinent data relative to product planning.

It is not only important to recognize a trend but to have the "timing" right and to have the information available enough *years* in advance for the company to design, tool, and otherwise be ready for the market with a new vehicle at the proper time – not too late and not too early.

It is also important to provide smooth continuity to retain public confidence in the company "image". Consumers like stability in the company that produces the vehicles they buy. They don't have faith in a company that jumps into one thing one year, only to abandon it for something different the next year. Consumers recognize progress as a smooth transition, not too fast, but continual improvement – evolution, not revolution. If a new vehicle of revolutionary concept is to be introduced the public must be "prepared" for it by proper and frequent publicity over a period of time. It must be shown to be the logical answer to changing times, the best solution, and the "in thing" for the avant-garde. No one wants to buy someone else's "experiment", so it must be "fully developed" in the eyes of the public as well as in fact. It must be a "trend setter", not a "flash in the pan".

HOW FORECASTING IS DONE AND TRENDS ARE DETERMINED

Forecasting and the study of "trends" is accomplished by the constant perusal of:

1 – Technical trade magazines
2 – Business and financial reports, even newspapers for economic influences.
3 – Government regulations and new laws – Federal and State.
4 – Competitor products
5 – Consumer spending patterns
6 – Trade meetings and conventions
7 – The results of "survey polls" to determine consumer preferences, etc.
8 – Anything that provides "intelligence" relevant to the problem.

This information is usually condensed by starting a "clippings file", with dates and references. It must be organized, filed, and cross-filed for prompt retrieval. Periodic review is not enough. The information must be studied and analyzed to determine patterns or trends. Then divide the trends into "long-term trends" of 10 to 25 years or more, and "near-term trends" of 5 to 10 years. There should be continuity and a smooth transition between the two.

TRENDS ARE NEVER STATIC – THEY CHANGE CONSTANTLY

Trends are not fixed. You do not analyze once and then say "well, that's the way it looks so let's file it for future reference." Every trend, long- or near-term, is in a constant state of flux and each factor affects other factors. Therefore, these studies require constant review and up-dating. A trend can accelerate or even reverse. Or it can become of no significance over a period of time. A new trend can overshadow it. Public interest in styles can change. New government regulations must always be considered. Sometimes conditions are in such a state of change and instability that trends are difficult to predict. This too, must be noted.

Economic conditions have a definite impact on sales and product design through buyers changing choice of type of vehicles. Inflation, taxes, imports, recession or depression, wars, population shifts, and changes in consumer attitudes all should be evaluated. Frequently certain influences are in conflict with other influences, such as public opinion vs government edict. It may take the wisdom of Solomon to predict the outcome. And, of course, our personal preferences and viewpoints should not influence the analysis. But they will! It helps to obtain opposite viewpoints for a balanced analysis, but this is not always possible.

RECOMMENDATIONS SHOULD ALSO BE MADE

After the analysis is complete recommendations to management should be prepared. These are seldom needed relative to long term trends unless requested by management. Short term trends are a different matter. Short term trends, 5 to 10 years, affect decisions that must be made promptly so recommendations should be made on:

1 – Technical research that should be undertaken in the engineering department to solve problems that loom on the horizon.
2 – Products that should be phased out and when.
3 – Products that should be modified and how.
4 – New products that should be developed and for what market and when.
5 – An analysis of how large the market will be for each new product. From this production quantities can be assigned.
6 – Feasibility studies for new designs.

LONG-TERM TRENDS – 10 TO 25 OR MORE YEARS

We now will attempt to provide a list of long-term trends as of this writing. Please realize this list is intended only as an example and is subject to frequent revision. No one should use this information for their own forecasting.

FUEL (as forecast during 1978 and 1979)

A – We are running out of fuel as we now know it. Petroleum supplies are being depleted. It is estimated that by the year 2020 (or before) natural oil as we now know it will be essentially used up as far as transportation is concerned.

B – Petroleum will continue to become more expensive as it is being depleted. By about 1985 mid-east oil production is expected to have peaked. After that, if not before, reduced production will be matched by a further rise in prices. (Shortages and price increases occurred again in 1979.)

C – Large oil companies and others are working hard to develop synfuel from coal, shale, and tar sands. This is long term and expected to be expensive.

D – Other companies are working on the production of alcohol based fuel from waste, etc. Although clean, it too will be expensive. It also provides only about half the miles per gallon but can be mixed with gasoline to stretch supplies.

E – Hydrogen from sea water is being studied.

F – The Federal government's Department of Energy (DOE) is sponsoring research in all manner of fuels and the conservation thereof.

G – Shortages, even gasoline rationing is a possibility at any time, and on short notice if imports of fuel are again restricted or severely curtailed. We now import approximately half our petroleum.

H – This nation and most of the rest of the world is progressing towards an "all electric" source of power as compared to its essentially petroleum base now. Wind, waves, solar, and nuclear may provide the sources.

I – The fuel shortages of the summer of 1979, the stepped decontrol of crude oil prices, rising prices, and a shift to smaller cars, and the discussions regarding a new energy policy has the public now concerned about what will occur in the future.

Reference 3, a 555-page book entitled "Energy in America's Future" indicates the nation can conserve energy and protect the environment without sacrificing economic growth, at acceptable – but higher – energy prices. It concluded that:

- The nation can learn to get more economic growth per unit of energy than in the past. An increase of 1.8% a year in energy use could support a rise in gross national product (GNP) of about 3.2% per year.
- Energy costs will rise unavoidably in the future as today's fuels run out and must be replaced by more expensive substitutes. Costs may level out at about double the prevailing price of 1978, in constant dollars.
- Slower rates of energy growth can be achieved while providing adequate protection for the environment, human health, and safety.
- Safe solar energy is more expensive and cannot alone compensate for the depletion of oil and natural gas that is already well under way.
- Nuclear power and synthetic liquid and gaseous fuels from coal will have to provide much of the nation's energy in the near future.
- Nuclear power's dangers are exaggerated in the public mind because of the failure to compare them with the dangers of other forms of energy. In that comparison, it said coal is a far more hazardous fuel, taking into account the accident rate in coal mining and handling, as well as its effect on air pollution.
- Nuclear energy may become more acceptable if stricter safety requirements are applied and if the plants are located away from densely populated areas.

VEHICLES

J – The National Highway Traffic Safety Administration (NHTSA) is now planning its second round of automobile fuel economy standards. These will be for 1986, 1987, and 1988. Late in 1979 the proposed standards will be issued and then finalized in 1980.

K – Vehicles must continue to become more efficient. This means smaller, lighter, less wind resistance, better power-trains, tires, etc.

L – Traffic congestion will not go away. Mass rapid transit is often mentioned as is car pools and more buses. But people still want the convenience of their own car.

M – As people become more affluent, families add to their stable of cars. Convenient small specialized urban-type vehicles may become highly desirable as second or third cars as traffic congestion and higher operating costs become more important than high-performance and excess interior space. Just as stylized pick-ups recently became the "in thing", this type of vehicle may become the "status symbol" on short notice.

ECONOMICS

The economy of this nation continues onward and upward, but never smoothly. It goes in a jagged sawtooth manner. It still has its ups and downs. Some companies go broke while others prosper. Economists come in all varieties from "eternal optimists" to the "doom and gloomers". For the long term period, nearly everyone agrees the economy is strong and will make many "adjustments" over the next twenty-five years. Long term growth seems evident. It is your company's job to decide how high the peak or high low the valley of the economy, where we are now, which way we are headed, how soon, and how the company market is affected.

NEAR-TERM TRENDS

This is always the critical period for management decisions. New product decisions must be made far in advance to initiate new models or completely new vehicles. Large automobile companies require from 5 to 7 years to go from the first drawing through to mass production of a completely new vehicle. The time periods break down somewhat like this:

A – One to two years in PRELIMINARY DESIGN to work up designs and alternatives for management to consider and eventually authorize for the next step. Feasibility studies are made. Three view drawings and inboard profiles are prepared, artwork, clay models, mock-ups, and specifications, performance estimates and cost estimates follow.

B – EXPERIMENTAL DEVELOPMENT of new ideas, components, etc., will require another year or two to develop in laboratories. Some of this may also be done by suppliers. Wind tunnel work on body contours needs to be done. Management decisions are required as to what can or cannot yet be used because the "state-of-the-art" is or is not sufficiently developed.

C – DETAIL DESIGN FOR PRODUCTION begins next. Prototype vehicles are built. This will take possibly a year or more when road testing is included. When finally approved the big gamble and large expenditures to produce come next.

D – MASS PRODUCTION requires facilities and coordination. Fabrication machinery must be chosen and ordered. Plant layouts for assembly lines according to proper assembly sequence must be worked out. Tools must be designed before they can be made – then tried. Then a pilot line is set up to slowly get production started and to determine problems and bottlenecks. Eventually, everything is working smoothly and mass production is underway. All this can easily require up to three more years – especially when a few changes are required.

These stages (A) through (D) can overlap slightly and sometimes be compressed with much overtime and/or luck.

A small company will not undertake as large a program as a big company, but all the elements are still there. Also, since production will be in lower quantities, less facilities are needed. Therefore, the time-span can be shortened – often cut in half or less.

Actual NEAR-TERM TREND predictions are a refinement of long-term trends and must blend with them. It is left to the reader to initiate his own near-trend predictions in consideration of how they may influence his own activities.

From the above timetable it is clear that management must watch NEAR-TERM TRENDS constantly to decide when a model (or models) should be started. Sometimes the time may be speeded up or slowed down due to economic or other conditions. However, it is as much a mistake to try to "force" as it is to try to "delay" a trend. Accept it, anticipate it if possible, and plan accordingly. For the sake of discussion and for examples in this book, we will "assume" the following NEAR-TERM TRENDS as of 1979. Five to ten years hence puts us to 1984 (5 years) and 1989 (10 years). Actually 1985, six years hence, is the year the *present* government regulations require the maximum miles per gallon for vehicle groups. This means continuing "higher efficiency" as our book contents indicate. Lighter, small vehicles with lower power requirements and better aerodynamics are factors that must be utilized to the fullest. The mass produced vehicles, both domestic and imported, are being completely redesigned from the ground up, all at great expense, to be completed for all models by 1985. Then, as mentioned in long-term trends, the regulations will be tightened further.

A small specialty company would never consider competing head-on with the large mass-production companies. The small company seeks out unique markets – often initiating new idea vehicles years before the mass-production companies could gear up. Additionally, quantities may be too small for the big companies but more than adequate for a small company. Electrics are an example. Small companies are getting into them now. The large companies expect this market to be large by the late 1980's. Some of the small companies who get into electric vehicles now will grow with the market and be medium to large companies by 1990.

MARKET ANALYSIS

Market analysis can be a rather tricky business. It is one thing to analyze the market today, and it can be something else to determine what it will be some five or more years in the future – when your product will reach the market. Many things can happen – both good and bad – some of which cannot be anticipated in an analysis.

Essentially the analysis should consist of:

A – Market potential. This is the total market for all competitors.

B – Your company's possible share of the potential market.

C – Your company's ability to reach its share of the market. This includes ability to design competitively, manufacture economically, organize a sales force including a nation-wide dealer organization with adequate promotional advertising, and finance the complete project until income balances outgo. All of this must be done in a progressive manner within the time frame allocated.

D – A step-by-step plan must be established for goals in the first year and for following years until the expected market share is reached.

E – Contingency allowances must also be made. Maybe the program will take longer than planned and require more money and more experienced people than planned. Will the market change during this initial period?

WHAT THE DESIGNER CAN DO ABOUT ALL OF THIS

If he is located in "Advanced Vehicle Development" and is well up in the organization, his efforts are *crucial* to the success of the program. His designs define the company product. Its manufacture, if to be economical, will require his best efforts at cost analysis and complete coordination with the manufacturing branch of the company. Although he will not be selling the vehicle to the public he must work closely with marketing people in both Sales and Advertising to be certain they fully understand the vehicles and can promote the features he designed into the vehicles. The designer must also explain and justify these same features to the financial people if they are to come up with the funds for the program.

The designs and the designer who coordinates and defends every feature of the designs is the keystone to the complete program. This is a demanding and rewarding job for a thoroughly experienced designer. It is not recommended for the weak at heart.

HOW THE MARKET ANALYSIS IS MADE

The market potential in the motor vehicle field can start with a review of monthly and annual sales as listed in WARD'S AUTOMOTIVE REPORTS for different types of vehicles. Other sources such as articles in various trade magazines are often reliable. Changes in volume from year to year will help to indicate trends. Any other factors such as economic conditions and competitor products will also influence the market. It is well to realize that the customer will not magically have the additional money and the desire to purchase your product as well as purchase whatever else he, the customer, planned to buy. It is an either/or situation. He must give up or forego something else to purchase your product. Therefore, your vehicle must be more desirable to him than whatever else he had in mind. Your vehicle is in competition not only with other vehicles but also with whatever else he is considering such as a boat, clothes, home improvements, entertainment, vacation trips or whatever, including possible savings or investing. Sale of your product comes from the *lack of a sale* of some other product or products.

Next it must be realized that this analysis must be made – not for today's market – but rather for the market when your product will actually be produced sometime in the future. This is why forecasting is so important along with a study of trends.

YOUR COMPANY'S SHARE

Rarely does any company have a market completely to itself and then only long enough for the competition to catch up. The originator of a new product may have the market to himself for a year or two – then – if in the interval he aggressively develops sales and product lines and keeps prices competitive he may possibly retain 50% of the market. Your analysis must not assume that automatically your company is capable of doing this. Is the product covered by a Patent? Are there other proprietary features? How can your product be made outstanding and remain so? Can it really compete pricewise? Can and will the manufacturing branch adequately tool up to achieve low costs? Does the financial branch have the capital funds needed or can these funds be obtained in time at acceptable rates or risk? What about other facilities and the needed experienced personnel that must be added? Just how aggressive can your company be – and as compared to other companies? Not only must the product be designed right – everything else must be right, too. These are hard questions and we must be realistic.

WHY SMALL URBAN CARS MUST COME ALONG BY 1985

Higher vehicle initial cost and higher fuel costs will, of course, be an inducement for the buyer to seek relief in a smaller vehicle. The manufacturers also have a very strong inducement to make them available by 1985. That is the year Corporate Average Fuel Economy (CAFE) must reach 27.5 mpg. So they simply must have some small cars for sale to balance the larger cars of lower mpg. How small?

Many people misunderstand the averages. If a large car averages 17.5 mpg and the fleet average must be 27.5 mpg, an equal number of small cars at 37.5 mpg *won't do it!*

An equal number of small cars *must average 64.2 mpg.* This is explained more fully in the chapter on Rules and Regulations. Then if sales quantities of large and small are unequal, this further alters the averages and must be included in computing the corporate average. From this it is evident the small urban car, with its higher miles per gallon, is required to permit some large cars to be available.

RECOMMENDATIONS – EXAMPLES

For our purposes in this book the product planning group will make the following recommendations.

1 – An ICE (Internal Combustion Engine) development program is beyond the ability (technically and financially) of a small company – along with body and chassis development at the same time. This may not be true for a racing project but is for a street vehicle that must meet government emissions and other regulations.

2 – Existing engines (foreign and domestic) and motorcycle engines provide sufficient variety for every size and type vehicle. Certain modifications can be considered for installation, racing, or other reasons but major changes should be avoided.

3 – The unique small city or urban vehicle should be considered. It may soon become quite popular as a second or third car for shopping, to and from work or school, and other stop-and-start driving. This vehicle may be thought of as an intermediate step between the motorcycle and the family car.

4 – The motorcycle market at one million annually should not be ignored. A three-wheeler under 1500 pounds is classed as a motorcycle and could make an ideal urban vehicle.

5 – An electric vehicle, possibly the small urban car, should be developed at this time. Electric components including improved batteries are becoming available in great variety and at reduced prices due to higher quantities. This will help the electric motorcycle also.

6 – The sale of plans, kits and parts for home builders is growing. Over 10,000 MECHANIX ILLUSTRATED readers ordered plans for URBA-ELECTRIC at $15.00 a set (reference 4). This market is worth considering by small companies.

7 – Feasibility studies should be authorized for the Advanced Vehicles Development Department for a new family of small vehicles, both with ICE and electric power. Due to the progress made in three-wheelers, these along with four-wheelers, should be analyzed also.

THE BELL CURVE AGAIN

It is strongly urged that these studies and eventual decisions be made promptly. Often, companies are reluctant to start new designs when existing sales are at their peak. A review of the bell curve phenomenon shows why this may be a mistake. Figure 3 depicts the loss of market and sales if the start is delayed until the present model peak is past. Figure 4 depicts how, by moving the start of the new product to the left, under the peak of the existing product, the loss of the market is greatly minimized.

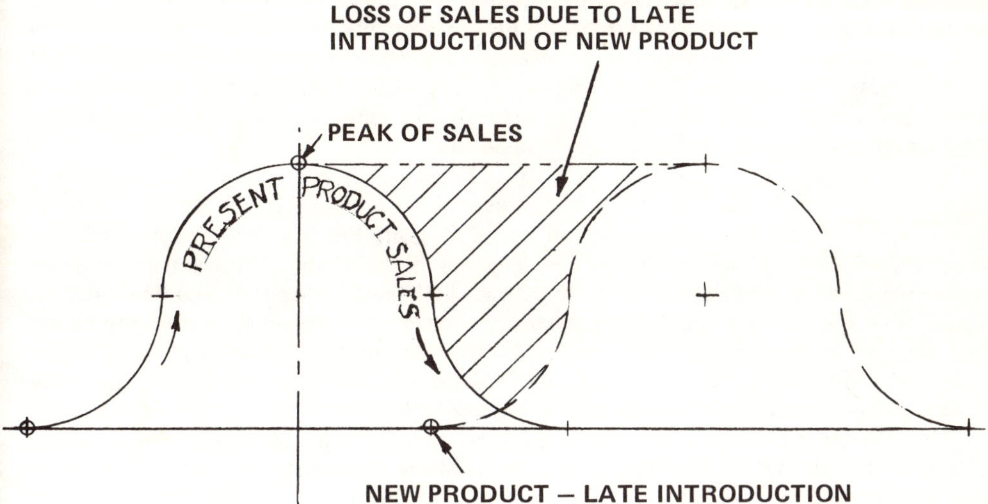

-- FIGURE 3 --

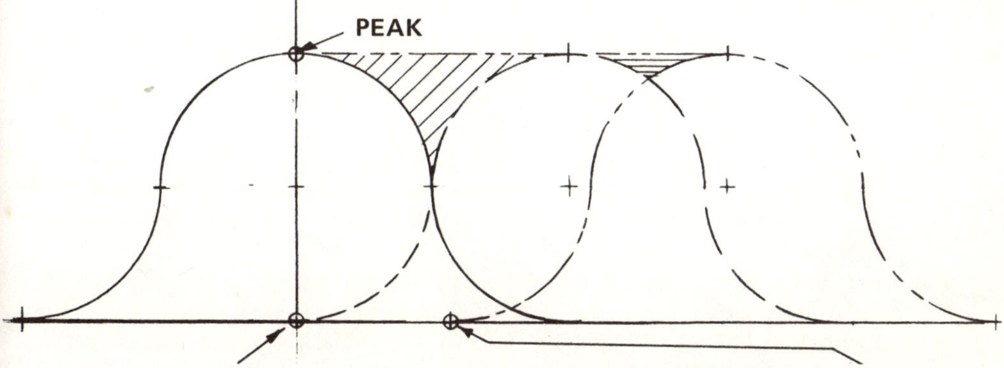

INTRODUCE NEW PRODUCT AT PEAK OF EXISTING SALES (OR BEFORE) TO REDUCE LOSS OF MARKET BETWEEN PRODUCTS PEAK SALES – HIGH START-UP COSTS OF NEW PRODUCT IS LESS OF A BURDEN WHEN BUSINESS IS STILL GOOD – SALES HIGH.

-- FIGURE 4 --

THE NEXT MOVE IS UP TO MANAGEMENT

Outstanding success of a product has a way of blunting the decisions that must be made by a management that becomes too conservative with its success. Remember – nothing lasts forever – and it is always easier to delay a decision than run the risk of making an error. The remarkable success of the Model T no doubt delayed the introduction of the Model A and at a great loss to the early FORD company. the VW Beetle success story somewhat paralleled the Model T and the Beetle replacements were introduced a bit late to keep sales near the peak. This shows one of many reasons why business has its ups and downs. Every company *must* do its research and development on a *continuous* basis so it can, at any time, move to maintain its momentum.

It may be argued that the new product will drain away sales from the older product which is still selling well. Possibly – but remember, the buying public has inertia in its buying habits. That's depicted in the bell curve. Sales are lost to the competition and are never recovered. Additionally, if the new product is less than perfect and needs some improvement, when introduced early there is a little cushion of time and company resources to bring the new product successfully through its "teething" period before the older product has become obsolete and no longer sells. Better to be ready with the new product(s) than "too little – too late", with losses that take the company to the brink of failure.

These recommendations have now gone to management. Management can sit on them and do nothing, disagree and reject, or decide to pursue the matter further. Since these recommendations cannot possibly provide all the answers, management will usually ask for feasibility studies to obtain more (and needed) information.

This brings us to the next chapter – ADVANCED VEHICLES DEVELOPMENT, the people who take the possible product from planning into feasibility studies, and then to the actual design.

Chapter IV

ADVANCED VEHICLES DEVELOPMENT – (1)

On an organization chart this function may be listed under a different or more exotic name. Essentially it is the first step in the engineering design of a new vehicle. For many years companies in different industries as well as the automotive industry called it simply "PRELIMINARY DESIGN".

Quite normally the "key" designers who begin with the feasibility studies usually proceed with the preliminary design of the vehicle and often stay with the project into detail design for production. Management prefers this in many companies to insure that the continuity of the original design concepts are never diluted or lost as the design is finalized.

FEASIBILITY STUDIES – FROM THE VIEWPOINT OF THE DESIGNER – MANUFACTURER – INVESTOR

Man can create anything he can envision – given enough facilities, funds, assistance, and time. That doesn't make it feasible, though. It may not even be desirable. Company managements need feasibility studies to help in deciding what to develop next, when is the best time to do it, and what will be needed. Good feasibility studies have a way of growing into management directives and become the *basis for specifications* to and/or by the ADVANCED VEHICLES DEVELOPMENT group. That is why the subject was put in this chapter. It should be realized that some studies or portions of a study may and should be prepared by other groups.

A single type feasibility study is not enough to indicate what can be accomplished for the reason that people have different interests. Each study must be from the viewpoint of a specific group of readers. The designer's interest is in function, appearance, style, performance, technical innovation. He feels that if he does a good job in these areas everyone will want his product. Maybe so, but this is just the beginning. Show the proposal to a manufacturer and he may have great interest as a customer, but none as a manufacturer. The problems and opportunities for manufacture simply have not been discussed. Show the proposal to a potential investor and he may agree he likes it, would like to have the product for his own use, but to invest in the venture? Again, his business interests haven't been touched. The risks, the competition, the size of the market, the potential for profit – these are of major interest to the investor.

NEW IDEAS AND INVENTING ARE NOT ENOUGH – THE DESIGNER MUST BE A BUSINESSMAN TOO

A designer must think like an entrepreneur – in fact, some designers become entrepreneurs when they go into business for themselves or eventually head up another company (or an Engineering Division). Specializing as a strong technical type is fine as long as you are willing to work on other people's ideas. All ideas have to be "sold" and technical analysis alone simply cannot do the whole selling job. The designer's viewpoint is well covered a bit later in this chapter and even later from preliminary design through details of components, performance, aerodynamics, etc., so will not be dealt with here.

The manufacturer concerns himself with the costs of labor and materials, and with processes and methods of manufacture related to the fabrication and assembly of all parts. If he is a sub-contractor his interest stops here, since he will be bidding to a prime manufacturer. As the manufacturer who brings it all together, his interests extend far greater to all the elements of research and development, prototypes and testing, capital funds needed to manufacture, facilities, personnel, marketing, and return to stockholders.

If no company exists to proceed with the designer's new idea, he, the designer, must approach outside manufacturers as indicated above, with his feasibility study complete for all interests.

If a new company is to be formed for the new idea (product) then venture capital is needed. This may come from investors in related fields or from venture capitalists who specialize in this type of investing. These people require both the designer's feasibility study and the manufacturing feasibility study along with another – the investor's viewpoint. How much money for how long a time at what degree of risk is of major interest to these investors as well as potential profit.

Since a new company with a new product involves more than ordinary risk, it is reasonable to expect higher than ordinary returns on the investment. A projected return of 25% or more per year is usually considered necessary to justify a higher risk investment. Further, it is not unusual for the first year or two to not yield any profits so the eventual return must compensate for this. The feasibility study must be realistic and must clearly set forth these factors. Two of the major elements of a feasibility stucy of interest to a manufactuer and/or investor are 1–CAPITAL FUNDS REQUIRED and 2–CASH FLOW ANALYSIS.

WHICH FEASIBILITY STUDY COMES FIRST?

The ADVANCED VEHICLES DEVELOPMENT group must do the first feasibility study. The designer must define the vehicle(s) rather completely with specifications, choice of major components, and sufficient drawings. This shows what the product(s) will look like, consist of, and what performance can be expected. From this data comparison with the competition can be made to determine potential market share and the quantity to be manufactured. The designer's feasibility study is the basis for all other feasibility studies. Nothing else can be studied until the product is defined. Then marketing, manufacturing, and financial studies can be made. These, however, are unique to each company because of varying facilities and ability, so are beyond the scope of this book.

HOW THE DESIGNER PROCEEDS

The first step is *not* to make an artistic sketch as so many people seem to think. The first step is to determine as near as possible "what the boss wants"! This is done through meetings and discussions so that a "set of objectives" or goals can be defined. These become an "engineering specification" eventually. *Then* the designer proceeds to work up his concept, or preferably several concepts of the vehicle(s), that will meet the proposed specification. Close coordination with management and sources of information is necessary as the concept developes. Problems, confusion, lack of information, misunderstandings, – all are rampant at this point. Changes are made and finally the ideas begin to firm up.

THE ENGINEERING SPECIFICATION

The first rough draft of an engineering specification, for our purposes, might be as indicated below. It will be "tightened" or more closely defined as work progresses, but the basic content cannot deviate from management's objectives without reasons, and without approval from management.

The ADVANCED VEHICLES DEVELOPMENT GROUP SHALL:

1 – Design a "family of small vehicles" sufficiently to establish feasibility of the concept of commonality of components and parts and to what extent this is practical.

2 – Show commonality in design to obtain:

a – Maximum production quantities and lower unit costs of components and parts on more than one vehicle.

b – Lower tooling costs by using the same tooling for items common to more than one vehicle. This will also reduce the need for other additional facilities and capital funds.

c – Greater variety of vehicles for larger market penetration.

3 – Design this family of small vehicles to carry two people and provide a separate compartment for items such as parcels or limited luggage.

4 – Include competition vehicles for racing, rallys, or other competitive events to promote this family of vehicles. These vehicles may include competition modifications to the basic stock vehicles and completely separate single-seat vehicles for all-out racing in club events.

5 – Limit its design studies essentially to small sports cars, competition cars, urban cars, and utility vehicles that incorporate commonality in design.

6 – Utilize existing fossil fuel powerplants, available in production quantities, that meet emissions laws, both exhaust and noise, and that may be installed with but minor changes.

7 – Utilize alternative electric powerplants, including hybrid designs in these vehicles, especially for the urban and utility vehicles. Again, available production components are desired.

8 – Include four-wheel and three-wheel (motorcycle) vehicles in the family of designs so long as commonality can be shown.

9 – Design these vehicles with emphasis on efficiency, excellent performance in category or class of vehicle, excellent handling such as in cornering, stability, and control, comfortable sports car "firm" ride, and with advanced styling that will endure.

10 – Consider the possibility that some of these vehicles may be marketed in KIT form, partially or completely assembled.

Upon completion of this study and review by management it will be decided which vehicle(s) will be developed first and which, if any, will follow in coming years. Facilities and capital funds required as well as market studies and possibly other factors must also be analyzed after the vehicles are defined.

HOW A DESIGNER DESIGNS

This is very baffling to most people and is of some concern to the student planning a career in this field. He would like to know how to go about it correctly. Here are some observations:

1st – He must understand what is wanted. *Not what he wants* – what the boss or the client wants or needs.

2nd – He must realize there are many solutions. He won't necessarily think of all the solutions so he must be open-minded about the ideas of others.

3rd – Ideas do not arrive in his or anyone else's mind in complete detail. Ideas need detail development to be made usable or so they can be evaluated. As one of my former bosses said, "An idea, like a newborn infant, comes in swaddling clothes". It needs nurturing and loving care to be brought to maturity. When mature, the clothes are no longer "swaddling". The idea, when properly dressed in a mature manner, is then ready for evaluation.

4th – Dr. Edwin H. Land, chairman of Polaroid Camera, at the 1978 annual stockholder's meeting spoke of how Polaroid's SX-70 was developed, (reference 5). Dr. Land mentioned that in a philosophic sense it took 40 years to develop. He then stated, "You start with a fantasy and visualize something as perfect -- and then with experiments you work from the fantasy back towards reality. The important thing is to keep your standards high, never to drop your original fantasy. Even if the goal of transforming the idea to reality appears insurmountable, it is much better to persuade people that there is a way to do it so they will remember that there is a way to do it in spite of the fact there is no (current) way to do it. All you have to do is decide to do it."

5th – Do not decide an idea is "good" or "bad". It may not be popular at the moment. Maybe it is not fully appropriate for the problem. Possibly it isn't mature enough for evaluation – yet. It may not be the "best" idea as presented. Others may not understand it. DON'T RULE IT OUT COMPLETELY. Even if not used, it may become a stepping stone to another more acceptable idea.

Here are some examples of ideas that fortunately were not permanently abandoned. (a) The rotary combustion engine. It was a new idea that was taken off the U.S. market shortly after its introduction. It is now back with detail refinements, maturity, and new popularity. (b) The electric vehicle has had several "rebirths". This time it looks like it is maturing, is timed right, and will remain and grow. (c) Front wheel drive has come and gone over the years. With better constant velocity "U" joints and the need for smaller cars, it has new purpose and is coming back strong. (d) Automotive aerodynamics, tried in the past, is needed now and is "in". (e) Three-wheelers. OK only for slow utility vehicles in the eyes of many, now offer additional advantages seldom understood before.

So, don't discard ideas. Look at them differently. Maybe with some further development, that is now possible they will solve your problem.

HOW A DESIGNER GETS IDEAS

1 – *He must be curious.* A curious person "finds out what makes something tick". He is interested in many things and in all things related to his job. He knows about competitive products and the latest state-of-the-art in his field. Research and development are of great interest to him.

2 – *He must be flexible – responsive to change.* The unexpected is a part of his way of life on the job.

3 – *He must be receptive to ideas.* A closed mind or a narrow mind is not receptive. He must be open.

4 – *He must be optimistic – even enthusiastic.* He must have these qualities with the innocence of a wide-eyed small child learning something new.

5 – *He must be a "diligent seeker"* in the inner quiet of his own mind. He doesn't weaken his efforts to "seek" by telling others what he is going to do in his thinking process. That is a form of "leakage of the mind". He conserves his energies to think rather than talk. The time to talk, if ever, is *after* the fact. He accomplishes first – then talks – maybe.

6 – *A designer "envisions".* He gets mental pictures of his ideas or designs. He thinks in pictures. He can mentally visualize an object, even in great detail, inside out if required. He can move parts around in his mind – make all kinds of mental changes. Then when he sees the solution he transfers the idea to paper in the form of a drawing or sketch so others can see and understand it.

7 – *He is grateful for all ideas – no matter what the source.* You, personally, don't have to consciously think up all the ideas that are used! The "NIH mentality" (not invented here) has long been observed in many offices. Avoid it like the plague. Jealousy and selfishness have no place in an engineering office or anywhere else for that matter. There are other good minds beside your own. Appreciate others. Give others the same respect you expect. You'll be surprised – it comes back in many unexpected ways.

There are also ideas that "come-out-of-the-blue" from your subconscious mind. Some call it the unconscious mind. Maybe it is something you "feel" – something intuitive. Acknowledge and be grateful to your subconscious just as we all should do to each other. The subconscious part of the mind is a willing and powerful helper if we (1) recognize it, (2) ask it for help, (3) accept what comes – even if it isn't what you thought you wanted. The subconscious, like a computer, has access to more information than your conscious mind. (4) Appreciate and thank it. This can be a mental "thank you" within yourself – but do it. The ideas of others need to be acknowledged and appreciated orally or written to the giver. Then there is no hesitancy to offer more ideas.

8 – *IDEAS COME IN ABUNDANCE – WITH EXPERIENCE* "like rain drops from heaven". Reading about it isn't enough. You must practice – do it all the time. Be curious – receptive – diligently seek – envision – in an optimistic enthusiastic manner – and be grateful. Quiet surroundings without distraction help.

These pointers are not original with the writer. The origins are lost in antiquity. Every major religion teaches these things in various ways. In the distant past the "heart" was what we now refer to as the subconscious mind. If you are non-religious, a study of psychology will provide much of the same material.

Of course, designers are not the only people who seek ideas. These pointers apply for anyone who is interested in creative work in any field.

Let's now move on to the next step. On the chart under ADVANCED VEHICLES DEVELOPMENT the first block is CHASSIS DESIGN – GENERAL ARRANGEMENT and CHOICE OF MAJOR COMPONENTS. We must now develop some layout drawings.

Chapter V

ADVANCED VEHICLES DEVELOPMENT – (2) CHASSIS DESIGN

GENERAL ARRANGEMENT AND CHOICE OF MAJOR COMPONENTS

In working out the chassis general arrangement and choice of components, tire size, based on expected gross weight, is usually the first tentative decision that should be made. Radials have less rolling resistance and are generally superior for our purposes. The smallest cross sectional size is 145 and is ample for the low weight we anticipate. Wheel diameter need not exceed 13 inches for an overall tire diameter of approximately 22.5 inches. The smallest size, 10 inch diameter, will provide a tire diameter of nearly 20 inches and will easily carry the load and weigh less, so also should be considered.

Next we should prepare a planview drawing somewhat as indicated in figure 5 to establish a first indication of tread, wheelbase (W/B), turning diameter and clearance for pedals, frame, and powerplant. This drawing shows rather typical dimensions for many small cars with 145-13 tires. The tread and wheelbase are also close to those of many small cars you might compare it with. The turning diameter of about 30 feet is as low or lower than most small cars have. In fact, 33 to 36 feet might be more nearly typical. Large cars normally exceed a 40-foot turning circle.

Now let's try again. This time let's use the smaller tire size, reduce tread and wheelbase, but keep the same turning diameter as shown in figure 6. The tread drops from 51" to 44" but surprisingly, we still keep 15½" clearance from inside the turned tire to the centerline (℄) of the car. The small diameter helps to reduce tread width while maintaining the same internal clearance. Incidentally, the early MG-TC had a tread of 45" with a larger tire and a slightly higher center of gravity so our proposed 44" tread, while narnow, might still be acceptable. A narrow tread permits some weight savings and a reduction in frontal area and air drag.

A short wheelbase further lightens the vehicle, permits a stiffer frame, and when properly balanced, the car can corner better. Of course, adequate length for occupants and powerplant comes first, but length beyond the minimum required serves no useful purpose. Reasonable increases in wheelbase while increasing turning circle may be acceptable as long as within 33 to 36 feet or so, without changing steering angle (27½ degrees).

THE INBOARD PROFILE AND "HARD POINTS"

Next an "inboard profile" preliminary design is begun. This might start out looking somewhat like figure 7. In this case the driver, engine, fuel, and basic roll cage frame are indicated. Other assumptions are made such as the location of the battery and reverse. Since a motorcycle engine is shown, the reverse is separate as described in chapter X – Power trains – ICE.

With this limited and tentative information a center of gravity (c g) position is roughly calculated and shown. With the c.g. position assumed we can then locate the wheels for a four-wheeler and for a three-wheeler.

The engine is most likely the first major component chosen after the tire size or even before in some cases. Horsepower, weight, size, cost, and availability are known or assumed at this point. Possible other engine choices are also considered and briefs are prepared to indicate the justification for the choice and the order of preferences.

The chassis designers will proceed to develop the complete chassis with the preliminary layouts that show the steering system, suspension members, drive train, and other major components. The location of these items, once fixed, become known as "hard points" and are rarely changed.

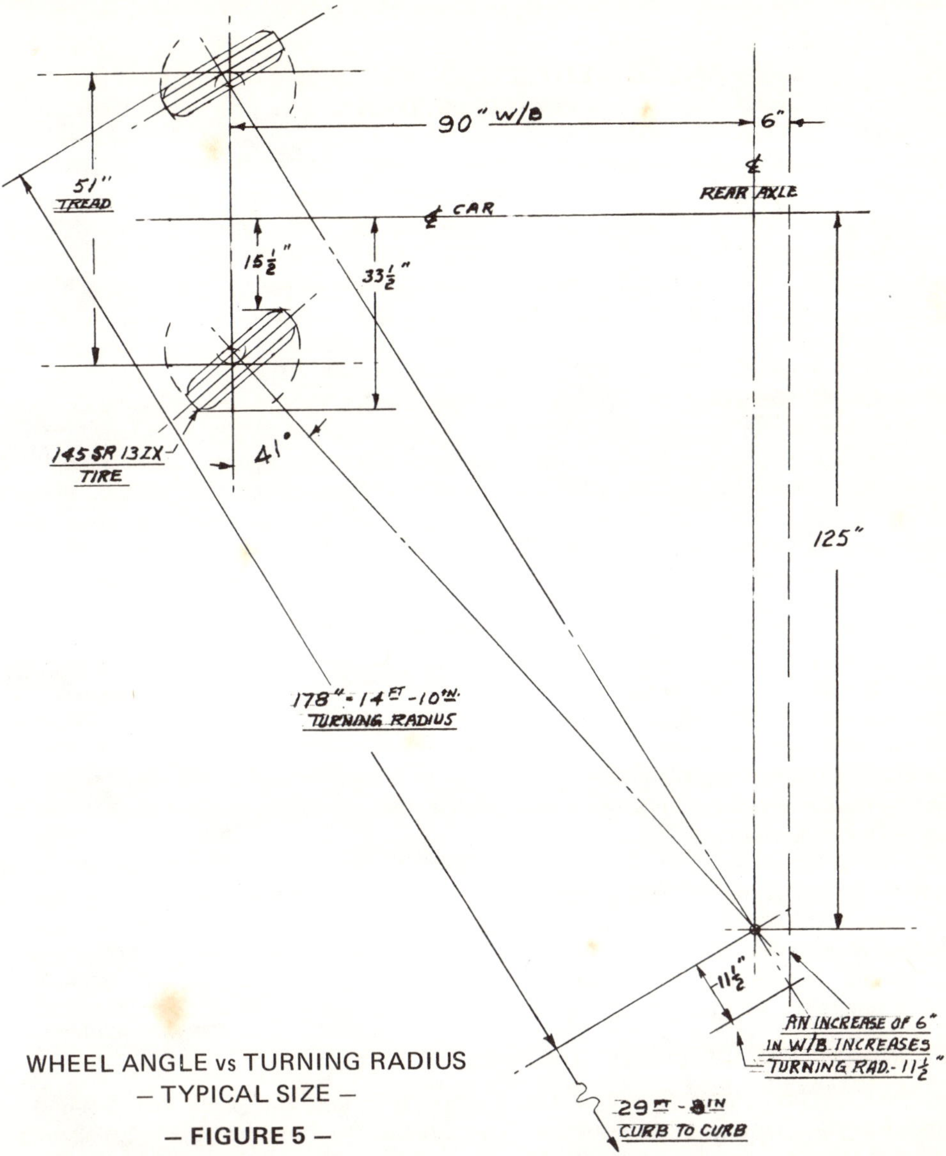

WHEEL ANGLE vs TURNING RADIUS
– TYPICAL SIZE –

– FIGURE 5 –

While this is being done the various staff specialists will be called in to assist with their recommendations, vendor information, test data, costs, weights, availability, etc. on all major components being considered.

COORDINATION REQUIRED

Coordination between everyone must be complete and constant through meetings and memos. Soon designers find themselves competing for vehicle space in critical areas. For example, in a typical new car the exhaust system with its catalytic converter, muffler, sensors, brackets, etc., requires more space than originally expected. The more sophisticated independent rear suspension (IRS) planned also needs more or different space and interferes with the exhaust system. The designers get togehter and see if they can resolve the problem between themselves. If not, they then report their mutual problem to their

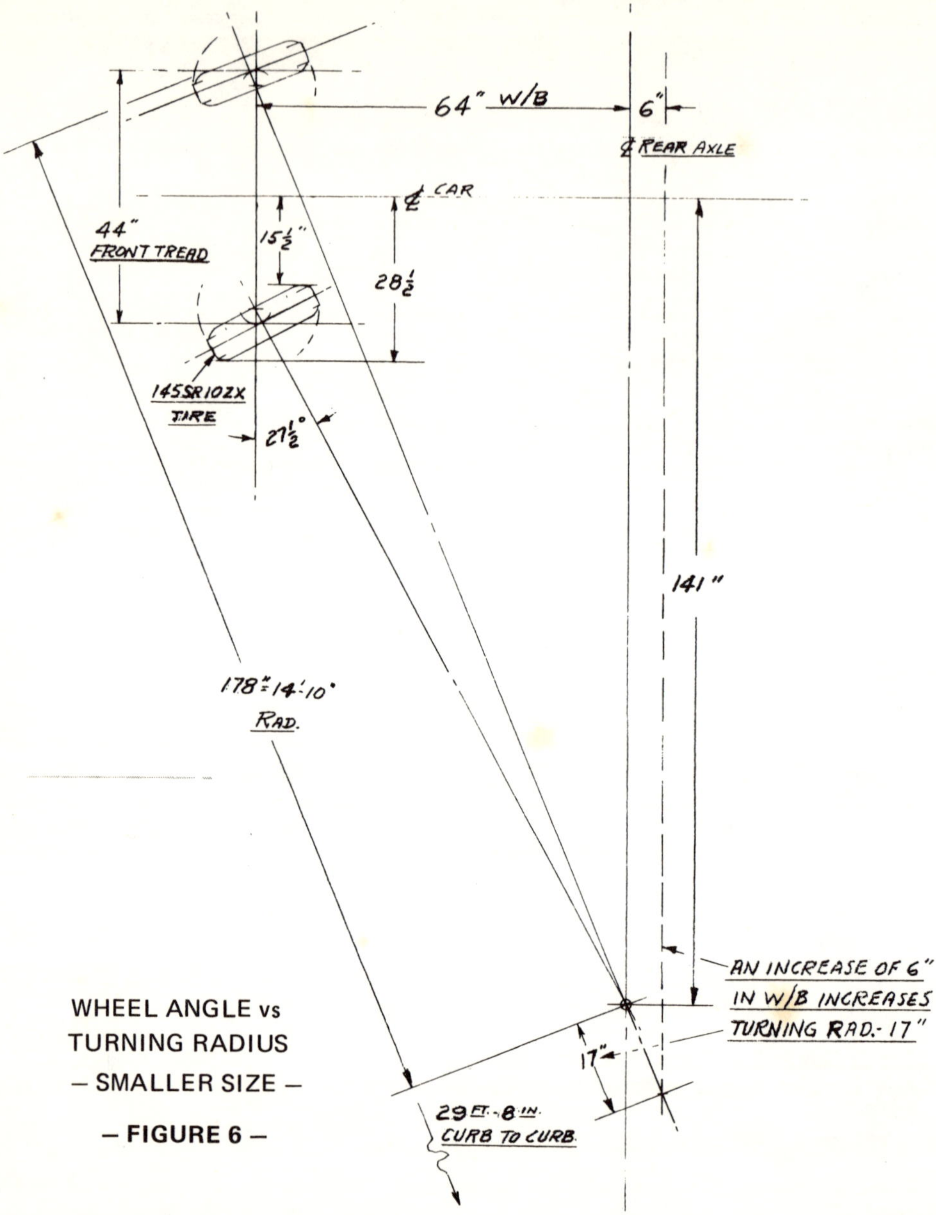

WHEEL ANGLE vs
TURNING RADIUS
– SMALLER SIZE –

– FIGURE 6 –

respective bosses. If the bosses still can't arrive at a decision they go higher to their bosses, and beyond until the problem is solved by one manner or another. Years ago the problem might have been handled by simply increasing the wheelbase to provide whatever space is needed for both designers. The decision would be made in minutes. Today the problem must be dealt with far differently. Any increase in wheelbase means added weight which might seriously affect the expected miles per gallon of fuel achieved and which must comply with the Federal CAFE regulations.

To adequately deal with the problem it may be necessary to reroute the exhaust pipe through a cross member in the frame. This may weaken the frame cross member so it

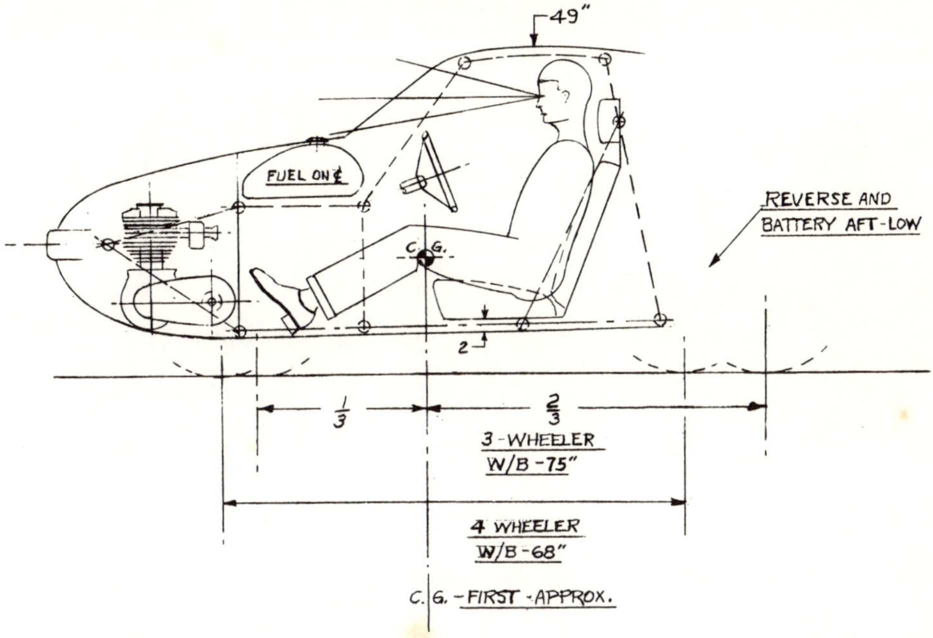

INBOARD PROFILE – PRELIMINARY DESIGN

– FIGURE 7 –

too must be redesigned to retain frame torsional strength – without adding any weight and hopefully no additional cost.

The designers must be very capable, innovative, and most cooperative.
As the chassis layouts proceed the body designers must also get under way.

BODY DESIGN AND STYLING

These designers can't wait until the chassis design is complete. They start with whatever information they can obtain – even though it is still tentative. This would be the inboard profile and the "hard points". Figure 7 might serve to get them started even before all the "hard points" have been established. Of course, they must then continue to fully coordinate their efforts with the chassis designers as changes occur and the design develops.

VARIOUS PROPOSED BODY LINES AND THE DEVELOPMENT OF A VEHICLE THEME – ITS PERSONALITY

A first attempt is made to indicate approximately what the vehicle might look like. Figure 8 provides the barest outlines of the four-wheeler and figure 9 indicates a start for the three-wheeler. I have never known of a case in which the original sketch ended up as the agreed-to design. It may provide some ideas and a start, but usually that is all.

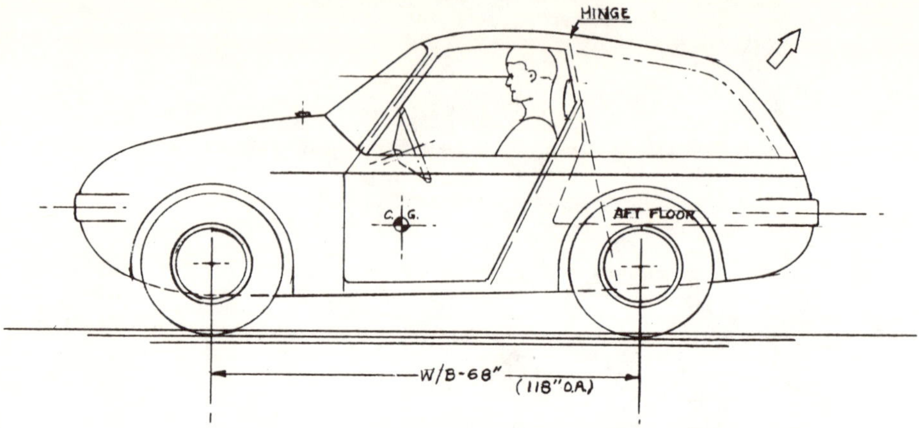

4 WHEELER – FIRST PROPOSED BODY LINES

– FIGURE 8 –

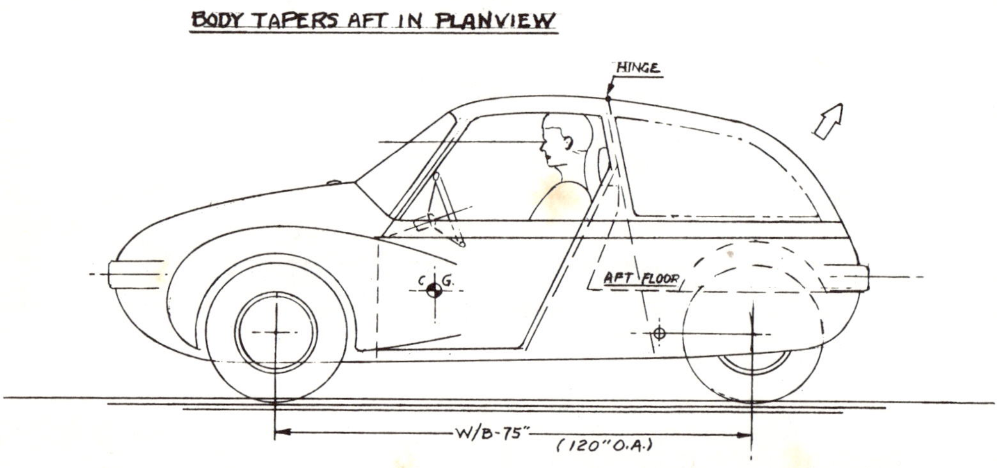

3 WHEELER – FIRST PROPOSED BODY LINES

– FIGURE 9 –

The designers may attempt a little larger vehicle such as shown in figure 10. This sketch uses the 13" wheels and a bit greater wheelbase to achieve a four passenger wagon. Most likely the management would reject it because it deviates too far from the specification and might not offer enough differences from existing vehicles now in production. It might, however, serve as a classroom subject for study or as a start for someone's design project. In any event, many sketches are needed along with discussions to develop an approved design.

The designer of the outward appearance of a motor vehicle is in a very powerful position. It is also a position of great responsibility. The automobile is generally recognized as an art form. It should have a personality. There should be an overall theme. Subtle differences in body lines, the proper use of molding and trim strips, color matching, and the overall "balance" of design, along with a "hundred and one" detail refinements from bumper to bumper (including the name), all properly blended and coordinated, go together to determine the theme of the vehicle. It must all be done in good taste and represent current thinking. It must be in vogue – not too advanced, and not a copy

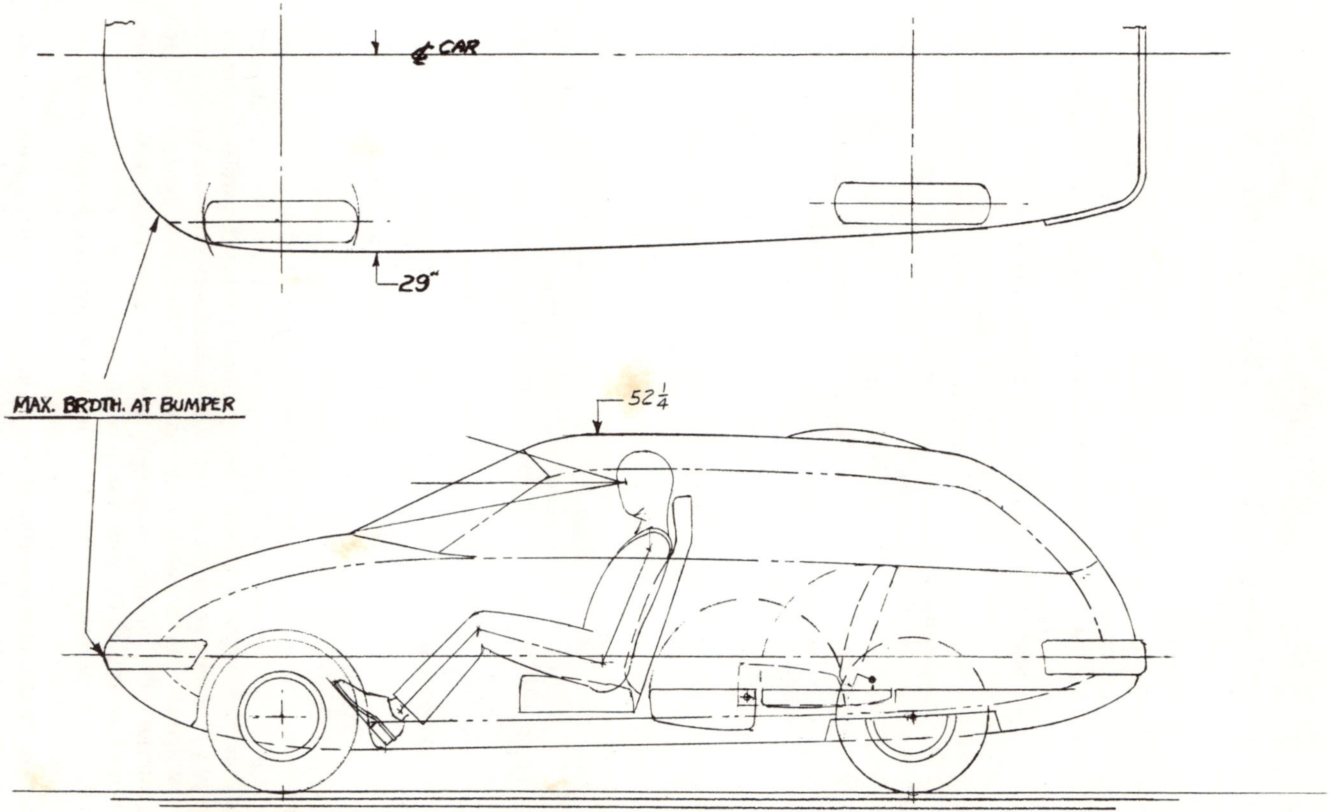

FOUR PASSENGER WAGON — 94 W/B — 155½ O.A.

— FIGURE 10 —

of something else. The old cliche, "I don't know anything about art, but I know what I like" applies to automobiles as well as to paintings. Every buyer has definite opinions even though his explanation is elusive. Naturally, the designer or "stylist", as he was formerly called, demands great freedom of thought – few constraints in developing his design. More often than not he is not an engineer. His background is in the field of art – the graphic arts. Today he most certainly should have some industrial design training. But even this is not enough for designing "higher efficiency" into motor vehicles of today and tomorrow. For example, it is generally agreed that it is more difficult to design a small vehicle that is beautiful, than a beautiful large vehicle. In both cases, the space for humans must remain about the same. Often people are even seated in a more upright position in the small car to conserve space. This plays havoc with the "fast" look. The boxy look of many of today's small cars is explained away with another cliche, "Form follows function", but that doesn't make the buying public any happier about it. Hopefully, since many new cars have the appearance, it is at least "in style" so therefore somewhat acceptable. The constraints of size limitation make the designer's job more difficult – no doubt about it.

Another constraint now being imposed is automotive aerodynamics in the search for greater efficiency. The old style designer is resisting it and he has direct access to management through his artistic renditions. Often management is sympathetic in feeling that it will not sell cars. Actually, nothing could be farther from the truth. Aerodynamics is probably the most misunderstood of all the disciplines in the industry. Possibly because we can't see the colorless air actually flowing around a surface or body, we tend to ignore or downplay it.

Established designers who should know better are on record for statements such as "Aerodynamics don't make a damned bit of difference on an automobile" and "you can't sell something that must look like a potato", and again "With aerodynamics won't all cars look alike?"

Obviously the time has come for a new breed of designers in the body design departments of some of the motor vehicle industry. Maybe managements needs a bit of updating, too. Or possibly they haven't been paying attention to the right people. It appears the old style designers and ultra conservative managements are kicking and screaming all the way to the new look – "The aerodynamic look" – in motor vehicles. Of course, aerodynamics make a big difference in the efficiency of cars – just as weight reduction and other factors. And a car doesn't have to look like a potato and all cars don't have to look alike. Gentlemen, it's time to stop fighting it – learn to use it – properly.

Admittedly there have been some so-called "streamlined" cars that were not fully accepted by the buying public. This was not the fault of aerodynamics. Something that isn't used properly or that is "immature" in concept, or that is not timely in its application, shouldn't be blamed for the failure of a vehicle – and then forever banned. Automotive aerodynamics are now "mature" and if skillfully utilized, can be an outstanding factor influencing efficiency, appearance, and the successful sale of vehicles.

One last statement about automotive "styling". There is a pattern, a continuous development of themes, year after year in certain cars. If you would like to study the history of automotive styling reference 6 may be of interest.

Now back to our examples and a bit more about how our designs might progress. After a few changes and many sketches and meetings, the four-wheeler might appear as shown in figure 11, with a racing version in figure 12. These designs are from the author's projects. Figures 13 and 14 are of the body mock-up in three-quarter front view and side view of the small racer. The fiberglass panels are complete but just resting on a wood mock-up structure instead of the finalized chassis, which is not yet fully developed.

A three-wheeler with engine in front might appear as in figure 15. A three-wheeler racer with engine in the rear is suggested in figure 16.

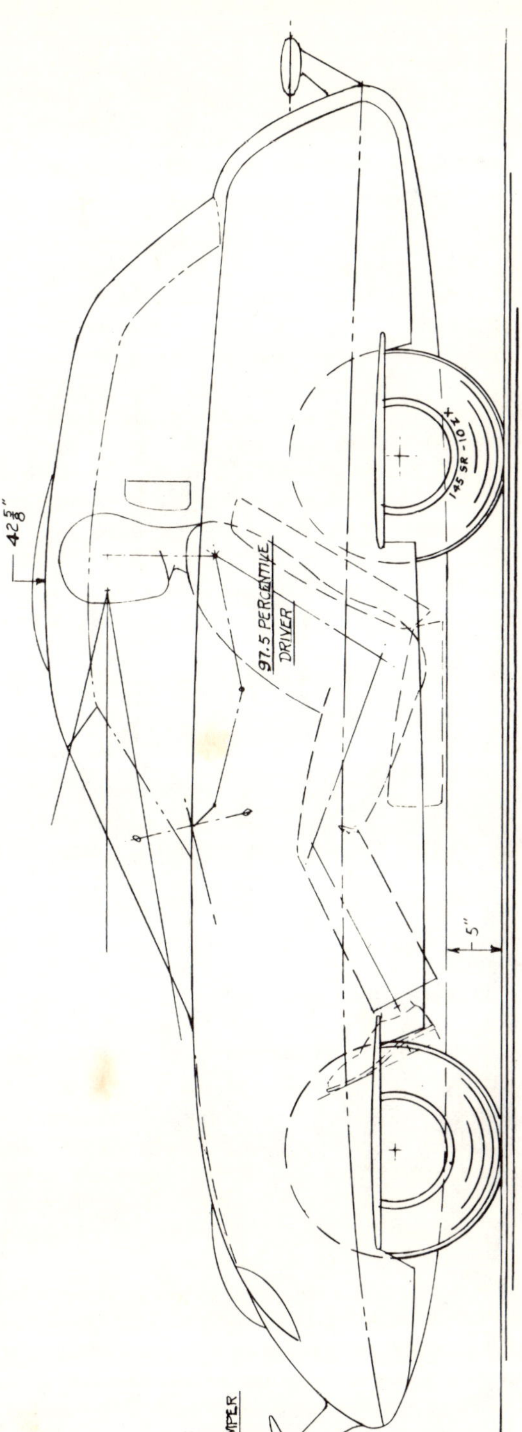

SMALL SPORTS RACER - G.T. – 66 W/B – STREET LEGAL

– FIGURE 11 –

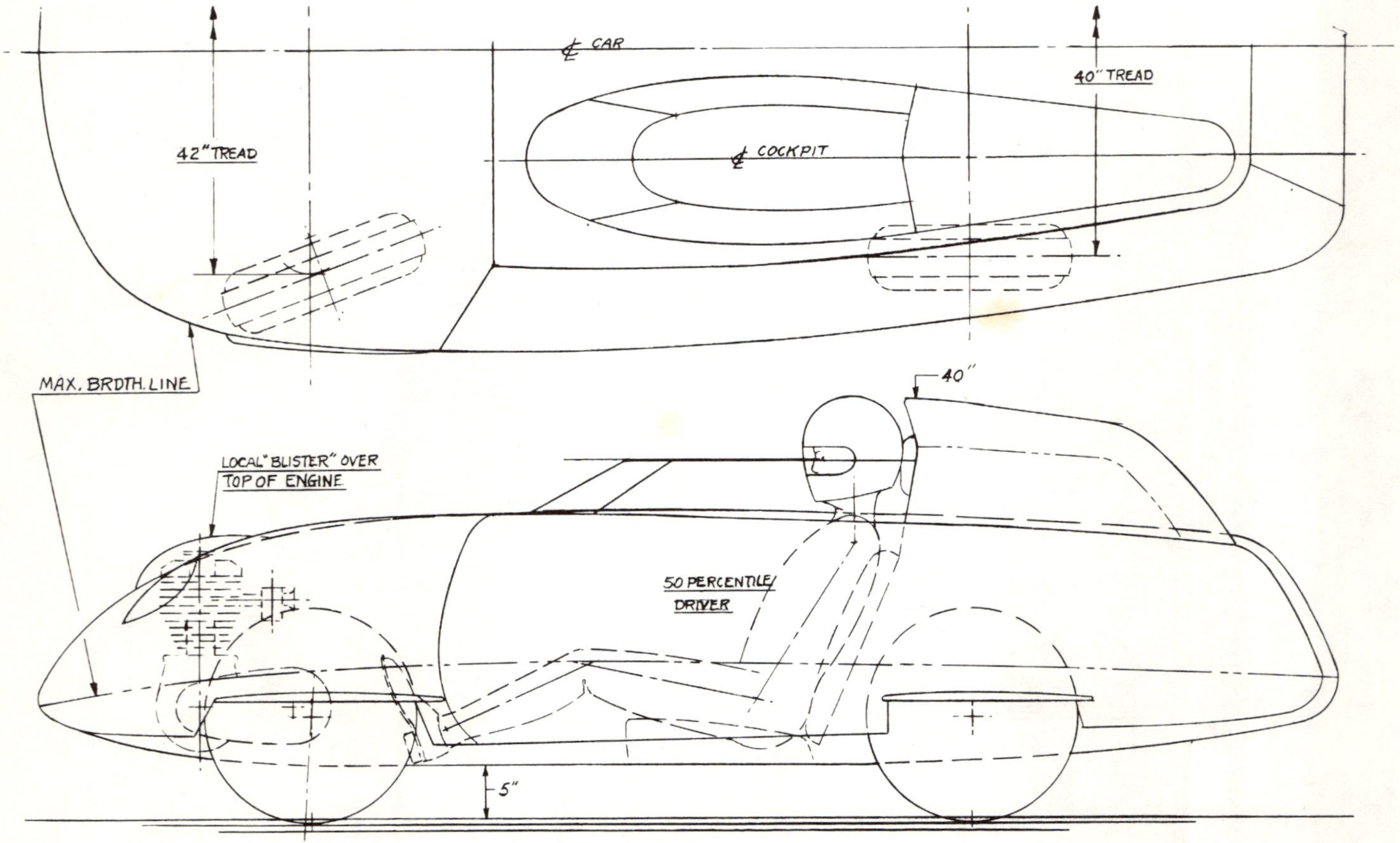

OPEN SPORTS CAR RACER – 64" W/B – 126" O.A.

– FIGURE 12 –

— FIGURE 13 —

— FIGURE 14 —

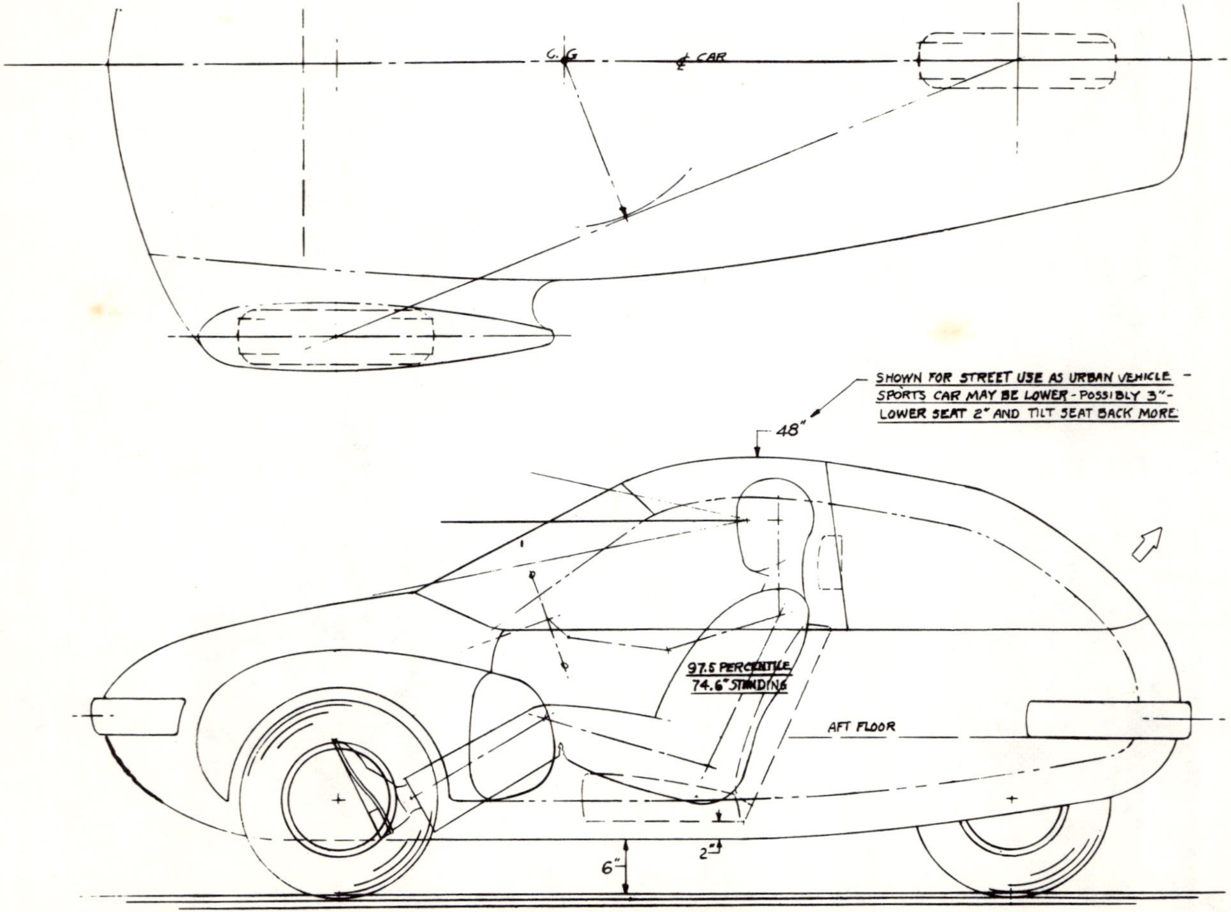

– 3-WHEELER – PRELIMINARY BODY LINES –

– FIGURE 15 –

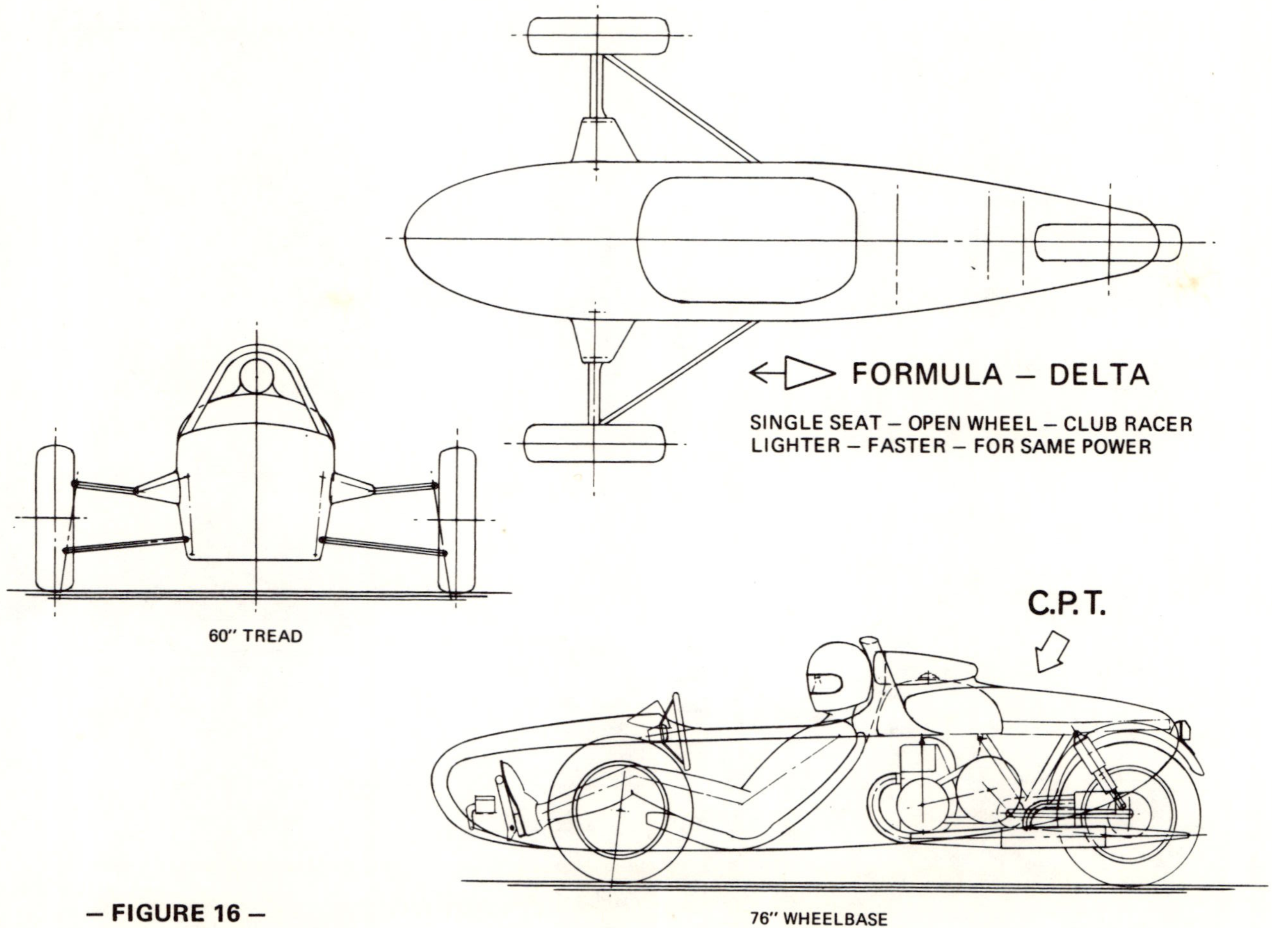

FORMULA – DELTA
SINGLE SEAT – OPEN WHEEL – CLUB RACER
LIGHTER – FASTER – FOR SAME POWER
60" TREAD
C.P.T.
76" WHEELBASE

– FIGURE 16 –

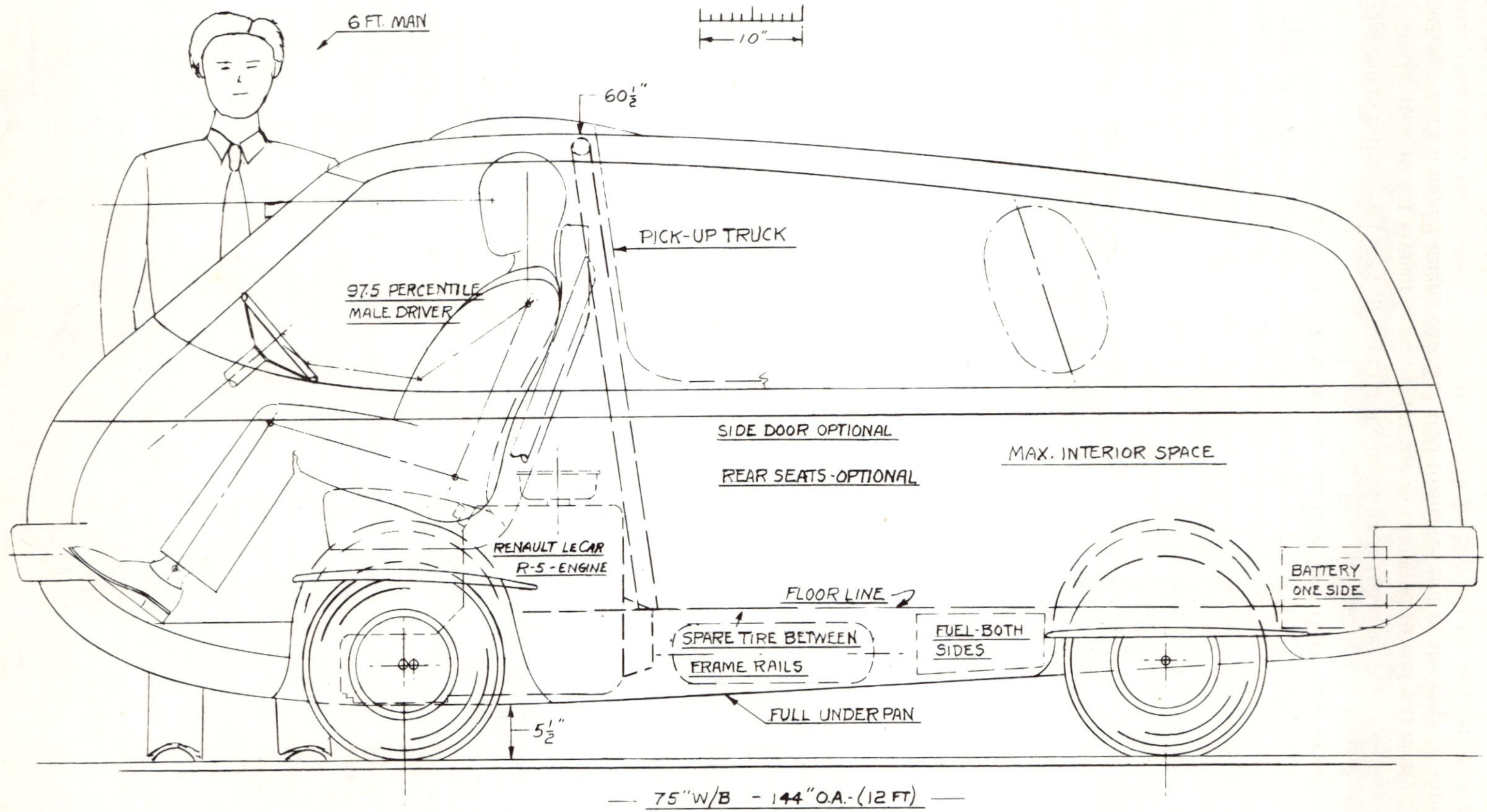

– AERO – MINI – VAN – FWD – C.O.E. –

– FIGURE 17 –

Commonality of parts in the case of the four-wheelers would include wheels, tires, suspensions, brakes, steering, reverse, instruments, basic engines, drive train, fuel tank, and body panels. Frames would be different and the "greenhouse", the (windshield, roof, and windows) would be additions. Seats and interior trim would also be different.

The three-wheelers could use many of the parts used on the four-wheelers.

If electric powerplants are to be considered the body designs need not change much. The chassis would have to provide battery space, preferably in front with motor and controls aft.

These designs, all of them, are not the only way to meet the specification. Proposed solutions can be endless, limited only by the imagination and ability of the designer.

For a change of pace and to keep the creative juices flowing let's now consider an Aero-Mini-Van with front wheel drive and with the cab-over-engine (C.O.E.). This might appear as in figure 17. This vehicle is small enough to still be classed as an urban car, yet provides the maximum amount of space on a very short wheelbase. Its simple body shell could (1) remain bare for lowest cost, (2) incorporate "deluxe" interiors for customers who so prefer, (3) be cut as shown for a pick-up truck design. Commonality of parts would remain high and a market analysis might be justified.

This design is a bit unique for a C.O.E. in that the roof line is but 5½" higher than the RENAULT "LE CAR" from which the chassis components are obtained. The six foot man illustrates the small size. Seat height is low enough to avoid climbing up into it. In fact, it should be very convenient for men and women who must enter and leave an urban vehicle often in the travels of a day. The short wheelbase and overall (O.A.) length of but 12 feet makes it very easy to park.

This design is shown with the smallest wheels – 145-10 tire size – with an overall diameter of 19.2 inches. The newer elliptical cross-section radials would remain nearly as small on a 12" rim and the Renault R-5 uses 145-13 tires, which are only slightly larger.

By careful design the weight can remain as low as the R-5 and aerodynamic drag need not exceed it. The drag coef. between 0.25 and 0.30 can be achieved by careful design. Fiberglass body panels would help to keep weight low. In this case the gear ratio in the transaxle need not be changed and performance may remain about equal to or better than the R-5.

The new transverse engined front wheel drive cars now in production in this country would also provide interesting powerplants for a Mini-Van such as this. It could remain FWD or the unit might be placed aft to drive the rear wheels.

In any event, an Aero-Mini-Van and/or pick-up with the newly available major components may eventually become a very useful as well as unique fun car at low cost for many people.

APPROVED LINES INITIATE THE MOCK-UP, CLAY MODEL, AND AERODYNAMIC MODEL FOR WIND TUNNEL TESTING

As soon as management decides which design is preferred, work can proceed with a full size mock-up. In fact, the mock-up begins before, at least in part, with the human factors portion such as seats, steering wheel, pedals, windshield, roof line, and other details as they become available. This is helpful to the designer, although it also puts some contraints on his design while assisting in selling the concept.

The clay model comes along with the sketches to provide a three-dimensional view of what the vehicle will look like. Few people can look at a three view drawing and fully comprehend the vehicle in all aspects. A model is most helpful. As changes are made the model can eventually be used to develop the body buck full size.

In the past an aerodynamic model for use in the wind tunnel was rarely considered. Years ago such a model might cost $15,000 to make, and in the eyes of some managements, only delay the project. Today the cost is higher and the need more urgent. Wind tunnel testing and refinements to design pay off in less power required, thus yielding better m.p.g. Critical areas can be studied and changed to achieve smoother air flow and less drag. Pitching moments, yaw, and other stability measurements can be determined, lift and induced drag investigated, as well as establishing the drag coef. for performance calculations.

The body designer needs this information to guide him in meeting his part of the specification. His favorite "treatment" of a headlight or wheelwell or something else just might not be acceptable in the wind tunnel, no matter how great it looks.

All these things simply must be worked out before *final approval* can be given if a more efficient design is to be achieved.

COMPUTER ASSISTANCE

In the past we had to prepare body drafts to extreme accuracy to achieve beautiful lines that could be transferred to the shop to create body panel dies and other tooling for production. This required many tedious manhours of effort and coordination. Today the approved "lines" can be taken off a model and reproduced to any scale, quickly, by computer. Even if the model lacks finely developed lines, the computer can go through a "smoothing operation" to average out errors such as humps or flats, to develop a perfect contour with absolute accuracy. This can be done in all directions. These contour lines are then defined as ordinate dimensions for use in making an accurate wind tunnel model, dies, and all shop tooling with everyone receiving exactly the same information. The shop computer controlled cutting tools create models, dies, all sorts of tools, all directly from the same computer tape.

Even the engineering department can use this computer data in an automatic drafting machine to make certain drawings. This not only speeds up the process but also eliminates much of the drudgery. Computer terms are described in more detail in chapter VII along with their use in other departments.

PROTOTYPE DEVELOPMENT

At one time a prototype vehicle was built to "try it out", see how it looked and handled, all with little thought about getting it into production at that point. Today such a vehicle would not be classified as a prototype. It would still be back in the "experimental" class. Maybe a "dream" car for auto shows, possibly a part of a racing program, but not a true prototype. The experimental cars, dream cars, and racers are still needed, but are rarely good enough to be the final step before production.

THE PROTOTYPE MUST SOLVE ALL MAJOR PROBLEMS AND PROVIDE A VEHICLE FOR HONEST EVALUATION

The prototype step comes before production. A true prototype vehicle may be "hand made" but it is built as near the production design in every detail as humanly possible. In that way all major problems can be better understood and resolved *before production* begins and at far less cost in both time and money.

No matter how small the company, how far behind schedule, or how much over budget, the prototype stage must never be eliminated. New companies or radically new projects often fail, simply because the prototype stage was rushed or not worked out in complete detail before production started. It is not unusual to build several identical prototypes to insure that all design problems are really resolved before the immense cost and effort of production is begun.

THE ADVANCED VEHICLE DEVELOPMENT DEPARTMENT IS ESSENTIALLY A SMALL COMPANY WITHIN A LARGE CORPORATION

This department is more than an experimental department. It is responsible for the vehicle design and development through all stages right up to turning it over to the engineering detail design for production department. It does not create all the detail design drawings, but it must create a complete vehicle that is ready to be fully detailed out. Think of it as a small company that develops a product but lacks the facilities to mass produce it.

IF IT CREATES THE CORRECT VEHICLE AT THE RIGHT TIME THE REST OF THE CORPORATION HAS A BETTER CHANCE TO ALSO BE SUCCESSFUL

If the Advanced Vehicles Development Department creates a vehicle that misses the market and doesn't sell well, the whole corporation suffers, no matter how great are its manufacturing and sales departments. This has happened to large automobile companies. In fact, no company can "bat a thousand" year after year.

One way to hedge its bets is to have a constant flow of new ideas with new vehicles "on stream" at all times for evaluation, but not necessarily scheduled for production. This makes it easier for management to choose and schedule models for production a year or two ahead of the normal planned schedule. Then, if an outstanding vehicle hits a year when the economy is also great (and maybe the competition is weak), it might be desirable to continue it into a longer run or just incorporate a "face lift".

Even if this is done it would be a mistake for management to cut back on efforts in Advanced Vehicles Development. Conditions in the economy, market forces, company profits, and the competition can change drastically in a matter of a few months. Then there will not be the usual long lead time needed for the development of a new vehicle. Management should have designs well along at all times for any eventuality. If the competition meets the new conditions quicker, another year may be lost and the corporation's financial strength and resources are weakened further. Sometimes it can become fatal if the dealerships also abandon the corporation.

What happens in Advanced Vehicles Development is always an indication of what most likely will occur in the corporation downstream a few years hence.

W-60

Chapter VI

ADVANCED VEHICLES DEVELOPMENT – PERFORMANCE CALCULATIONS (3)

Top speed, gradeability, acceleration, and fuel consumption are the primary performance calculations and will be discussed here. Essentially we are dealing with two factors (1) the force required to push (or pull) the vehicle, and (2) velocity (V). The force required to push or pull the vehicle is the sum of (a) rolling resistance, (b) wind resistance, (c) powertrain losses between engine and drive wheels, (d) if on a grade the additional force needed to overcome the grade, (c) if accelerating, the additional force needed to accelerate at the required rate. Sea level and "standard" air pressure and temperature are assumed, along with still air and a smooth hard level road surface.

EXAMPLE – SMALL SPORTS CAR RACER

Chapter V, figure 12 depicts the small racer we will use for our example. Its weight is low, its frontal area is minimal, and its drag coef. is quite low. The G.T. version of figure 11 would be heavier, have greater frontal area and a higher drag coef. so would have greater power requirements for equal performance. An electric would be entirely different with much greater weight due to batteries and corresponding reduced performance, although frontal area and drag coef. might be similar to the G.T. vehicle.

ROLLING RESISTANCE

Rolling resistance varies considerably based on the type of tire, new or worn, air pressure, miles of warm-up to stabilize the tire materials, and road surface as explained in chapter XII and reference 68.

Dr. Hoerner, in his book, reference 7, provides this formula which I have used for bias ply tires and believe to be good.

$$\frac{\text{lb. resistance}}{\text{100 lb. load}} = 0.5 + \frac{15}{P} + \frac{0.0035V^2}{P}$$

V is velocity in mph. P is tire pressure in pounds per sq. inch (PSI)

For convenience I have prepared a chart, figure 18, based on 35 psi and for a vehicle weight of 1,000 pounds. Any other weight can be used with a multiple to obtain rolling resistance.

In our example we will use the data provided by reference 68 in chapter XII. Table I provides the calculations and figure 19 shows these values plotted on the lower curve.

WIND RESISTANCE OF SHAPE – FRONTAL AREA AND DRAG COEF.

Determining wind resistance is a bit more complicated. First we must determine frontal area. Figure 20 is a scale drawing of half of the body at the maximum cross section including that portion of the wheel exposed below the body. Each square equals one square inch, so we need only count the squares within the contours. The areas are tabulated for separate portions as indicated, totalled, and converted to square feet.

The drag coef. was determined from wind tunnel data provided by the Aerophysics Department, Mississippi State University, and discussed in detail in chapter XV – Automotive Aerodynamics. The shape, as tested, had an estimated drag coef. of 0.2060. With certain improvements this could be brought down to about 0.1794, it was estimated. Since the body molds have already been completed not all the changes could be made. Additionally, we must increase the value to include internal wind resistance for engine cooling, ducting, air inlet, and exhaust system.

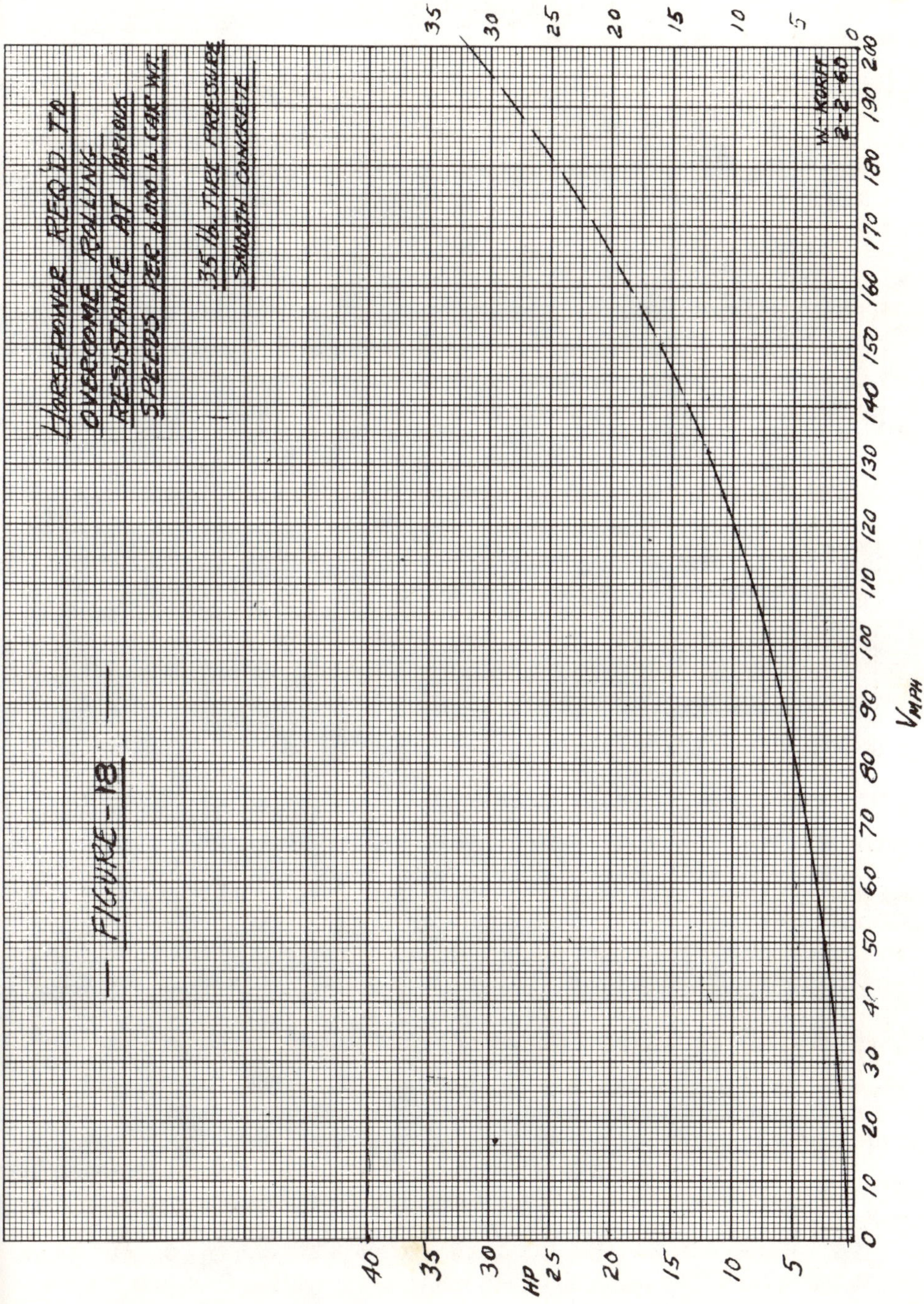
— FIGURE-18 —
HORSEPOWER REQ'D. TO
OVERCOME ROLLING
RESISTANCE AT VARIOUS
SPEEDS PER 1,000 lb. CAR WT.
35 lb. TIRE PRESSURE
SMOOTH CONCRETE
W. KORFF
2-2-60
HP
V MPH
0 10 20 30 40 50 60 70 80 90 100 110 120 130 140 150 160 170 180 190 200
0 5 10 15 20 25 30 35 40

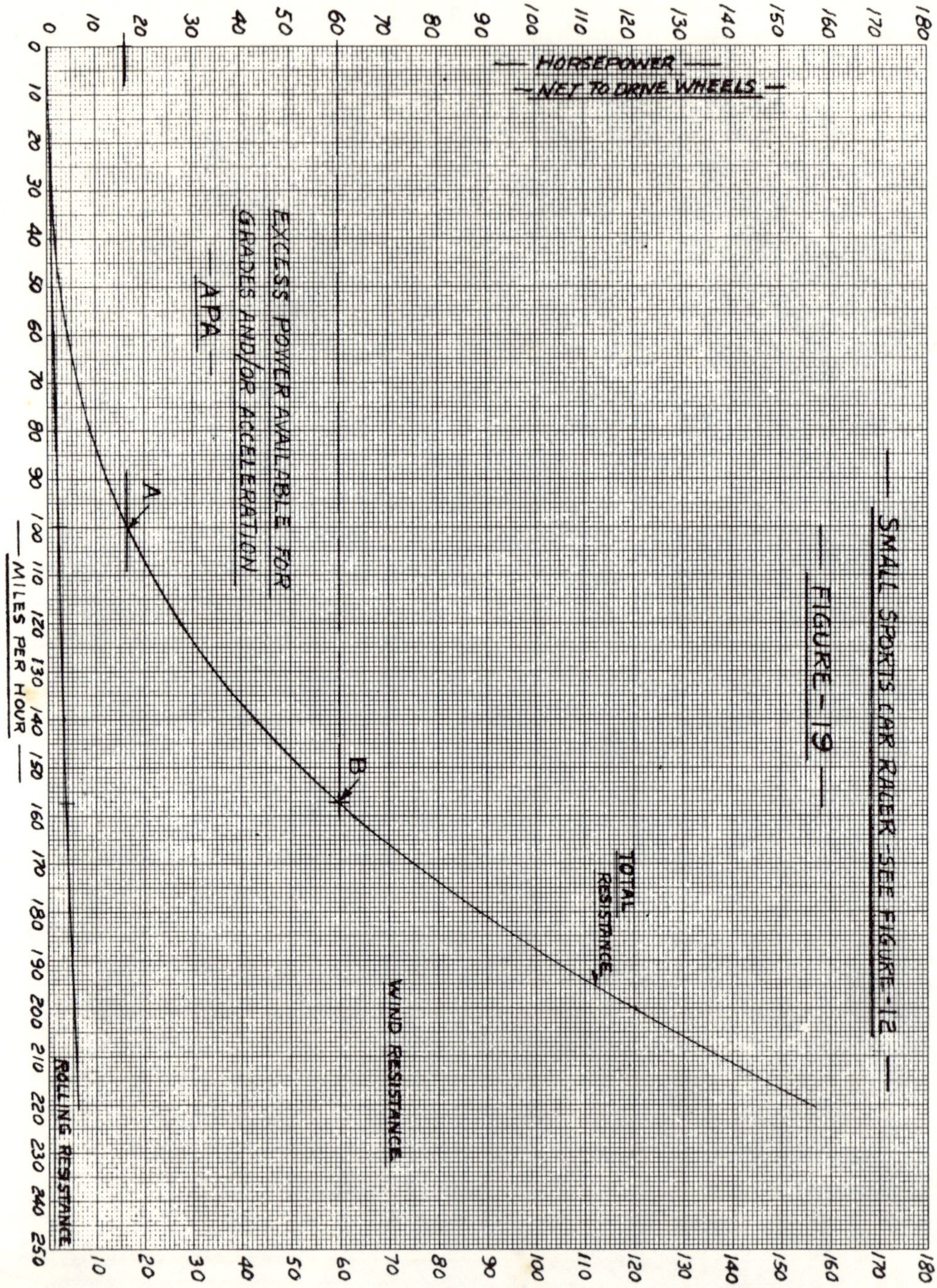
SMALL SPORTS CAR RACER - SEE FIGURE-12
FIGURE-19
HORSEPOWER
NET TO DRIVE WHEELS
EXCESS POWER AVAILABLE FOR GRADES AND/OR ACCELERATION
APA
A
B
TOTAL RESISTANCE
WIND RESISTANCE
ROLLING RESISTANCE
MILES PER HOUR
0 10 20 30 40 50 60 70 80 90 100 110 120 130 140 150 160 170 180 190 200 210 220 230 240 250
0 10 20 30 40 50 60 70 80 90 100 110 120 130 140 150 160 170 180

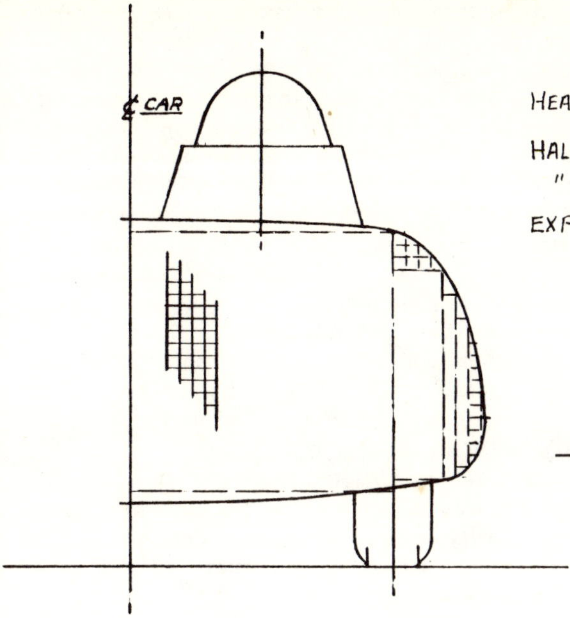

— AREAS —

HEADREST & WINDSCREEN	-	135 SQ. IN.
HALF BODY - (SHOWN)	-	587 " "
" " (NOT SHOWN)	-	587 " "
EXPOSED TIRE (SHOWN)	-	38 " "
" (NOT SHOWN)	-	38 " "
TOTAL	-	1,385 SQ. IN.
EQUAL	-	9.62 SQ. FT.

– FRONTAL AREA – RACER –
– FIGURE 20 –

TABLE I

CALCULATED ROLLING RESISTANCE OF TIRES

VEHICLE Small Sports Car Racer – Figure E-8

ASSUME
(1) Radial tires – 32 psi – fully warmed up – new.
(2) Smooth roadway.
(3) Gross Vehicle Weight (GVW) = 840 lb. (chapter XVI)
(4) Ref. L-1, chapter XII provides coef. of R.R. to 60 mph – extrapolate beyond.

FORMULA (A) $C_f \times GVW = R.R.$ in lb. **(B)** $\frac{lb. \times V}{375} = HP$

EXAMPLE – 40 mph .0080 x 840 = 6.720 $\frac{6.720 \times 40}{375} = 0.7168$ HP

V–mph		C_f		Pounds		Horsepower
40	–	.0080	–	6.720	–	0.7168
60	–	.0084	–	7.090	–	1.1343
80	–	.0088	–	7.392	–	1.5769
100	–	.0092	–	7.728	–	2.0608
120	–	.0096	–	8.064	–	2.5805
140	–	.0100	–	8.400	–	3.1360
160	–	.0104	–	8.736	–	3.7274
180	–	.0108	–	9.072	–	4.3546
200	–	.0112	–	9.408	–	5.0176

Note: These HP values are plotted on the lower curve of figure 19.

As a first approximation and without detail redesign and testing, we assume a drag coef. of 0.22 can be achieved with everything included. Table II provides the formula for obtaining the wind resistance in pounds and horsepower for any velocity (V), the calculations from 40 to 200 mph, and the basic data to be used in the formula. These values were then added to those for rolling resistance to obtain total resistance in horsepower as shown in figure 19. This is the power required at the drive wheels on a smooth, hard, level roadway with no wind in any direction.

The drive train behind the crankshaft includes transmission gears, propeller shaft, differential gears, half shafts to the wheels, bearings, seals, and lubricant at each unit and might easily account for 10% power loss. This loss must be included to obtain the power required at the crankshaft for any chosen velocity (V) in mph.

To get up to this speed requires more power than to maintain it once there. Further, the roadway is rarely ideal for long stretches and completely windless conditions occur briefly at certain times of the day in test areas. For this reason it is practical to either increase the horsepower requirements a bit, or lower the estimated maximum speed slightly for so-called "average" or typical conditions.

In most parts of the country a 15 mph wind is considered about average. Therefore, if driving into it, it becomes a head wind and we can figure the additional drag directly. If it is a side wind the resultant of forward velocity and side wind can be determined. This quartering wind may increase the drag coef. at least 10% for good shapes, and much more if the body shape has sharp edges to disturb air flow.

Note that tailwinds, which help rather than hinder, occur only 25% of the time because they are in the aft quadrant only.

As data comes in on design changes, both improvements and losses can be re-evaluated for more accurate calculations.

TABLE II
CALCULATED WIND RESISTANCE OF SHAPE

VEHICLE Small Sports Car Racer – Figure E-8
FRONTAL AREA – 9.62 sq. ft. (from figure F-1)
COEF. OF DRAG (C_d) – 0.22
FAC_d = 9.62 x 0.22 = 2.1164

FORMULA

(A) $\frac{FAC_d \times V^2}{391}$ = lb. drag at V **(B)** $\frac{\text{lb.} \times V}{375}$ = HP req'd at V

EXAMPLE (A) $\frac{2.1164 \times 1600}{391}$ = 8.66 lb. at 40 V **(B)** $\frac{8.66 \times 40}{375}$ = 0.9237 HP at 40 V

V(mph)		V^2		Lb. Drag		Horsepower (HP)
40	–	1,600	–	8.6605	–	0.9237
60	–	3,600	–	19.4860	–	3.1178
80	–	6.400	–	34.6418	–	7.3903
100	–	10,000	–	54.1279	–	14.4341
120	–	14,400	–	77.9441	–	24.9421
140	–	19,600	–	106.0906	–	39.6072
160	–	25,600	–	138.5674	–	59.12207
180	–	32,400	–	175.3740	–	84.1797
200	–	40,000	–	216.5120	–	115.4728

EXCESS POWER AVAILABLE FOR GRADES AND ACCELERATION

To show how this works let's assume our racer uses a motorcycle engine of 66.6 H.P. First, we deduct 10% for drive train losses to net 60 H.P. at the wheels. Referring to figure 19 the absolute terminal maximum speed would then be at "B" and comes out 157.5 mph. Since it requires more power to get up to speed than to maintain it, this value could only be reached if we had a slight downhill run at it (or a tailwind) before stabilizing out on the level with no wind. Actually, about 150 mph might be a better or more practical value. Then at this point "B" or thereabouts, there is no more power available for grades and/or acceleration.

Now, let's assume we are at point "A" on the curve with a speed along the straight of 100 mph. At point "A" on the curve only 16.5 H.P. is required for wind resistance and rolling resistance. The remaining 43.5 H.P. is available for grades and acceleration. Of course, we don't always have exactly a maximum of 60 NET H.P. on tap unless an automatic transmission keeps the revs. up or the driver has a close ratio transmission of about 5 speeds or more and is always in the proper gear. For our purposes now we will assume this is true. With 43.5 H.P. available for grades and acceleration, let's now move on to how these are calculated.

GRADEABILITY

This is essentially the additional power required to *lift* the vehicle at a specified speed. This is equal to the percent grade multiplied by the gross vehicle weight (GVW) divided by 100. For example, our GVW is 840 lb. A 10% grade will increase the "push" required by 84 lb. Therefore, to convert this "pounds push" to horsepower, we again use the formula (B) from Table I. This is:

$$\frac{\text{lb.} \times \text{V}}{375} = \text{HP} \qquad \frac{84 \times 100}{375} = 22.4 \text{ HP at 100 mph.}$$

Clearly it takes more power (at 100 mph) to climb a 10% grade than to just maintain the speed on level ground. This shows the importance of keeping the weight low. It also shows why a heavily loaded truck must slow down greatly to negotiate long grades.

Our little racer could either climb a steeper grade at 100 mph, or it could accelerate while climbing this 10% grade.

If the maximum grade a truck or car could climb at a crawl, with no wind resistance to be included is to be calculated, this formula from reference 8 is useful.

$$G = \frac{T \times r \times e}{R \times W} - 1.5\%$$

T is maximum engine torque in lb./ft.; r, the total reduction ratio between crankshaft and drivewheels; e, efficiency of transmission in percent; R, effective drivewheel radius in feet; and W the gross vehicle weight (GVW) in pounds. The 1.5 percent represents the grade equivalent of the rolling resistance. G is gradeability in percent.

This formula can also be used for small industrial vehicles and electrics of low power to determine the maximum grade that can be negotiated. Of course, with sufficient power wheelspin can become the limiting factor. If all wheels are not drivewheels, the grade is then limited by the percentage gross vehicle weight on the drivewheels and the coef. of adhesion of the tires.

Let's now go on to acceleration.

ACCELERATION – TWO DIFFERENT KINDS

As far as the engine is concerned, the added power required to accelerate is equivalent to climbing a grade and we will determine it on this basis.

First let's realize we are accelerating masses in two different ways. Moving the vehicle down the road at an increasing rate is the obvious way – linear. However, additionally we are accelerating in rpms everything that *rotates* such as the crankshaft, flywheel, clutch, transmission gears, propeller shaft, differential gears, half shafts, road wheels and all engine accessories. This, too, takes power.

The equation for linear acceleration in pounds force is:

$$F = \frac{W}{g} a$$

W is the GVW of the vehicle in pounds, g, the constant of gravity (32.16), and a, the acceleration in feet per second. From this it is evident that the force needed to accelerate at a certain rate varies directly with the vehicle weight. However, this equation does nothing about the angular acceleration of the rotating parts. Angular acceleration is accomplished with torque and varies directly with the moment of inertia of the rotating parts.

To determine the various moments of inertia of all rotating parts is quite a job. We are indebted to Mr. P. M. Heldt (ref. 8) for doing just that many years ago. In the case of a small foreign car of 2860 lb. he found the moment of inertia of the rotating parts to be equal to 11.2 percent additional car weight, in direct drive. In another case the value went up to 18.75 percent. An American car had a value of only 7.5 percent. In intermediate gears the ratio is much higher.

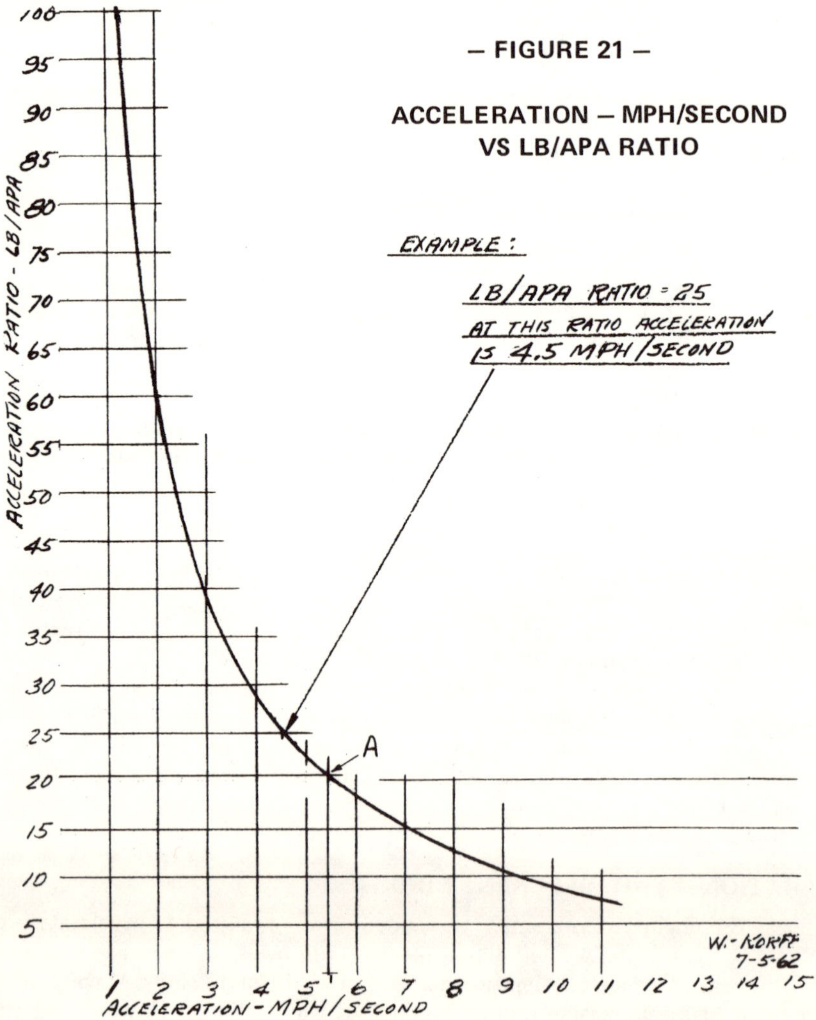

In another example he found the value to be 9% additional car weight in direct drive, 17.8% in second, and about 45% in low gear.

Today's cars have smaller, lighter wheels but larger tires. Some eliminate the propeller shaft by having the engine and transaxle directly connected at either the front or rear of the car. Higher strength gears are loaded heavier so can be smaller for the same power and thus have a bit lower moment of inertia. In any event, the designer can individually determine the moments of inertia of all rotating parts of his vehicle or apply a "fudge factor" to Mr. Heldt's values for a quick assessment of the differences in power trains. I do the latter.

If, for example, the equation for linear acceleration is used on our little vehicle of 840 lbs. the figures might come out like this. The equivalent weight is 840 GVW plus 9% more for accelerating the rotating parts for a new total of 840 + 75.6 = 915.6 lb. If we want an acceleration of 2.5 mph/second this comes to 3.67 feet per second^2. The values in the equation then become $\frac{915.6}{32.16} \times 3.67 = 104.5$ lb.

This is the force on the circumference of the driving wheels and may be converted to torque or horsepower.

As mentioned earlier, acceleration is equivalent to climbing a grade. As Mr. Heldt points out, linear acceleration at the rate of one mile per hour per second calls for the same propelling force and axle torque as climbing a 4.56 percent grade. If we then add for rotating parts the value goes up to about a 5% grade in high gear. In intermediate gear the value is equivalent to a 5.36% grade and in low gear, a 6.6% grade. However, since torque goes up in lower gears the values might be 5% in high (as before), 3.3% in intermediate, and 2.4% grade in low.

Years ago I simplified the problem a bit for my own calculations by constructing a curve as depicted in figure 21 based on the previously mentioned equation for linear acceleration and converting the values to horsepower and mph per second.

ACCELERATION IN MPH/SECOND VS LB/APA RATIO

In this curve we introduce a new value – "acceleration ratio – lb/apa". The pounds (lb) value should be the gross vehicle weight (GVW) plus an additional percentage for the inertia of the rotating parts. APA stands for Available Power for Acceleration. The APA, in horsepower, is obtained from your own chart constructed for your own vehicle similar to figure 19.

In our example, at point A, we have an APA value of 43.5 HP. Assuming a LB weight of 840 + 75.6 = 915.6 lb. – 915.6 divided by 43.5 = 21.05 (LB/APA). This is at "A" on the curve and shows an acceleration of 5.42 mph/sec.

This curve, (figure 21), was used to determine acceleration rates for the GOLDENROD world land speed record car to plot time and distance to see how fast it would be going by the time it reached the measured mile. Chrysler Engineering calculated acceleration in greater detail and with proper power curves for these engines at altitude, corrected efficiencies in each grear ratio, and allowances for shifting time. By comparison the author's simplified approach was optimistic in speed and distance by about 4% for the Bonneville course. This is the only direct comparison of this method with others, so I do not know if this degree of accuracy will hold for other cars. I feel that it is adequate for first approximations at least while in the Advanced Vehicles Development performance calculation stage.

PLOTTING SPEED AND DISTANCE IN SECONDS (FOR PASSING)

In highway travel as well as in racing, it is desirable to know the ability of a vehicle to pass others from certain speeds and not necessarily from a standing start. This can be done with the help of LB/APA ratios and table III.

First, along a straight line, mark off equal steps to represent seconds, possibly up to twenty seconds. Let's now assume we want to start the plot at 40 mph to find out how far and how fast we can go in twenty seconds. From table III we can show that, at 40 mph we travel 58.8 feet in one second. Going back to our curve for power required (figure 19 in our example) we can find out how much APA we have. Then, with this we determine LB/APA ratio as before and, referring to figure 21, the acceleration in mph/sec. is obtained. One second later the vehicle is going 40 mph plus that rate for a new higher mph. This new higher mph is then checked out on table III to determine the dis-

TABLE III

DISTANCE TRAVELED PER SECOND AT VARIOUS SPEEDS

MPH	Distance traveled in 1 sec.	MPH	Distance traveled in 1 sec.	MPH	Distance traveled in 1 sec.	MPH	Distance traveled in 1 sec.
40	58.8	81	119.1	122	179.0	163	239.7
41	60.3	82	120.6	123	180.5	164	241.2
42	61.8	83	122.1	124	182.0	165	242.7
43	63.2	84	123.6	125	183.4	166	244.2
44	64.7	85	125.1	126	184.8	167	245.7
45	66.2	86	126.4	127	186.3	168	247.2
46	67.7	87	128.1	128	187.8	169	248.7
47	69.1	88	129.4	129	189.3	170	250.2
48	70.7	89	131.1	130	190.8	171	251.7
49	72.2	90	132.4	131	192.2	172	253.2
50	73.4	91	134.1	132	193.6	173	254.7
51	74.9	92	135.4	133	195.1	174	256.2
52	76.4	93	137.1	134	196.6	175	257.7
53	77.8	94	138.4	135	198.1	176	259.2
54	78.1	95	139.8	136	199.6	177	260.7
55	80.6	96	141.3	137	201.1	178	262.2
56	82.1	97	142.6	138	202.8	179	263.7
57	83.5	98	143.7	139	204.1	180	265.2
58	85.0	99	145.2	140	205.6	181	266.7
59	86.5	100	146.7	141	207.0	182	268.2
60	88.0	101	148.2	142	208.4	183	269.7
61	89.5	102	149.7	143	210.0	184	271.2
62	91.0	103	151.2	144	211.4	185	272.7
63	92.4	104	152.7	145	213.0	186	274.2
64	93.9	105	154.2	146	214.4	187	275.7
65	95.4	106	155.6	147	216.0	188	276.4
66	96.8	107	157.2	148	217.4	189	278.7
67	98.3	108	158.6	149	218.0	190	279.6
68	99.8	109	160.2	150	220.4	191	281.7
69	101.4	110	161.4	151	221.9	192	282.6
70	102.7	111	163.2	152	223.4	193	283.4
71	104.2	112	164.6	153	224.9	194	285.2
72	105.7	113	165.1	154	226.4	195	286.4
73	107.2	114	166.7	155	227.9	196	287.6
74	108.7	115	168.3	156	229.4	197	288.9
75	110.2	116	170.0	157	230.8	198	291.0
76	111.7	117	171.5	158	232.2	199	292.2
77	113.2	118	173.0	159	233.7	200	293.4
78	114.7	119	174.5	160	235.2		
79	116.1	120	176.0	161	236.7		
80	117.6	121	177.5	162	238.2		

tance traveled in the second interval of one second and tabulated on a separate sheet of paper. These steps are repeated for each second for as long a time or distance desired. At any point it is possible to also see how fast the vehicle is going.

HORSEPOWER CORRECTIONS FOR TEMPERATURE AND ALTITUDE

Of course, our vehicle is not operating at laboratory conditions. (S.A.E. Test code J816 b) Under hood temperatures are usually higher than 85 degrees and many areas are more than 500 feet above sea level. So our engine is really producing something less than its rated horsepower even if perfectly tuned.

Existing cars, almost without exception, permit the carburetor to suck in hot turbulent air from around the engine. The fuel/air ratio is determined by weight: weight of fuel and weight of air. Hot air is lighter so at higher inlet air temperatures the mixture is air-starved and power drops off. How much? One percent with every 10 degrees F rise in temperature. It is entirely possible that under hood temperatures reach 185 degrees F and more in certain air cooled engines with the carburetor quite near the hot cylinders and with only a slight movement of air or a baffle at best. This is 100 degrees warmer than standard outside air at 85 degrees F. This temperature rise therefore causes a loss of power amounting to 10%. It obviously pays to direct cool air to the carburetor air inlet or air cleaner.

Altitude causes a power loss of 3½% for each 1,000 feet. The Bonneville Salt Flats are at 4,228 feet, so account for a loss of 13.05% power. Air resistance is also less as discussed under aerodynamics, but not enough to compensate for power loss.

Many race courses are at about 1,000 feet, so you can expect a loss of 1¾% quite normally. Mountain driving in the West is frequent and the altitude is often six to eight thousand feet, thus losing 19% to 26% in addition to losses due to under hood temperatures.

Air turbulence could account for some starvation and amount to possibly 2 or 3 percent. Picking up inlet air in a slightly pressurized area and smoothing the air flow would be helpful.

All these losses are more serious than at first realized. While the reduction in power is measured as a percentage of maximum engine power, the loss comes directly off the *excess* power available for grades and acceleration so in reality gradeability and acceleration suffer much more.

FUEL ECONOMY

With rising fuel prices and CAFE economy ratings, everyone is now interested in fuel economy. By plotting power required at chosen speeds, choice of gear ratios, and with excellent curves of fuel consumption for a given engine, it is possible to predict fuel usage fairly accurately. Steady state (no grades, no acceleration), reasonable speed on the low side with higher tire pressure, will yield the lower power required figure, and then if the gearing is "tall' for low engine rpm and more efficient fuel consumption (like an overdrive), excellent results are obtained. Grades and power to pass can then best be handled by downshifting along with more fuel.

100 (OR MORE) MILES PER GALLON ON THE FREEWAY?

People formerly asked, "Wot'l she do?" meaning top speed. Nowadays this most likely refers to miles per gallon. "100" seems to be a magic number and all sorts of schemes and inventions show up to reach it. There really isn't any mystery about it, just efficiency, or higher efficiency.

A G.T. as shown in our figure 11 is about as small, lightweight, and of minimum drag as is reasonably practical for two people. At a freeway speed of, say 60 mph, it may only require about 6 horsepower. Then, with overdrive, or otherwise "tall" gearing for improved engine operation, fuel consumption can be optimized. An efficient small gasoline engine, properly tuned, might then achieve around 80 miles per gallon.

If a small diesel is used, miles per gallon would improve some 25% to the magic 100 miles per gallon. With an "ADIABATIC" diesel as described in chapter X, the figure would go even higher! These values assume "steady state" driving only. Grades and passing require downshifting or preferably turbo charging for a burst of more power when needed.

From an investment standpoint such a high mileage vehicle is quite desirable because it will easily repay the initial cost of the car in savings on monthly fuel bills over a period of years.

Such a vehicle could be quite a challenge to the electrics planned for the mid 1980's.

ELECTRONIC TRIP COMPUTOR'S FUEL ECONOMY FIGURES

Mr. Robt. W. Irvin, Associate Publisher/Editor of AUTOWEEK, reference 9, drove a Cadillac Seville over 300 miles in Michigan and reported the following fuel economy read-outs on the car's electronic trip computer.

1) – When you start the engine on a cold day and drive off, the fuel economy might be only 5 mpg.

2) – As the engine temperature climbs to 180, the fuel economy improves to about 14 mpg on the highway.

3) – However, it is possible with this gauge to show the car developing 50 mpg. All you have to do is get behind a semi-tractor-trailer and draft him. It was no big deal to get 35 mpg this way, and even the 50 mpg figure was possible, if you wanted to drive like Richard Petty at Daytona.

From Mr. Irvin's experience two things clearly stand out.

a) – A cold engine and cold tires, as with short stop-and-go city driving, yields very poor mileage. This is the logical best use for electric vehicles or very small engined urban cars.

b) – Wind resistance at highway speeds uses up a lot of fuel. Major improvements in automotive aerodynamics can greatly improve fuel economy on the highway.

Chapter VII

DETAIL DESIGN FOR PRODUCTION

The DETAIL DESIGN FOR PRODUCTION GROUP follows the original concept as delineated by the ADVANCED VEHICLES DEVELOPMENT GROUP. No matter how great the concept we still haven't got a vehicle ready for production until *all* the details of the design are worked out *on drawings.*

As noted earlier it is possible to build an experimental vehicle or for that matter any "one off" vehicle with very few drawings. Parts are made without special tools at great expense and are hand fitted without assembly fixtures. All this requires many manhours but is still less expensive if only one or possibly just a few vehicles are to be built.

Mass production is an entirely different story. To better understand this, let's now define the different stages of vehicle design from concept to production.

EXPERIMENTAL – PROTOTYPE – AND PRODUCTION STAGES EXPLAINED

The first step in developing a new concept is to experiment. The experimental stage is carried out by the Advanced Vehicles Development Group. Experimental means, loosely speaking, exactly what the word indicates. We are experimenting, trying something different. Therefore, it may be assumed we do not know the outcome for certain. Otherwise it would not be an experiment. We must be prepared for failures and make changes until we have a machine that accomplishes its intended purposes. When the experimental vehicle is satisfactory to management, we are ready for the next stage – prototype design.

PROTOTYPE means the first model or example for later production articles. The prototype may look like the experimental vehicle and may perform like it, but there the similarity ends. The experimental version may have parts that are marginal and changed so often that they should be redesigned. Some materials may not be adequate for mass production because of difficulties in fabrication as learned in the experimental vehicle. The purpose of the prototype design is to overcome these problems before production begins. More changes must be made. But we are no longer experimenting. We are eliminating all aspects of the design that would delay or interfere with production or be too costly. Once prototyping is finished, we have a vehicle that should not require any more changes, or at least only very minor changes that can be incorporated with slight or no risk during the production stage.

The PROTOTYPE DESIGN stage may begin in AVD (Advanced Vehicles Development) and may even be completed there, or it may be started and completed in DETAIL DESIGN FOR PRODUCTION. Often it begins in AVD and makes a smooth transition along with assigned key designers to DETAIL DESIGN FOR PRODUCTION when management so authorizes the production stage.

PRODUCTION means the development stages are completed. It is no longer an experiment and the prototype now shows us how it will be when mass produced. Officially, management authorizes the budget and line design engineers to establish a full blown DETAIL DESIGN FOR PRODUCTION GROUP for this project.

HOW MASS PRODUCTION IS ACCOMPLISHED FROM DETAIL DRAWINGS

This is when the inexperienced get into trouble. They usually underestimate what is required to get into production. They often think they can still experiment a little in this stage, or, due to a tight budget and limited time, possibly skip over the prototype stage. To do so may bankrupt a company. Let's see how mass production is accomplished.

Excellent tooling is an absolute requirement, to reduce the labor content of manufacturing, to achieve a competitive selling price, and to manufacture at a rapid pace within limited space. Tooling cannot be designed without detailed drawings of each part of the vehicle. Actually, the production shop does not work from engineering drawings as does the experimental shop. Production employees work only with specific items from previous departments and with assigned tools to accomplish certain operations, according to the Manufacturing Planning Department's "operation sheet" for that "station".

The "operation sheet" spells out in exact detail, step by step, precisely what should be done by the employee at that station. Thousands of "operation sheets" are needed to proceed from cutting raw materials, through all processing, fabrication, sub-assembly, testing, and final assembly. Yet no "operation sheet" can be planned by a manufacturing planner until he has a detailed engineering drawing of the part with a materials list, the specifications of the materials, processing required, and tolerances of all dimensions. Then, and only then, can the manufacturing plan proceed. The planner can then decide what tools, both standard and specially designed, are needed and can inform the tool designers. The planner can then also begin the preparation of the "operation sheets" and the sequence of assembly. All this requires a tremendous amount of coordination. It all begins with the detail drawings of each part from the engineering department. And remember, if anything is changed, the whole process must be carefully reviewed to see what was affected. Always expect a "domino effect."

HOW MANY ENGINEERING DRAWINGS ARE NEEDED?

It has been estimated that a typical so-called full size car has approximately 15,000 parts. Each part must have a drawing. Next, there must be assembly drawings and installation drawings showing how the parts fit together. In order to make some of these drawings, the designers must make "layouts" and other drawings to investigate the effect of tolerance extremes (stack layouts) during manufacture to see if the parts can still be made to fit together when tolerances are at the limits, instead of at the mid point. As the project progresses changes start to creep in. So, much of the previous work must be done over.

As can be seen, the "detail design for production" stage of a new vehicle requires a lot of engineers and draftsmen (or drafters) to develop the detail design of each part. Actually, the drafting time required to make the final drawing that is used is a small part of the total engineering effort required. In the aerospace industry this may be as low as 5% for a new complicated design. For automobiles it may be about 10% – rarely more than 25%. The remaining time goes for investigations, creative thinking, proposed layouts, alternative designs, and coordination. This makes it difficult to always come up with good engineering estimates of how many manhours are needed for a new project. It also points up the need for experienced, qualified designers. Motor vehicle design is approaching the sophistication of aerospace design in many respects. What makes the problem even more misunderstood is that the final solution often must be reduced from a complicated design, through several steps to simple, but effective design because in technology – SIMPLICITY IS THE ULTIMATE SOPHISTICATION!

Remember also, there are many constraints on a design. Briefly these are:

1) – Target weight (maximum) for CAFE fuel economy.
2) – Federal emissions and safety requirements.
3) – Function, performance, and style within the original concept.
4) – Target cost for each part and for the final vehicle.

After the parts are designed for production, requiring many months of work, the detail designs must be carefully coordinated with the engineers in the Manufacturing Division and the buyers in the Procurement Division. When mass production is very high, as for example, a million vehicles per year of a specific model, then production is essentially a continuous process instead of a batch process. Maximum economies are then possible.

Tool set-ups are infrequent. Everything can be planned thoroughly with complete tooling at every step.

EFFICIENT LOW PRODUCTION – THE INTERMEDIATE STEP

Only the largest automobile manufacturers really reach the high quantities involved in continuous production and even then not for all models.

Designing a vehicle for efficient low production is in many ways a greater challenge. The designer must know more about the manufacturing process to reduce costs through proper design for a low total quantity of vehicles. Since parts will be fabricated on the batch process, tools will be set-up and removed in stamping presses and machining operations more often, and this increases costs. Since nearly every part will be made by running off a certain quantity, then stopping to change tools to make a different part (the batch process), costs to make each part are higher to include the set-up costs. Then that quantity of parts must be moved to storage until used in the next step.

This moving to storage requires effort and additional cost. The storage area requires additional space and inventory control at further cost. Then, when that batch of parts is to be used it must be moved on to the next station, again with a cost for moving. When the designer understands all this, he then sees the importance of designs with fewer parts to be manufactured by his company. He also learns that there is merit in using many parts from other models, and in obtaining high production parts from outside sources. In many ways he is finding out about the advantages of an "assembled product" to reduce cost.

It takes great skill to design a product that looks and is completely new and advanced, yet low in cost and thoroughly reliable through the use of many existing components and standard parts available from outside sources. Don't "re-invent the wheel" is still good advice.

HOW MANY DETAIL DRAWINGS FOR A LOW PRODUCTION, SIMPLE URBAN VEHICLE?

Again, this needs additional definition. How low is the production and how simple is the vehicle? Answers vary greatly and only specific quantities and specific designs provide decent answers. To give some indication the following are from the author's design files.

THE MINI-K THREE-WHEELER

This small three-wheeler was designed for low production and to be sold in KIT form. It is no longer available. Its simplicity and general appearance are evident in figure 22. An exploded view of the chassis shows the parts by number in figures 23 and 24. To adequately define in detail all chassis parts for manufacture, along with assembly drawings to show how they go together, required approximately 250 drawings. The home builder used the assembly drawings only. This does not include any drawings of the engine, transmission, shock/springs, electrical system including lights, horn, switches, and instruments. The body is a relatively simple, box-like design without doors, windshield, and top, and is of wood construction for the convenience of the home-builder. About 25 drawings were needed to define all parts of the body. Needless to say, changes and alternative designs represented some of the total number of chassis drawings. This is unavoidable in any project. Now for some more complicated designs.

A SMALL, SIMPLE, OPEN ROADSTER

This vehicle would use the chassis for the racer in figure 12 (see page 36), with rudimentary fenders, a windshield, soft top, simple doors and the minimum number of body parts. About 300 detail, assembly, and installation drawings are needed for the chassis and about 50 more for the body are required to reasonably define this vehicle's parts for low production manufacture. Again, this does not include engine or other purchased parts. The body remains quite simple, although it has more parts than the MINI-K.

MINI-K THREE-WHEELER

— FIGURE 22 —

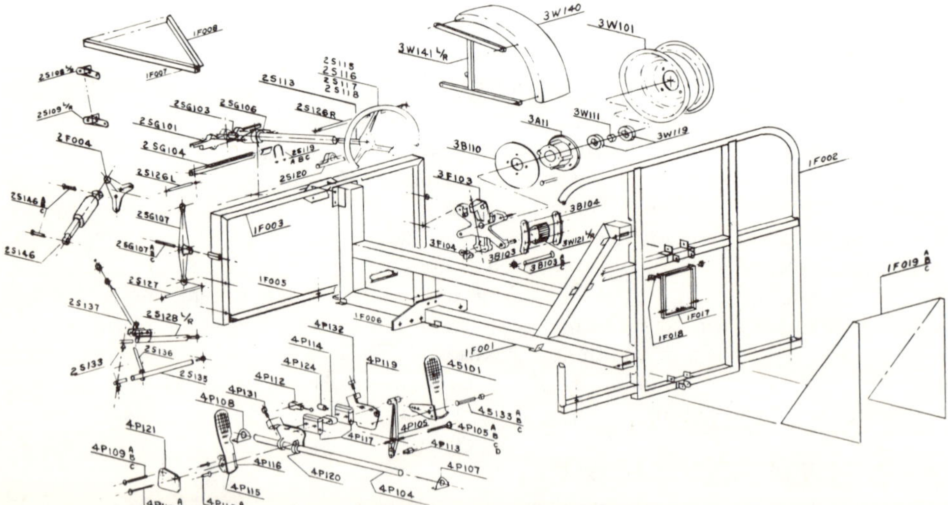

EXPLODED VIEW
CHASSIS FORWARD – MINI-K

— FIGURE 23 —

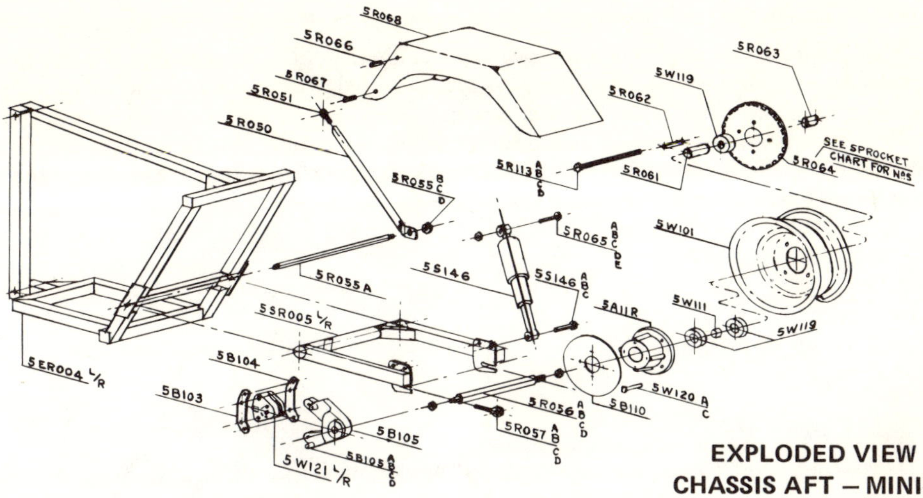

EXPLODED VIEW
CHASSIS AFT – MINI-K

– FIGURE 24 –

THE OPEN SPORTS CAR RACER

This design, another feasibility study in chapter VI, is shown in figures 12, 13, and 14. The chassis, mentioned above, fitted with streamlined body, was designed and a model of the body tested in the wind tunnel as described in the chapter on aerodynamics. Full size loft lines of the body were made. Clearances at wheel wells with tire bump and rebound were investigated. Body attachment points to the chassis were detailed in. Windscreen and fairing behind the roll bar were designed also. All this and many more items were investigated in full size layouts. All together, about 75 to 100 drawings were made, but the effort was quite time-consuming.

THE SPORTS CAR - GT (based on the open sports car racer)

This feasibility study is shown in figure 11 of chapter V. This is a further refinement of the open sports car racer. Bumpers, doors, windshield, roof, and the interior design must be added. Another hundred or more drawings would be required in addition to those already prepared for the open sports car racer. Inexperienced designers lacking "know-how" in body design could easily develop twice as many drawings, attempting to achieve a design for production.

Larger and more sophisticated vehicles require many more drawings. This brings us back to the functional chart of chapter II. Note that DETAIL DESIGN FOR PRODUCTION is divided into three groups as follows.

THE THREE BASIC GROUPS IN A DETAIL DESIGN FOR PRODUCTION PROJECT

A – POWERPLANT INSTALLATION

Even though the engine, transmission, and accessories are purchased there remains a great deal of design work in developing the installation for a specific chassis. Radiator size and location, exhaust routing, instruments and controls, clearance problems in the frame and at the steering gear, and hood line clearance are but a few. Engine mounts for vibration control, proper gear ratios and driveline details are a few more. Complicated emissions controls at the carburetor and in the exhaust system add more problems. If more than one engine size or type is to be offered as an option, then it all has to be done again. Optional accessories such as turbo-charging further complicates the installation. A specialized group of designers are needed with the know-how for this aspect of vehicle design. They work only on the models of the assigned project, but seek and obtain assistance and advice from staff powerplant engineers and the testing laboratories.

B – CHASSIS DETAIL DESIGN

This group of designers are also specialists in their field. They develop the design of the suspension system, frame, steering, brakes, and final drive train, using as many standardized units as possible. They must coordinate their efforts with both the powerplant installation group and the body detail design group to resolve interferences and attachment points. Deflections under road shocks and vibration from the power train require a great deal of attention to keep them isolated from the body and its occupants.

Optional equipment for the chassis will include heavy duty suspension, brake options, and tire sizes and types.

Federal safety requirements involve the bumpers, front and rear, and other elements of the chassis.

This group of designers also receive assistance from staff specialists in frame design, suspensions, and from the testing laboratories.

C – BODY DESIGN

This group is often the largest because it works on several different body types for the same chassis. These would include two doors, four doors, hatchback, notchback, and a variety of interiors from standard to deluxe GT sports designs with bucket seats.

These designers also seek and obtain a great deal of assistance from staff groups such as aerodynamics, production design, weight, structures, stress analysis, materials and processes, and the testing laboratories. Design details of vehicle elements are discussed in these other chapters under the appropriate headings.

The BODY DETAIL DESIGN GROUP also works very closely with the designers and stylists in ADVANCE VEHICLES DEVELOPMENT, to be certain the original concept is not lost in productionizing the design. The mock-up and the experimental vehicle are referred to continuously.

Body designers must also coordinate their efforts with the chassis and powerplant designers on matters of interference, controls, instruments, attachments, access, and whatever else becomes a problem.

Manufacturing planners and tool designers also have questions and coordination meetings with them are needed.

OTHER PROJECTS

The DETAIL DESIGN FOR PRODUCTION function is not limited to a single project in a large company. Each vehicle model name has a separate project. To balance out the engineering manpower, these different projects are usually staggered a bit so their peak manpower requirements do not all occur at the same time. Designers may then be transferred from a finishing project to a beginning project – often designing the same portions of each vehicle.

HOW THE COMPUTER SERVES ENGINEERING AND MANUFACTURING – TOGETHER

NC – ADM – CG – CAD – CIM – DELPHI METHOD. Sounds a bit intimidating? These are only a few of the initials and names used in this business. The list is almost endless and continues to grow year after year. What do they mean and what do they do? We will find out shortly, but first let's see how the job gets done without these aids and what the problems are.

In drafting a curved line such as an auto body contour or an airfoil section for a wing on an airplane, the draftsman must obtain a smoothly blended curve without humps or flat spots. The curve or loft line is made full size on a large sheet of aluminum painted white and laid out in grid lines ten inches square in an air conditioned room, so that tolerances due to temperature are not introduced. Descriptive geometry is used to develop designs three dimensionally with sections as needed. When the "lines" are complete they must each be copied accurately and dimensioned for use in designing tooling and later in making the tools and eventually the part. At each step slight errors or tolerances in measurement can and do occur. This then requires more time-consuming coordination with engineering and some compromising and further blending to achieve good tools and satisfactory manufactured parts. The part is close, but never *exactly* what was drawn in engineering. The need is to improve accuracy and compress time. That is what the computer does in serving engineering and manufacturing in an integrated manner. At first it appeared that the computer would primarily take the drudgery out of drafting. Now we know it also takes much of the drudgery out of tool design, tool making, and part fabrication as well as improve accuracy and compress time.

In the early 1960's the cost of going to computers in this work was very expensive, difficult to incorporate, and the errors of learning made it hardly worthwhile even for a large corporation. The Air Force helped the aircraft/aerospace industry get into it because the long-term benefits were so great. This became the first generation effort. Later, after things got rolling, we moved into the second generation phase with much improved methods and integrated procedures between engineering and manufacturing. Now, the third generation not only consolidated all this, but with further improvements in electronics have cut costs greatly, and we have learned how to use all this in even more applications. It is no longer only for the large companies but is finding uses in smaller companies. Suppliers quickly moved in and we have a "snow-balling" effect that should not be ignored. Now for the meaning of those initials and a bit more of what is accomplished.

NC means Numerical Control. Through NC we can take any curved line for example, and with a few dimensions to define it and with the aid of the computer, develop numerically all the dimensions required, with accuracy, out to as many decimal points as desired. If there is any unevenness, NC can perform a blending or smoothing function to achieve the degree of perfection desired – all very quickly.

All points on the curve can be located three-dimensionally in space relative to any other surface or point – also very quickly. These numbers then may be used also for tool design, tool make, and for the manufacture of parts. There is no error possible in transferring this information from engineering to other departments in the company, because everyone uses exactly the same numbers.

ADM means Automatic Drafting Machine. This machine takes the NC data and makes a drawing with extreme accuracy, to any scale, and will add as many views or sections as desired. And it does it quickly.

CG means Computer Graphics. It looks a bit like a TV set in that it will graphically depict an object on its screen from the computerized data fed into it. In fact, if so programmed, whatever is depicted may be rotated to a different view, enlarge portions for greater detail, or reduce size for adding related parts. The observer/designer may use an electronic wand to add to, erase data, or otherwise alter what is depicted. Simultaneously another observer in the shop or elsewhere, with a similar machine, can see what is happening. They can talk back and forth with their CG machines to discuss changes or to convey relative information of the design depiction. Drawing changes can be made promptly.

CAD means Computer Aided Design. This is a more general term embracing essentially what was mentioned in NC along with ADM and CG machines.

CAM means Computer Aided Manufacturing – also a more general term embracing NC as used throughout manufacturing.

CIM means Computer Integrated Manufacturing in which computer data of all kinds is properly integrated into all departments or functions of manufacturing.

DELPHI METHOD is a systematic technological forecasting technique for obtaining intuitive concensus of a panel of specialists or experts who are unknown to each other. Certain design specialists or engineers may be called upon to participate or to provide inputs. One forecast is that in 1990 fifty percent of the work force on the shop floor in manufacturing will be highly skilled and trained engineers and technicians keeping automated, robotized and computerized plants operating. This came out of the Third Annual Industry Briefing on the Air Force Integrated Computer Aided Manufacturing (ICAM) program as reported in IRON AGE magazine, July 31, 1978.

Students and others interested in the use of computers are urged to take courses and read up on the subject. Excellent books are available, computer leasing services and computer manufacturers provide instructions and practical information so that you may keep up to date. College level courses are taught for those who wish to specialize in this field.

Chapter VIII

HUMAN FACTORS

"Creature comfort" helps a great deal in selling a motor vehicle. The designer must recognize the comfort and convenience of the driver and his passengers is often of more immediate importance than the technical features and specifications of the vehicle. Ease of entry and exit, leg room, head room, a comfortable seat, unobscured visibility, controls within easy reach and easy to operate, nothing to bump a knee on, are just a few of the things that must be considered.

A FULL SIZE MOCK-UP OF THE INTERIOR WITH DOORS AND WINDSHIELD IS A NECESSITY.

A full size mock-up in great detail is the only sure way to obtain a satisfactory arrangement. The mock-up may be made of wood framing, plywood, cardboard in no-load areas, and miscellaneous other materials as needed. It is important that it be strong enough to be usable for seating and it should be adjustable for modifications. It should be tried by different sized people and their comments noted. A mock-up is not a "one-shot" thing. It should be built early in a project, incorporate changes to chassis and body body developments as they occur, and grow in detail refinements. Later it can even be used to work out the ideal placement of instruments, switches, and levers relative to the steering wheel. Door controls and foot pedals with either seat or pedal travel should be included.

A starting point is needed. This is usually the seat bottom and back, with everything else coming in place relative to the seat. Simple drawings are needed to show how the mock-up is to be built in the shop. The following drawings may be used as a starting point. Your application most likely will be slightly different, so do not hesitate to make your own changes as you see fit.

SUGGESTED SEATING ARRANGEMENT – FOUR PASSENGERS

Figure 25 shows in side view how the driver and rear passenger are arranged in a small car. Note that the rear passenger's feet, at least the toes, are placed under the front seat to conserve space, overall length, wheelbase, and weight. The driver is shown as a 97.5 percentile male and the rear passenger a 50 percentile male. Fifty percentile means that, of the male population, half are shorter and half are taller. In the case of the 97.5 percentile male, only 2.5 percent of the male population are taller. All the remaining are shorter. Dimensions of each portion of the body are shown as an aid in working out proportions. There are many excellent references on seating and passenger car systems. References 10, 11, and 12 are noteworthy.

If your designs are for snowmobiles or heavy machinery the operator will be wearing heavier clothing and this must be allowed for. The S.A.E. Handbook, part II, section 41, Construction and Industrial Equipment, is most helpful in working out these arrangements.

BACK-TO-BACK ARRANGEMENT

Certain vehicles such as four-passenger golf cars and in-plant vehicles find the back-to-back seating arrangement advantageous. This is illustrated in figure 26. Often the rear seat can be folded down to provide cargo carrying space. Realize again, these dimensions are only suggested and may be altered for specific applications as dictated by a proper mock-up.

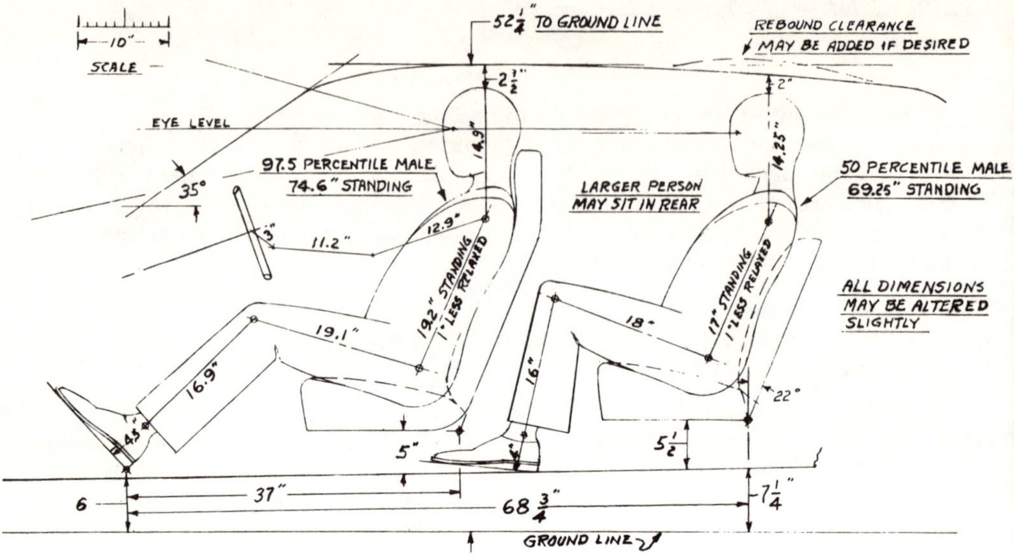

SUGGESTED SEATING ARRANGEMENT – 4-PASSENGERS
– SMALL VEHICLE –

– FIGURE 25 –

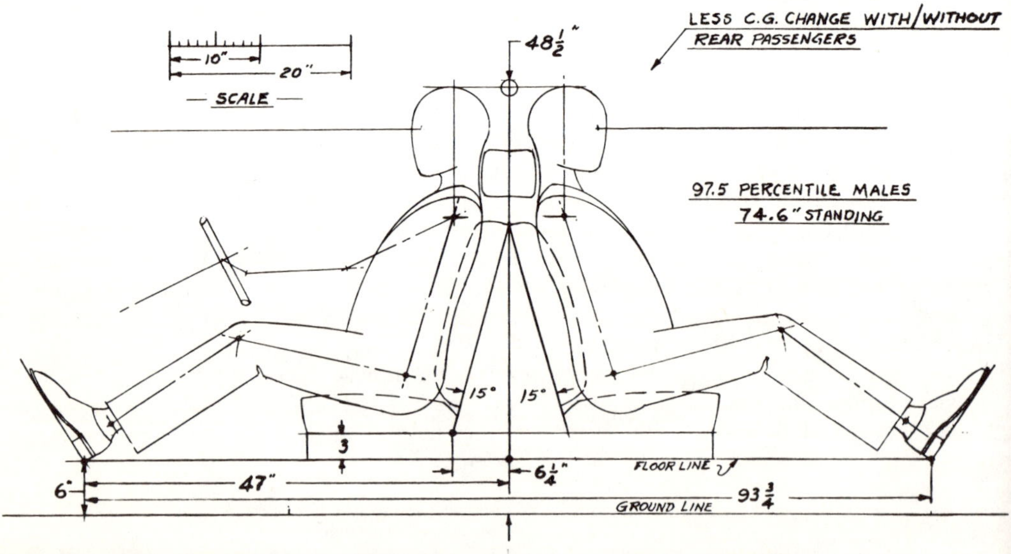

BACK-TO-BACK ARRANGEMENT
– FIGURE 26 –

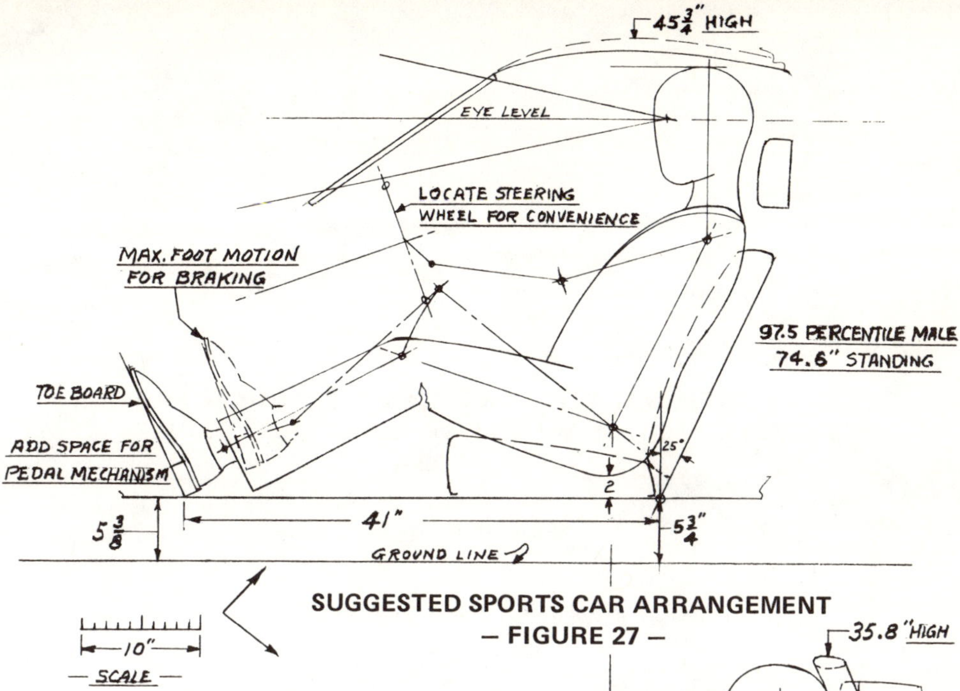

SUGGESTED SPORTS CAR ARRANGEMENT
– FIGURE 27 –

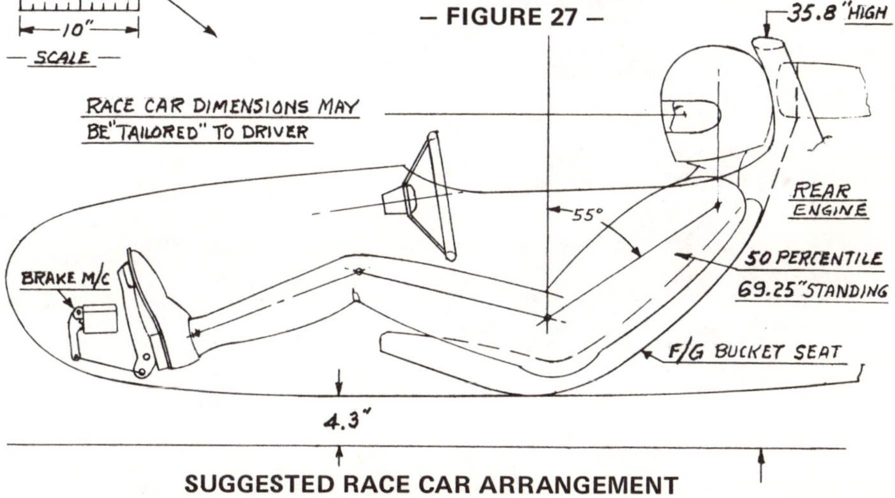

SUGGESTED RACE CAR ARRANGEMENT
– FIGURE 28 –

SUGGESTED SPORTS CAR ARRANGEMENT

Figure 27 provides dimensions for sports car seating. Note the seat back is at a greater angle and the seat bottom is on the floor. The roof line is about as low as it is reasonably convenient to make it. A similar sketch with more details in the front view, such as the centerline frame and body width, is provided in figure 29. Figure 30 shows a cross-sectional view to provide the minimum dimensions for sports car entry and exit. These dimensions, especially the head room, have been reduced in a few cases, but this should never be done without a careful mock-up development. Obviously, sports cars are not for everyone.

SUGGESTED RACE CAR ARRANGEMENT

Figure 28 depicts the driver in a reclined position as may be done with a rear engine design. Very few front engine front drive powertrains would permit a driver position this low.

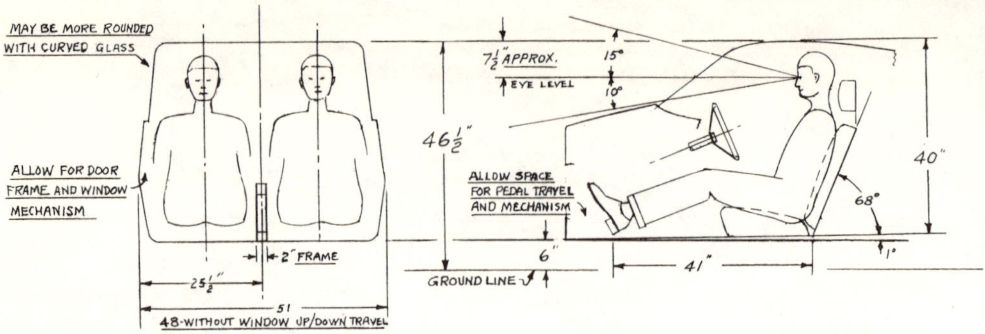

6'2" MAN SEATED IN SPORTS CAR – LOWEST PRACTICAL POSITION
– PEDALS OR SEAT SHALL ADJUST FOR SMALLER DRIVER –
– FIGURE 29 –

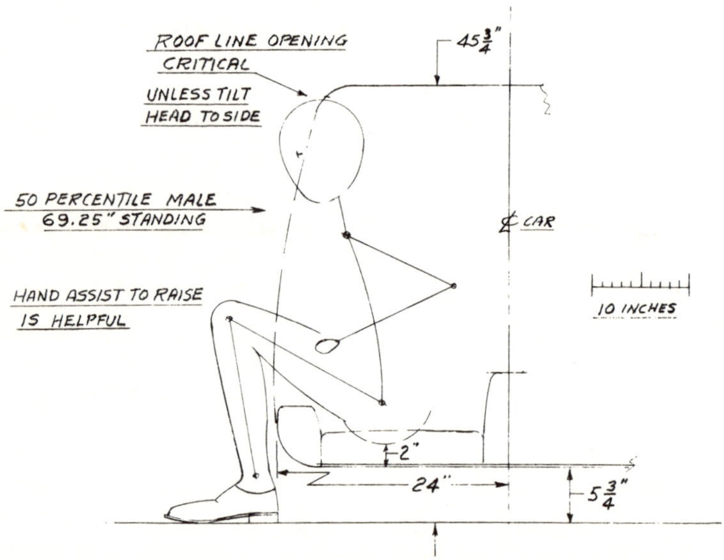

MINIMUM DIMENSIONS FOR SPORTS CAR ENTRY/EXIT
– FIGURE 30 –

LEG AND FOOT SPACE

It is quite common for inexperienced designers to be a bit scimpy on leg room or distance from the driver's heel to the seat back. Be sure you don't make this mistake. Seat or pedal travel should be at least 4 inches and preferably 6 inches. Figure 31 shows pedal location and minimum foot room (width) for clearance. Bear in mind these are the absolute minimum and should be increased unless the distance between the front wheel well and the transmission simply cannot be increased. This is more acceptable if only two pedals are required (throttle and brake) as with an automatic transmission.

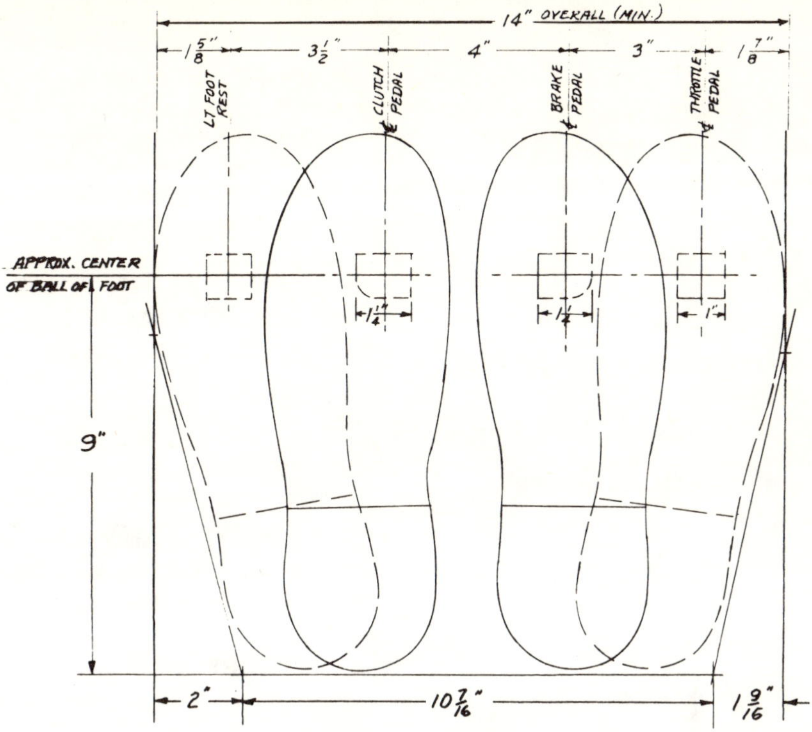

PEDAL LOCATION AND MINIMUM FOOT ROOM FOR CLEARANCE – BETTER TO EXCEED THESE WIDTHS

– FIGURE 31 –

Figure 32 shows the pedals and feet in side view. For safety the throttle idle position is forward of the full on brake position. If the driver's foot overlaps the throttle it will not be depressed. The pedal travel shown is for small vehicles and is usually greater for larger, heavier vehicles.

DOORS

Doors are not depicted because most of the many different types are shown throughout the book, especially in the last chapter. The conventional swinging door is simple and it is difficult to improve on. Many people are fascinated with gull-wing doors. They are nice but usually are heavier, more expensive to make, and the window glass cannot move up and down, at least not fully. Sliding doors have been improved considerably for vans, postal trucks, and utility vehicles. They also have an advantage for small urban vehicles when parked near other cars.

CARGO CAPACITY SHOULD MATCH THE MISSION

Unfortunately many cars are designed or "styled" to look great with little or no thought as to required space for items the owner intends to stow and transport. "Looking great" may favor a short or non-existent trunk space which may be totally impractical for the vehicle's intended use.

Needs also differ with vehicle types. Obviously, a race car only must provide for the driver, and a pick-up truck is expected to have a bed for bulky items too large to put in the trunk of a passenger car. In between these extremes the needs should be considered for each application.

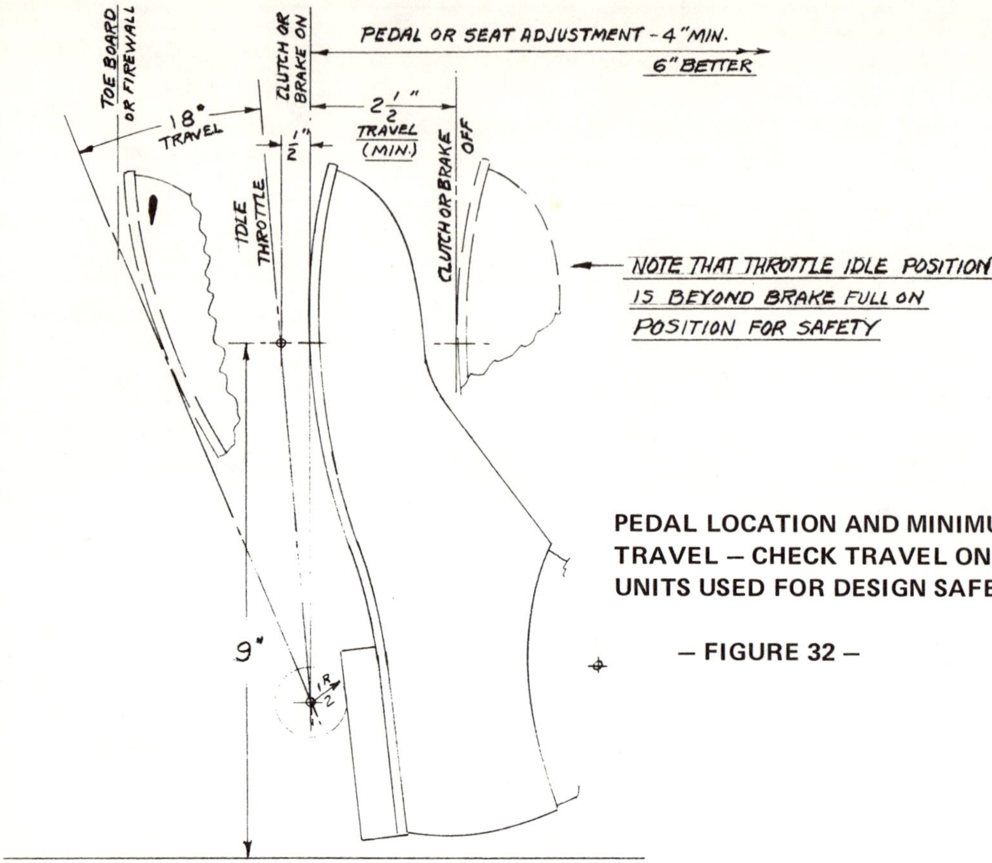

PEDAL LOCATION AND MINIMUM TRAVEL – CHECK TRAVEL ON UNITS USED FOR DESIGN SAFETY

– FIGURE 32 –

SPORTS CAR

Consider the cargo needs of the typical owner of a sports car. In addition to daily chores such as to and from work and shopping, a sports car is expected to take its owner and a friend on overnight trips to sporting events at distant places. This means luggage for both, and often special sports equipment. Typically, small sports cars only provide nooks and crannies for small soft bags that hardly serve as luggage. Anything additional must be strapped on outside and be subjected to weather conditions or stowed between the passenger's legs. Clearly, more thought should be given this problem by the designer of the car.

STATION WAGONS

In this case the owner needs more seats for children or friends and relatives. Usually the rear seats should also fold to provide a large, open area for bulky items at times. Many salesmen use wagons to carry their displays and other business equipment. Wagons come in different sizes, depending on chassis dimensions and will continue to be popular with a large group of customers.

One of the problems, especially with a small wagon, is access openings for bulky items. The rear door must be carefully designed to permit access of the largest item that interior space will accept.

ECONOMY SEDANS

These little econo-boxes usually are on a small chassis and are expected to provide space for four people only. If anything else is to be carried it must be put on a seat or the package shelf. Rear seats should fold flat for added space, and a rear door or window lift provided.

URBAN CARS

These are the smallest, most economical vehicles, and are designed specifically to be most convenient in traffic and for parking. Rarely are they designed for more than two people, and they are used for short distances – to and from work and for shopping. The urban car must carry groceries and shopping parcels in a decent sized space with easy access. This means more than a parcel shelf.

Since urban cars must remain small and interior space is important, another concept might be the MINI-AERO-VAN with C.O.E. (cab over engine) as shown in figure 17 (on page 40). This provides maximum interior space, easy entry and exit, minimum wheelbase for agility, and short overall length for easy parking.

Chapter IX

FEDERAL AND STATES RULES AND REGULATIONS EXHAUST AND NOISE EMISSIONS – SAFETY – MPG (CAFE)

The designer has a great, even awesome, responsibility to conform to all the rules and regulations of the Federal government and the States, in the design of a motor vehicle. The manufacturing division also must deal with rules and regulations concerning safety and the environment in producing the vehicle, or for that matter anything else. Since these rules and regulations change, become more complicated, and generally more restrictive, this chapter will only provide an indication of the problem. The subject should be reviewed constantly and updated to remain current.

The author makes no value judgments of his own, but will attempt to provide relevant information at this time. Complete, updated rules and regulations should only be obtained from the Federal and States originating sources.

Pollution, health, smog, and emissions are terms that are often used to describe this problem. To some degree these words are even used interchangeably, but they should not be. The general public associates the deterioration of health with all pollution, smog, and emissions, but this isn't true according to recent information. Let's look at each, particularly in relation to health, and put vehicle emissions in proper perspective.

Pollution is a general term and may be applied to air, water, or in fact, nearly anything. Health may or may not be affected by certain forms of pollution as we shall soon see.

Smog is a more restrictive term applied to what happens to air, photochemically, and primarily from the ingredients nitrous oxide (NOx) and hydrocarbons *in the presence of sunlight.* In chapter XV, under physical properties of air, we read that air naturally contains 78% nitrogen and about 20% oxygen. *It is well known and yet suppressed that we cannot reduce hydrocarbons because vegetation emits, in the evening, about four times the hydrocarbons generated by vehicles, storage tanks, and other man-made sources.*

Areas with very little sunlight, and all areas at night, do not produce smog (but have pollutants). Windswept areas, even with bright sunlight, do not have smog. It is blown away.

The principal fixed sources of NOx in most smoggy areas appear to be large powerplants generating electricity, and oil refineries. *They increase the concentration of NOx.* It has been reported (reference 13) *the DeNOx systems to reduce NOx by as much as 90% in these large powerplants are available now and can be installed within 2 years!*

As to mobile sources, automobiles with EGR (exhaust gas recirculation) only reduce NOx by 30% and this is accomplished by reducing the oxygen in the engine's cylinders with which the nitrogen combines. This also reduces power and performance. It also may not be a major smog factor, as studies mentioned later seem to indicate.

As to hydrocarbons emitted by automobiles, this is handled by the catalytic converters and again, does not seem to be a major factor in smog as our references point out.

Emissions are thought of mostly in relation to cars, but as we shall see, almost everything and every creature, including people, emits. Emissions may be good or bad and each should be properly evaluated, not just assumed to be dangerous.

Now that we have a handle on the terms pollution, smog, and emissions, let's find out what recent studies bring out, and later how we might possibly put it all in perspective.

SEVERAL VIEWPOINTS ON THE EMISSIONS PROBLEM

EXHAUST EMISSIONS

In an effort to provide another side of the automobile exhaust emissions problem as it relates to human health, we refer to a book "Air Pollution and Human Health", written by Lester H. Lave, head of Carnegie-Mellon University's Economics Department, and Eugene P. Seskin, senior research associate at Resources for the Future. This comprehensive 367 page book was published by the John Hopkins University Press. The book is based on their 10-year study of the pollution levels in 117 major metropolitan areas with the death rates in the same periods.

The story was reported in Automotive Industries, April, 1978, page 23, by Joseph M. Callanhan and titled "Car Smog Role is Debunked". (Reference 14). Lave and Seskin *concluded that particulate and sulphur oxides emitted by stationary sources were related to the deaths and to illnesses,* but that the *hydrocarbons, carbon monoxide and oxides of nitrogen produced by cars have no effect on public health* – – They said back in the 1960's, when only a limited amount of auto emissions was being controlled, *mobile sources* only contributed about 4% of the nation's toxic pollution. However, this small percentage will also be further reduced by the recent findings that nature produces at least 95% of the three major automotive pollutants.

Both Lave and Seskin served on the 1974 National Academy of Sciences Committee on the cost benefits of automobile pollution control. The study was financed by the Resources for the Future, supported by the Ford Foundation.

Among other things, the study indicates that the country has been and is spending far more for automotive pollution control than it's receiving in benefits.

In further support for the Lave and Seskin study, we need only to recall the London "Peasoupers", the polluted fogs of the past 600 years that killed and sickened millions. Coal fires were still burning in the city at the end of World War II. In 1952 the disastrous dense yellow sulphurous smog with its choking funes, tragically caused more than 4,000 deaths among the elderly and those suffering from breathing and heart conditions. As a result of this tragedy, the city of London decreed its whole area a smokeless zone. *The problem was solved without the slightest modification to motor vehicles.*

For information on natural air pollution see reference 15. Its sources include: Stanford Research Institute, Naval Research Laboratory, Air Control Assn., Science and Environmental Science and Technology.

The report mentions the following among many others.

1– Some 87% of 202 million tons of reactive hydrocarbons are generated each year by natural sources, usually pine and other trees.

2– Seepage from petroleum, natural gas, and tar deposits also make a substantial contribution, especially in California with its major faults in terrain. (This seepage is also hydrocarbons).

3– Oxides of nitrogen are 98% natural and originate in biological action and organic decomposition.

4– Carbon monoxide is 93% natural, coming from honey locust trees, alfalfa, swamps, sea waves, forest and prairie fires.

5– Carribbean sea surface water was found to be supersaturated with carbon monoxide.

6– Dr. Chambers of the University of Texas School of Public Health asked "Why, then, have we put ourselves in the position of embarking on a multibillion dollar transformation of our transportation and Energy consumption systems without solid assurance that it is mandated by solid evidence?"

Now let's go back to 1940, the year the writer moved to the Los Angeles basin, before smog was discovered or even noticed. Of course we had fog, which was rarely objectionable except along the beach roads when it got too heavy. Before World War II the main industries in Southern California were tourism, movie making, and aircraft factories. It was generally considered a vacation land. We had cars and traffic but limited manufacturing and related industries. Air quality was no problem except for the smudge pots in the orange groves in the winter. WW II years caused a tremendous influx of people (and their cars) from 1941 to 1947 at the end of the war. Of course, industry had expanded with the war effort, including industries that used oil, gasoline, and petrochemicals as well as those involved in the transfer and storage of these items.

The great increase in service stations, with which it seemed nearly one on every corner dispensing gasoline fumes with every sale, added more tons of emissions to the air, causing smog. They still do. According to reference 16, about 30% of all hydrycarbon emissions in the basin come from gas evaporating out of the car fuel lines and tanks. The service stations in the region will soon all be fitted with special nozzles to trap gas vapors while refueling. All this *before* it gets into the vehicle's engine.

Before ever getting to the service stations, the fuel must be transferred by trucks from storage facilities, with additional losses to the air as anyone can notice when in the vicinity of the tank farms and refineries. Similar losses occur all through industries who use fuel – so it isn't always the automobile.

Industry continues to grow and grow and people are still coming by the thousands to fill those jobs. Many other parts of the nation experienced similar growth of industry and air quality began to suffer.

In 1951, Dr. Arie Jan Haagen-Smit of Caltech published a controversial paper on the subject of how photochemical smog is produced. He reproduced it in his laboratory, analyzed auto exhaust, and found its contents similar. Like all first experiments, it could not possibly include all other sources of similar emissions. The matter soon became political and the automobile was the convenient whipping boy. It was only necessary to attack 3 or 4 large Detroit auto manufacturers. To attempt to attack thousands of manufacturers in hundreds of industries in every major city was not nearly as easy, would have been politically unpopular, and of course, there was not yet sufficient evidence. The great petro-chemical industry, based on oil, was also growing at this time. There was also an emerging suspicion that industry in general was a culprit, but it was felt the car was the one to clean up first.

In 1963 the Clean Air Act became law and the Environmental Protection Agency (EPA) was established to implement the law. The PCV (Positive Crankcase Ventilation) was the first emissions control item added to auto engines. We all have 20/20 vision in hindsight, but wouldn't it be interesting to try to see where our air quality would now be if industries had been required to clean up first instead of cars. Maybe also see what could be done to clean up some of nature in parts of California at least.

Mr. Lave indicated that the major defect in the Government's auto emissions control program was that it *assumed* that all automotive emissions were toxic. But this is not true. The major point of the argument in "Air Pollution and Human Health" is that the country has been and is spending far more for automotive pollution control than it is receiving in benefits. Even back in 1976, Janet E. Dickenson, chief air pollution analyst for the Southern California Air Pollution Control District, said the district staff did not recomment adopting the State's ARB 1977 stricter standards. (Reference 17)

SOUTHERN CALIFORNIA AIR BASIN IS UNIQUE

The California South Coast Air Basin is unique in many ways – no doubt about it.

1– We have lots of sunlight to photochemically produce oxidant (also called ozone).

2– The air sloshes back and forth from the mountains to the sea with the ocean breezes, both on-shore and off-shore.

3– Inversions occur. This is a layer of hot air that acts abnormally to effectively put a lid on the cooler air under it to trap pollutants. Then a build-up of smog occurs.

4– There has been and continues to be a great increase of industry in the southern part of the state, along with more vehicles and people to commute greater distances to and from work. In 1960 there were 3.1 millions of vehicles. In 1969, the figure rose to 4.05 millions. The south coast air basin in 1978 had 6.3 million cars and 10 million people. (Reference 16). There were 11.74 million cars licensed throughout California in 1977.

Additionally, throughout the nation including Southern California, it takes about 10 years to see auto emissions controls make an impact, since it takes this long for the older used cars with lesser controls to no longer be in use. Reference 16 is very comprehensive in reporting the excellent progress that has been made in air quality. Much has already been done on stationary sources as well as on mobile sources. If no stricter controls are imposed and existing controls continue to be enforced, the steady improvement will continue. Automobile exhaust, even the non-toxic portion, is amazingly clean right now. Many authorities now believe that further efforts on cars cannot possibly yield benefits that are worth the effort and cost to the Nation and to the car buyers.

"FOR SOME REASON, ALL OUR CONTROLS, ALL OUR TECHNIQUES – THEY JUST AREN'T WORKING!"

That statement was made by Dr. Thomas F. Heinsheimer, Vice Chairman of the Board of the South Coast Air Quality Management District (AQMD) in reference 18. Dr. Heinsheimer, an aerospace corporation scientist, also asks, "Are we better off now than 10 years ago? I'm not sure. We don't know how bad it might be if we hadn't made all those rules. *But what if our basic understanding of the physics and makeup of smog is wrong?* (Italics added by author) and "We'd better start understanding what we are doing."

These statements were made after 30 frustrating years since the first smog agency was established in 1948. September 15, 1979 had 102 degrees heat which tied a 70-year record for that date, but failed to blow the lid off the Southland's 9-day-old smog layer.

GREATER PERSPECTIVE IS NEEDED TO SOLVE THE POLLUTION PROBLEM

It is gratifying to note that after decades of blaming the automobile for most, if not all our pollution, the problem is beginning to be put in proper perspective. Reference 19, "WHAT THE POLLUTION AUDIT FOUND" brings this out very nicely. Patrick R. Zimmerman, assistant environmental scientist, Washington State, Pullman, WA., made this impressive study and two other related studies under research grants from the Environmental Protection Agency (EPA). He is now doing a fourth study. Quoting from the reference: "Zimmerman's studies of natural pollution are unique because he is the first person to measure all the naturally-produced hydrocarbons in the air. Being the first scientist to make actual field studies on the subject, he has developed a sound system for measuring the natural pollutants for the entire nation". The reference opens by stating, "A few months before the Clean Air Act of 1970 was enacted by Congress, the National Air Pollution Control Administration (predecessor to the EPA) announced to the nation that, "In 1968 approximately 32 million tons of hydrocarbons (HC) were emitted in the United States. Of the sources surveyed, motor vehicles were the largest single emitters of HC, accounting for 15.6 million tons per year." By comparison and again quoting, "The study (Zimmerman's) showed that 87.3 million tons a year, or 80% of the hydrocarbons in the nation in 1976, came from a wide variety of trees, plants, fallen leaves, nearby soil and marine areas. These sources weren't even mentioned in the 1970 government inventory of the nation's air pollution."

The study mentions the need to identify specific species since all of these do not participate equally in forming ozone.

The references also describe the method used to make the audit and several other related items of interest.

To continue with our sub-title subject – the need for greater perspective – the writer would like to suggest:

A– Messers Lave, Seskin, Chambers, Zimmerman, and others have done excellent jobs of learning much about the natural pollution of trees, plants, soil, and marine areas. Now studies are needed in other areas of nature and man's environment as proposed by the following.

B– Wild life should be studied. Birds, animals, insects, – all are emitters. How much deserves an approximation.

C– Domestic animals. Farmers are well acquainted with the pollution in the farm yard. Cattle feedlots and stockyards near cities are emitters. Shouldn't this, too, be evaluated?

D– And then we have ourselves! The human body "burns" to stay alive. Intestinal gas is well known, and trips to the bathroom (which is vented into the atmosphere), occur many times day and night, especially during TV commercials. Breathing is continuous, bad breath is a form of pollution and breath purifiers add another pollution ingredient. The skin is always emitting, day and night, even with daily bathing, as anyone can notice in a crowded bus or building without ventilation.

People consume about 2000 calories of "fuel" per day per person, plus what we drink. Children less and hard working men more. Multiply this by a population of 220 million. Next, we don't eat all our food raw like wildlife. First we process some of it at a factory by cooking, frying, boiling, or whatever. Then we package it. Then store it, transport it to markets, shelve it, haul it home and reprocess it over our stoves and finally wash it down with all kinds of liquids from "soft drinks", coffee, tea, wine, beer, and "hard liquor". Boy – do we "emit" – even *before* we nourish ourselves!

E– Should our kitchens and restaurant's kitchens be equipped with some kind of catalytic converter for the exhaust gases like an automobile? We stopped using weekend outdoor incinerators decades ago.

F– Smoker's pollution – how much – is it a significant factor?

G– Our pets, dogs and cats, like humans, are also emitters and require "fuel" in the form of factory processed food. What is their fair share?

H– Dry cleaning of clothes at the cleaners or at home is also a factor, as are our sprays including hair sprays and household bug sprays, cleaning solutions, and the like.

I– Our homes, apartment buildings, offices, and factory buildings are heated and air conditioned, both of which require energy and emit. Paint and roofing materials emit from the sun's rays.

J– The great concentrations of humanity by the thousands in high rise apartments and office buildings clustered together create concentrated pollution. Think what is vented to the air and through the sewage systems. Shouldn't this be measured or estimated to find out how great a factor it is?

K– Possibly the great concentrations of humans is a mistake. Our crowded cities may be efficient but environmentally wrong. Nature's wildlife doesn't concentrate to the extent that we do. *Quite possibly the major problem is concentration of pollution. Dispersal of people may make a big difference.*

L– Maybe we should determine the preferred limits of concentration in our large cities. If they are environmentally obsolete we might have a better, healthier, safer, and more enjoyable life by dispersing a bit. This might even reduce the related bureaucracy that goes with bigness. (Government pollution?)

M– Government control of factory pollution has come a long way over the last decade but the cost to industries is running into the billions of dollars – all of which must be passed through to the general public in higher costs for products. In an effort to reduce these costs without reducing effectiveness, the "bubble concept" has come into being and apparently was conceived independently, but at about the same time in industry and in the EPA. According to reference 20, "A Brave Experiment in Pollution Control", Mr. John Barker, an environmental engineer with ARMCO, and Douglas M. Costel, head of the EPA, and his staff got the idea about the same time. Rather than reduce pollution "stack by stack", the "bubble concept" would tell a company how much its overall pollution must be reduced, and allow the company to decide how it shall be accomplished and on which facilities. Another part of the concept is to allow a newcomer to "offset" the pollution it will create by cleaning up pollution produced by facilities already in the area. Many questions still remain such as: all pollution is not equal and cannot be treated equally.

Again, as with people, possibly dispersal of industries may be as effective as any other concept and should be evaluated. Possibly the worst polluters could operate with accepted safety in remote areas, isolated from other polluters including nature, in some cases.

N– In trying to put a handle on the total pollution problem it is suggested that five things need to be determined.

1– Percentagewise, how much pollution is created by wildlife, domestic animals, humans and how we live? If plant life contributes 80% of the hydrocarbons then industry and our vehicles end up with a very low percentage after the creatures take their share.

2– Since all pollutants are not equal, a weighted value for each needs to be developed.

3– Since concentrations of pollution appears to be a major factor, what can be accomplished by dispersing the sources of pollution rather than, or in addition to, reducing emissions?

4– Health is our primary concern. The varieties of pullutants should be compared and listed on a scale of, say 1 to 10, to decide, healthwise, where our efforts should be expended. Maybe we've been wasting time and money on the wrong things.

5– What can be done about nature's pollutants? Maybe we could accomplish more faster by looking into these sources. At least in concentrated areas of the nation.

These thoughts (A through N) are suggested to help get this pollution problem into proper perspective. We've already made the automobile the whipping boy. Now let's find out more about the overall picture.

A few parametric evaluations by qualified University people could quite possibly become the basis for funding from government or private agencies, to bring the parts together in studies that are aimed at the *whole* problem.

Possibly this may initiate a thesis or two to help advance some student's academic degree.

SAFETY – ACCIDENT AVOIDANCE AND ACCIDENT PROTECTION

Now let's attempt to put safety in its proper perspective. For many years we've known that it's the "nut behind the steering wheel" who is primarily responsible for automotive safety. The National Safety Council reports that 83% of all motor vehicle accidents are caused by driver error or improper driving, rather than by vehicles or the road. (Reference 21). Let's now break it down a bit more. It's beginning to look like the car owner and/or driver should be blamed for even more than 83% – in fact, nearly all of the accidents that occur when his or her responsibilities are fully indicated.

After all these years drivers under the influence of alcohol still account for about half of the fatal accidents. Drugs, more prevalent than ever, account for an unknown additional number of fatalities. The insurance companies recognize that youth (not all) also are involved in a disproportionate share of fatal accidents and are charged more for their insurance premium accordingly. Inexperienced drivers and/or thrill seekers and those careless people who are accident prone do their share also. Then we have those owners who refuse to maintain a car and its tires. And last, many people have never learned about "defensive driving." Learning to drive a car is not enough. Everyone should know how to drive under adverse conditions such as in the rain, on ice, when a tire blows out or goes flat, and when someone else makes a mistake in their car. Drivers must learn to anticipate trouble; that is part of defensive driving. It simply isn't possible to design a mechanism that is always "fool proof" and that will last forever without maintenance. Actually, the owner/driver has far more control over the proper safe use of a vehicle than the designer and manufacturer.

Designers and manufacturers have done a great deal on their own to constantly improve vehicles. Safety glass, better tires, stronger bodies, vastly improved handling, better brakes, deep dish steering wheels, just to name a few, are features that came along without government rules. Racing showed the way with roll-bars and cages, safety belts and harnesses, although those came from aircraft originally. Incidentally, no aircraft passenger is provided with a "passive restraint." If a belt is sufficient in a private or commercial airplane – why not in a car?

No doubt industry will continue to invent and devise additional items on its own. Nothing is beyond improvement. The industry has a good record on its own, based on data shown in ACCIDENT FACTS, (reference 22). It states: The death rate per registered vehicle has been declining rather uniformly except during the depression and WW II years. In fact, the death rate decrease from 1946 to 1961 was better than for the years 1964 to 1973. In 1974 the oil embargo and the 55 mph speed limit is credited for further improvement.

According to the Motor Vehicle Manufacturers Association, traffic fatalities in the U.S. are now the lowest of any surveyed nation, at 3.5 persons per 100 million vehicle miles driven in 1976. The figure dropped to 3.27 in 1978.

Other countries are compared as follows: Belgium 12.6; France 10.1; West Germany 8.4; Japan 8.0; Netherlands 6.9; Finland 5.9; Australia 5.9; Canada 5.6; United Kingdom 4.5; and Norway 4.4.

So, in fact, our record in this country is still pretty good for whatever reasons and even with our faults.

The reason for the above discussion of emissions and safety is to indicate *what and who is primarily responsible for the situation.* In the matter of safety-Accident AVOIDANCE may be as much or more important than Accident PROTECTION after it occurs. Again, let's get our priorities right and work on the most important problems first, and put the responsibility where it belongs. It appears the government, Federal and States, should really concentrate on the "nut behind the wheel" rather than the vehicle manufacturer and the designers.

MILEAGE STANDARDS (CAFE)

The Corporate Average Fuel Economy (CAFE) standards were set because of the alarming depletion of this nation's petroleum and to reduce our increasing dependence on foreign oil. Over half (52%) of U.S. petroleum consumption supports transportation – 29% is used in private automobiles (reference 23). The resulting massive downsizing of automobiles is so great that it must be taken in several steps over a period of years. The astronomical cost is running into the tens of billions of dollars for redesign, new tooling, and new facilities; all of which must be passed on to the car buyer in higher prices. Increased efficiency is another way to achieve improved mileage with reasonable downsizing. Now, after 4 years and less alarm, it appears there will be ample petroleum to the year 2000 and beyond, barring wars and embargoes, according to the respected Petroleum Industry Foundation (reference 24). The revolution in IRAN and reduced oil exports is a case in point. The more restrictive vehicle emissions controls will require more unleaded fuel and this has posed a problem for refiners of gasoline, but even that should cause only temporary shortages, according to "Supplying Gasoline for Tomorrow's Cars" (reference 25). The study actually predicts a slow decline after 1985.

By now large cars are referred to as "gas guzzlers" and are somewhat unpopular, although there is a definite need for some large cars and station wagons in large families, and for some businessmen. Therefore, a few are permitted under the CAFE standards.

THE RULES AND REGULATIONS – SEE HOW THEY GROW

In reference 26, Richard C. Gerstenberg, former General Motors Chairman, was quoted as follows: "Government today has something to say about how we design our products, how we build them, how we test them, how we warrant them, how we repair them, the compensation we pay our employees, and even the prices we charge our customers".

The reference further quotes Detroit as claiming it is spending almost as much time describing how they meet federal standards as they do in devising the methods to accomplish the task. "In 1972, the federal regulations covering certification of our cars for federal emissions took 140 pages", said S. L Terry, Chrysler Corp. - - - "In 1975, the regulations took 242 pages. Since then the number has jumped to 500 pages. In addition, it takes 167 more pages for the government to explain what they want us to do. In 1972 it took 774 pages for us to respond to those guidelines in our applications. In 1975, it took 8,758 pages".

G.M. estimated it spent $1.3 billion to meet all government requirements in 1974, excluding the actual cost of necessary new hardware, which amounted to $270 per car. It is left to the reader to estimate the cost for 1980 and the following years. These figures do not include the cost, payrolls, etc., of the people in the bureaucracies in Washington, which run into more millions and is paid for out of taxes.

There are now two dozen federal agencies regulating motor vehicle manufacturing and use and nearly 20 congressional sub-committees considering legislation that has a direct effect on the automobile and its industry.

From the above growth pattern it is easy to see this chapter could not possibly provide up-to-date and complete material on Rules and Regulations. The rules and regulations are a growth industry with diminishing returns that must be dealt with. Company funds must be provided to comply.

If yours is a small company with limited resources for these matters do not despair, there are ways to get started as you will learn later. However, your company's long-range plans must include funds for this effort.

FEDERAL EMISSIONS STANDARDS – AUTOMOBILES

The next big drop in emissions standards occurs in 1981 and is as follows: (ref. 27).

HC – Hydrocarbons – From 1.5 to 0.41 grams per mile
CO – Carbon monoxide – From 15 to 3.4 grams per mile
NOx – Oxides of nitrogen – From 2.0 to 1.0 grams per mile

Additionally, carburetors must be tamper-resistant, emissions controls for high altitude cars must be improved, emission systems warranted for 50,000 miles.

FEDERAL NOISE STANDARDS – AUTOMOBILES

All U.S. cars and light trucks built since 1975 have been designed to meet an 80 dBA standard. This probably will remain, since above approximately 35 mph tire noise becomes predominate and therefore the reduction of engine noise becomes not very cost-beneficial (ref. 27).

FEDERAL MOTOR VEHICLE SAFETY STANDARDS (FMVSS)

These standards are issued by the National Highway Traffic Safety Administration (NHTSA). At this writing they consist of the following:

TITLE	STANDARD NUMBER
Accelerator Control Systems	124
Brake Fluids	116
Brake Hoses	106-74
Brake Hoses, Hydraulic	106
Brakes, Air Systems	121
Brakes, Hydraulic	105-75
Brakes, Hydraulic Service, Emergency & Parking	105
Brakes, Motorcycles	122
Bus Window Retention and Release	217
Child Seating Systems	213
Controls and Displays, Motorcycles	123
Control Location and Identification	101
Door Locks and Door Retention Components	206
Exterior Protection	215
Flammability of Interior Materials	302
Fuel System Integrity	301-75
Fuel Tanks, Fuel Tank Filler Pipes and Fuel Tank Connections	301
Glazing Materials	205
Headlamp Concealment Devices	112
Head Restraints	202
Hood Latch Systems	113
Impact Protection For The Driver From The Steering Control System	203
Lamps, Reflective Devices, and Associated Equipment	108
Mirrors, Rearview	111
Motorcycle Helmets	218
Occupant Crash Proection	208
Occupant Protection in Interior Impact	201
Reflecting Surfaces	107
Roof Crush Resistance	216
Seat Belt Assemblies	209

TITLE	STANDARD NUMBER
Seat Belt Assembly Anchorages	210
Seating Systems	207
Side Door Strength	214
Steering Control Rearward Displacement	204
Theft Protection	114
Tires, New Pneumatic.	109
Tires, New Pneumatic for Vehicles Other Than Passenger Cars	119
Tires, Retreaded Pneumatic.	117
Tire Selection and Rims	110
Transmission Shift Lever Sequence, Starter Interlock and Transmission Braking Effect	102
Truck – Camper Loading	126
Vehicle Identification Number.	115
Warning Devices	125
Wheel Nuts, Wheel Discs, and Hub Caps.	211
Window Systems, Power Operated	118
Windshield Defrosting and Defogging Systems.	103
Windshield Mounting	212
Windshield Wiping and Washing Systems	104

Amendments and additions are always underway and are listed in the S.A.E. "Automotive Engineering" on the Legislative Log page of each issue. The FEDERAL REGISTER, issued daily by the Department of Health, Education, and Welfare (HEW) is the official source of all such information.

"NHTSA reveals five-year plan" (ref. 27 & 28) provides an indication of what is coming. Some 13 on-going proceedings will be dropped from active consideration. Mileage standards for passenger cars and light trucks through 1988 model year will be established by August 1981, and finalized by August 1982. Better side impact protection for occupants, improved braking requirements and reduced pedestrian fatalities are being worked on.

Manufacturers must provide data and tests to obtain Vehicle Certification in the assigned classifications.

CORPORATE AVERAGE FUEL ECONOMY (CAFE) STANDARDS IN THE 1980s

New car fuel economy standards are required by the Energy Policy Conservation Act of 1975. The National Highway Traffic Safety Administration (NHTSA) sets the standards. They are as follows:

20 miles per gallon in 1980 (for cars)
22 miles per gallon in 1981 (for cars)
24 miles per gallon in 1982 (for cars)
26 miles per gallon in 1983 (for cars)

The 27.5 mpg figure for 1985 was arbitrarily set by congressional decree and is subject to review and possible change by NHTSA. This review is due in 1979 (as this is being written).

NHTSA's announced aim in this area is to give auto makers four to six years lead time between proposed standards and effective date.

NHTSA's fuel economy standards for light trucks and vans are due in 1979 also.

Corporate Average Fuel Economy (CAFE) standards are based on a weighted average of all cars sold by each corporation. A few large less-efficient cars sold must be balanced by smaller, more efficient models sold to achieve the average. If only one model is sold, then it must equal or better the figure shown for that year, or obtain an extension. The EPA does the testing to determine mpg in conjunction with their emissions testing. Some changes are being considered as this is being written.

The CAFE standard is actually more difficult to balance than it first appears. For example, if a large car gets 17.5 mpg, it will consume 0.5714 gallon for 10 miles of travel. To balance this out there remains only 0.1558 gallon for 10 miles for the small car, so that they both average 27.5 mpg. This means the small car must average 64.2 mpg. This is what the CAFE standard means.

This is why U.S. cars must continue to become smaller, lighter, more aerodynamic, and more efficient in every way. Even while meeting tougher emissions and safety standards.

The electric car and its hybrid versions can effectively improve corporate averages, and this may be an important factor in the renewed interest in electrics for the mid-1980s and beyond.

MOTORCYCLES AND THREE-WHEELERS (UNDER 1500 LB.)

EXHAUST EMISSIONS – 50 cc DISPLACEMENT AND OVER – 1980 MODEL YEAR

No crankcase emissions are permitted.
HC – 5.0 g/km
CO – 12.0 g/km
NOx – (deleted)

Manufacturers with total U.S. sales of less than 10,000 vehicles per year, will not be required to submit vehicles for certification testing by EPA, but will be required to affirm in writing to EPA that their motorcycles conform to the applicable standards as confirmed by emissions test data generated by or for the manufacturers . . . . .

The above information and much more is covered in the FEDERAL REGISTER, Vol. 42, No. 3, Wednesday, January 5, 1977. This is under title 40, chapter 1, page 86. It continues for 40 pages of fine print to describe test procedures, among other things.

FEDERAL REGISTER, Vol. 42, No. 208, Friday, October 28, 1977 (6560-01) should also be reviewed for additional technical amendments and corrections. It continues for four pages.

NOISE EMISSIONS – ALL STREET MOTORCYCLES (PROPOSED)

1980 – 83 dB (A)
1982 – 80 dB (A)
1985 – 78 dB (A)

Mopeds – 1980 – 70 dB (A)

OFF-ROAD BIKES

170 cc or less	Over 170 cc
1980 – 83 dB (A)	1980 – 86 dB (A)
1982 – 80 dB (A)	1983 – 82 dB (A)

California noise emissions are more strict as described later under state requirements.

SAFETY

Refer to the previously listed Motor Vehicle Standards. Certain standards apply to motorcycles as indicated. Additionally, NHTSA has new long-range plans for rule making running through 1983. As this is being prepared these items are not yet finalized.

Three-wheelers, as motorcycles, need comply with fewer standards, thus simplifying the job considerably.

TEMPORARY EXEMPTION FROM COMPLIANCE WITH SPECIFIC SAFETY STANDARDS (ALL MOTOR VEHICLE TYPES)

It is possible to appeal to NHTSA for a one- to three-year duration exemption from compliance with a specific safety standard, pleading low-quantity production and financial hardship prohibiting demonstration of compliance. As many as 2500 vehicles per year of a car model have been permitted exemptions.

USE OF SUPPLIER'S APPROVED COMPONENTS FOR COMPLIANCE AT MINIMUM EXPENSE

No designer designs all components, nor does his company manufacture all components used in a motor vehicle. This is fortunate in many ways, not the least being the obtaining of Federal approvals. The designer need only choose components, parts, and other items that have been approved through the efforts of the outside supplier/manufacturer. Generally these are products manufactured for original equipment manufacturers (O.E.M.) such as headlights, tail-lights, turn signals, switches, windshields, windshield wipers, brakes, wheels, tires, seat belts and harnesses, and bearings to name a few.

The list now has been enlarged to even include engines, transmissions, and accessories, with complete emissions controls. A complete powerplant can be installed with but minor changes to avoid altering emissions levels. It is necessary to avoid increasing vehicle weight and/or wind resistance, preferably reducing them so less power is required. Then with proper final drive ratios, usually the same or numerically lower ratios (taller gearing) the vehicle will achieve the same or better emissions than when installed in the vehicles first designed for.

Three-wheelers might also use motorcycle powerplants under similar conditions.

O.E.M. components represent a major portion of a motor vehicle and should be utilized fully. Trade magazines such as AUTOMOTIVE INDUSTRIES, June issue, provide O.E.M. suppliers in great detail. Other trade periodicals and supplier catalogs are useful also.

PERFORMANCE STANDARDS FOR ELECTRIC AND HYBRID VEHICLES

The Department of Energy (DOE) on May 30, 1978 announced release of a set of minimum performance standards for electric and hybrid vehicles to be used in a nationwide government/private demonstration project (ref. 29). These regulations were adapted as proposed except for certain modifications made effective July 3, 1978. For further information contact Mr. Chas. E. Pax, Office of Assistant Secretary, Conservation and Solar Applications, 20 Massachusetts Ave., N.W., Washington, D.C. 20545, or phone 202-376-4893. The final performance standards were also published complete in ELECTRIC VEHICLE NEWS, August 1978. Briefly, the standards are as follows:

Fact Sheet: **DEPARTMENT OF ENERGY ELECTRIC AND HYBRID VEHICLE (E&HV) PERFORMANCE STANDARDS**

Summary: Public Law 94-413, Section 7, as amended, calls for DOE to establish performance standards that specify the minimum criteria for the Demonstration Project. These standards were published in the FEDERAL REGISTER on May 30, 1978. The standards, highlighted below, *will be revised periodically* as the E&HV state-of-the-art improves. For the complete standards refer to the FEDERAL REGISTER, Document No. 78-14887. The test conditions and procedures for determining levels of performance are stated in SAE J227a.

Definitions: Electric Vehicle (EV): a vehicle which is powered by an electric motor drawing current from rechargeable storage batteries, fuel cells, or other portable sources of electrical current, and which may include a nonelectrical source of power designed to charge batteries.

Hybrid Vehicle (HV): a vehicle propelled by a combination of an electric motor and an internal combustion engine or other power source and components thereof.

Personal-Use Vehicle: a vehicle designed to carry 10 persons or less and which is designed and primarily used for transporting the vehicle operator and passengers.

Commercial Vehicle: a vehicle other than a personal-use vehicle.

STANDARDS	PERSONAL USE	COMMERCIAL
Forward Speed Capability	80 km/h for 5 minutes	70 km/h for 5 minutes
Range	50 km (EV) and 200 km (HV) on SAE J227a/C cycle	50 km (EV) and 200 km (HV) on SAE J227a (a/B cycle
Acceleration	0 to 50 km/h in 15 s.	0 to 50 km/h in 15 s.
Gradeability Limit	20% grade for 20 s. either backward or forward	20% grade for 20 s. either backward or forward
Battery Recharge Time	10 h. from 80% discharge	10 h. from 80% discharge
Battery Life	75% of range after 12 mos. or 15,000 km; 100% of acceleration and gradeability	75% of range after 12 months or 15,000 km; 100% of acceleration and gradeability
Emissions, Safety and Crashworthiness	Federal Motor Vehicle Safety Standards and some safety standards appropriate only to E&HVs	Federal Motor Vehicle Safety Standards and some safety standards appropriate only to E&HVs

VEHICLE CODE – STATES

Up to now we have only considered Federal Rules and Regulations as they apply to a manufacturer of motor vehicles. The ultimate buyer or customer must still obtain a license in his own state in order to operate the vehicle on public streets, highways and secondary roads. Each state has its own vehicle code published in book form by the State Department of Motor Vehicles. It is usually revised and updated to some degree about every other year. A copy can be obtained for a small fee from a Department of Motor Vehicles (DMV) office, or by writing to the main office at the State Capitol.

It is important that a designer be quite knowledgeable about the sections dealing with motor vehicle design, construction, and features. Specific rules deal with headlight height, candlepower and beam pattern, stop and tail-lights, turn signals, brake design for safety, including parking brake, exhaust pipe location, emissions, and features too numerous to mention here. Usually, these items conform to the standards set forth in the S.A.E. Handbook, but they should be studied carefully.

It should be noted that racing cars do not use public roadways, so are exempt unless otherwise noted. Racing club rules apply in this case.

NEW VEHICLE STANDARDS SUMMARY – FEDERAL AND CALIFORNIA

To avoid seeking related data in different places, the Federal and California standards have been combined in the following summary. Reference 30 is the Federal source and reference 31 is the State of California source for those who seek additional information.

PASSENGER CARS

YEAR		STANDARD	TEST PROCEDURE	NON-METHANE HYDRO-CARBONS (4)	CARBON MONOXIDE	OXIDES OF NITROGEN (7)
1980		Calif.	CVS-75	0.39 gm/mi (0.41 gm/mi)	9.0 gm/mi	1.0 gm/mi (1.5 gm/mi)
		Federal	CVS-75	(0.41 gm/mi)	7.0 gm/mi	2.0 gm/mi
1981	(5)	Calif. (A)	CVS-75	(0.41 gm/mi)	3.4 gm/mi	1.0 gm/mi (1.5 gm/mi)
		Calif. (B)	CVS-75	0.39 gm/mi (0.41 gm/mi)	7.0 gm/mi	0.7 gm/mi
		Federal	CVS-75	(0.41 gm/mi)	3.4 gm/mi(6)	1.0 gm/mi
1982	(5)	Calif. (A)	CVS-75	0.39 gm/mi (0.41 gm/mi)	7.0 gm/mi	0.4 gm/mi (1.0 gm/mi)
		Calif. (B)	CVS-75	0.39 gm/mi (0.41 gm/mi)	7.0 gm/mi	0.7 gm/mi
		Federal	CVS-75	(0.41 gm/mi)	3.4 gm/mi(6)	1.0 gm/mi
1983 &		Calif.	CVS-7	0.39 gm/mi (0.41 gm/mi)	7.0 gm/mi	0.4 gm/mi (1.0 gm/mi)
		Federal	CVS-75	(0.41 gm/mi)	3.4 gm/mi	1.0 gm/mi

(1) Passenger car as defined in Title 13 of the California Administrative Code means any motor vehicle designed primarily for transportation of persons having a capacity of twelve persons or less.

(2) Hot 7-mode test.

(3) Hydrocarbon emissions from 1975-76 limited production vehicles may not exceed 1.5 gm/mi.

(4) Hydrocarbon standards in parentheses indicate total hydrocarbons. When applicable, manufacturers may elect to certify vehicles to either the non-methane or total hydrocarbon standards.

(5) For the 1981 model year manufacturers may choose options A or B separately for their gasoline and diesel product lines. The option chosen in 1981 must be retained for the 1982 model-year.

(6) Federal carbon monoxide standards can be waived to 7.0 gm/mi for 1981-82 by the EPA Administrator after a public hearing.

(7) Manufacturers selecting the oxides of nitrogen standards in parentheses must certify vehicles for 100,000 miles instead of the normal 50,000 miles.

gm/mi – grams per mile.
CVS-72 – constant volume sample code start test.
CVS-75 – constant volume sample test which includes cold and hot starts.
7-mode – 137 second driving cycle test.
PPM – parts per million.

OTHER REQUIREMENTS

Effective 1978, evaporative emission standards are 6.0 grams per SHED test for the 1978-79 model years and 2.0 grams per SHED test for 1980 and later model years.

Beginning with the 1980 model-year, the maximum projected emissions of oxides of nitrogen measured on the federal Highway Fuel Economy Test shall be no greater than 1.33 times the applicable passenger car oxides of nitrogen standards shown above.

SHED Test (Sealed Housing Evaporative Determination) – A method for measuring evaporative emissions from motor vehicles.

Reference 32 is the source for California noise emissions standards and the California Highway Patrol checks compliance.

NOISE STANDARDS

FEDERAL (Office of Noise Abatement and Control – ONAC) EPA

1978 – Medium and heavy-duty trucks	83 dB(A)
1982 – Medium and heavy-duty trucks	80 dB(A)

All U.S. cars and light trucks since 1975 were designed to meet an 80 dB(A) standard although no such Federal standard has been issued.

CALIFORNIA (Reference 32) and California Highway Patrol booklet: "Motor Vehicle Noise Laws"

		Speed Limit Zone	
		Less than 35 mph	More than 35 mph
(1)	Any motor vehicle with a mfgr's gross vehicle weight rating of more than 10,000 pounds and any combination of vehicles towed by such motor vehicle	82 dBA	90 dBA
			More than 45 mph
(2)	Any motorcycle other than a motor-driven cycle	77 dBA	86 dBA
(3)	Any other motor vehicle and any combination of vehicles towed by such motor vehicle	74 dBA	82 dBA

MOTORCYCLE LIMITS

For purposes of section 27200, the following noise limits shall apply to any motorcycle, other than a motor-driven cycle, manufactured:

(1)	After 1969 and before 1973	88 dBA
(2)	After 1972 and before 1975	86 dBA
(3)	After 1974 and before 1981	83 dBA
(4)	After 1980 and before 1986	80 dBA
(5)	After 1985 and before 1990	75 dBA
(6)	After 1989	70 dBA

For purposes of section 27200 the following noise limits shall apply to any motor vehicle with a gross vehicle weight rating of 6,000 pounds or more manufactured:

(1)	After 1967 and before 1973	88 dBA
(2)	After 1972 and before 1975	86 dBA
(3)	After 1973 and before 1978	83 dBA

For purposes of section 27200, the following noise limits shall apply to any other motor vehicle, not specified in this article manufactured:

(1)	After 1967 and before 1973	86 dBA
(2)	After 1972 and before 1975	84 dBA
(3)	After 1974	80 dBA

The above information is not complete but serves to indicate, in part, the noise standards in California for certain years and for certain types of motor vehicles in certain speed zones. The complete data is too detailed for inclusion here. Designers and manufacturers are urged to seek complete updated data at the time of need, and from references, both Federal and State.

SPECIALLY CONSTRUCTED VEHICLE (EXAMPLE – CALIFORNIA)

A "Specially constructed vehicle" is a vehicle of a type required to be registered under this code not originally constructed under a distinctive name, make, model, or type by a generally recognized manufacturer of vehicles. The writer has registered two experimental vehicles under this definition. Among other things, it is required to obtain a brake and light certificate from an official State Certified Testing Station and the vehicle is subject to careful inspection before being registered. The builder must also take all his receipts or bills of sale for chassis, body, and engine to the local Department of Motor Vehicles (DMV). Chassis and engine numbers are checked against a list of stolen cars by the California Highway Patrol office, who also checks for safety glass, the brakes and lights certificates, and whatever else is required, such as smog control devices that were in effect the year the engine was manufactured. Other states may have similar or otherwise regulations for these vehicles. A summary of the inspection requirements for California, CHP form 885, can be obtained by sending a self-addressed stamped business size envelope to: California Highway Patrol, 1325 Arden Way, Sacramento, CA 95813.

"Specially constructed vehicle" usually includes cars built from a kit, and home constructed vehicles using various components and a set of purchased drawings. For example, EV FOCUS, May 1, 1978, mentioned that more than 10,000 sets of plans for the URBA ELECTRIC described in MECHANIX ILLUSTRATED, Feb. 1977 issue, were sold. When the buyers complete their electrics, they will obtain their vehicle license under this classification in their own states. Dune buggies, three-wheelers, electrics, and unusual vehicles of all kinds that are home-built, may be licensed under this classification providing they comply with the State Code relating to vehicle design features and safety.

The National Highway Traffic Safety Administration (NHTSA) now considers any kit car customer who completes his car and puts it on the street, to be a manufacturer. According to Z. Taylor Vinson, NHTSA Senior Staff Attorney, "The definition of "manufacturer" described in Section 102 (5) of the National Traffic and Motor Vehicle Safety Act includes "any person engaged in the manufacturing or assembling of motor vehicles, whether or not for resale, while those importing motor vehicles for purposes other than resale would not fit the definition.

"In any event, the prohibitions of the Act (Section 108 (a) (I) (A) are not directed specifically against manufacturers but against any "person" who, among other acts, "introduces into interstate commerce" a motor vehicle that doesn't meet all applicable Federal motor vehicle safety standards. We believe that a car driven solely intra-state, but with inter-state traffic, would be in "interstate commerce". This means, in effect, that any time a motor vehicle leaves private property, it will be in interstate commerce".

The above and much more was reported in reference 33. As the reference mentions, "However, this interpretation of the law hasn't been tested in court, and most kit cars are exempt from Federal regulations because the regulations apply only to new cars and most kit cars are officially considered used cars"

In a letter to California Congressman Robt. Badham, Joan Claybrook, Administrator of NHTSA stated, in part, ... "The options that are open to the kit purchaser in completion of his vehicle with new or used equipment, are so numerous as to defy rational attempt at regulation under current procedures. While vehicles resulting from certain combinations of equipment might be regarded as "new" on balance, it appeared to us more likely than not that the installation of used equipment by the kit purchaser would create a vehicle that could not totally comply with every standard applicable to new vehicles, and that common sense requires it be treated as "used". Also, as a policy matter NHTSA does not initiate enforcement action against an individual who assembles a single vehicle for his own use". She further states, in a letter to Senator Warren Magnunson, "Obviously, a hybrid vehicle consisting of both new and used parts may be physically incapable of meeting the complete range of new vehicle standards, but the number of these vehicles on the roads has, in the past, been negligible, and the agency had not considered their regulation a priority matter".

HIGHER EFFICIENCY – THE WONDERFUL FALL-OUT FROM FEDERAL REGULATIONS

When emissions controls were first installed as add-ons, power and gas mileage dropped and performancs suffered. Then safety regulations added weight and more cost to make matters worse. Then the fuel crisis hit, causing mandated higher miles-per-gallon of gasoline. Conventional large cars became known as "gas guzzlers" and "dinosaurs". Small foreign "econo-boxes" were in. But many buyers learned that was not necessarily the best solution for them in this vast country, with greater distances often to be covered and with their need for more interior space.

As the industry struggled mightily, new approaches to the problems had to be found - - fast! Survival was at stake in some cases. In seeking better engineering solutions, considerable help was evident in studying the sophisticated methods pioneered in the aircraft and aerospace industry. Electronic controls on the engine improved its efficiency. Detailed weight and stress analysis led to much redesign, better use of existing materials and often use of higher strength materials of less weight, transmission improvements, tires of lower rolling resistance, front wheel drive to save space and weight, all became of immediate interest. Aerodymanics suddenly became more than of passing interest. Even electric vehicles, commercial and private, provided some alternatives. So much was going on, it all had to be computerized for faster answers. Higher efficiency became the name of the game, and it was force-fed at great expense to remain in the game. Although the buying public ultimately must pay through higher prices, the pay-back will eventually make it all worthwhile.

COMPETITION EVENTS

The rules and regulations for competition events are set by the governing body in charge of the event. Rules change and are updated from year-to-year so will not be covered here. The best advice is to join the club sponsoring the event, to learn what is going on and how to participate. For example, the Sports Car Club of America (SCCA) is well established and has many classes of cars identified by engine displacement and separated into stock and modified divisions. Additionally, there are Formula cars – open-wheel racers of different sizes.

Safety requirements for vehicle and driver, inspections, and driver testing are fully covered in the club membership and rule book.

Local or regional competition events, controlled locally, often have their own rules.

Motorcycle racing is also controlled by its own club, as is drag racing, Bonneville Salt Flats events, and stock car racing.

A new type vehicle such as the Formula Delta three-wheeler described in the Advanced Vehicles Development chapter would be allowed to run in a certain group, provided car and driver pass inspection and are accepted. Trophies are not normally awarded unless at least three similar vehicles are competing in the same race or event. If the vehicle type becomes popular and owner/drivers request it, a separate class will be officially established. This has been done many times, such as with Formula Ford, Formula V and super Vs. And as vehicle types become obsolete and unpopular, their classes are dropped.

There always seems to be a desire for low cost vehicles to provide club racing for more owner/drivers.

Chapter X

POWER TRAINS – INTERNAL COMBUSTION ENGINES (ICE)

Engines are mass-produced in great variety throughout the world and are available for every possible application. To further understand just how much variety there is, the reader is referred to the Engineering and Statistical issue of AUTOMOTIVE INDUSTRIES, April issue, each year. (Reference 14).

In view of the need for sophisticated emissions controls, testing, and certification, the designer and his company can save a great deal of time, money, and effort by using an existing powerplant with its needed accessories, essentially as is with only minor modifications such as rerouting portions of the exhaust system. Fuel injection and turbo-charging are also available on nearly all sizes of engines for increased horsepower. Diesels are available in many sizes.

Engine design is beyond the scope of this book and experimental developments will not be included, because vehicle designers are primarily interested in powerplants that are readily available. A general overview with advantages and disadvantages of each type, as installed, is intended.

AIR-COOLED VS. WATER-COOLED

The air-cooled engine has great appeal because of its simplicity. Elimination of the radiator, water pump, and plumbing reduces weight, complexity, and some cost. But times are changing. The air-cooled engine is not as quiet as a water-cooled engine, and noise pollution is a problem. First, clearances in an air-cooled engine are greater. This causes internal rattling noises until operating temperature is reached. There is also a noticeable power loss when cold. Motorcycle owners seem willing to tinker and go through a short warm-up period that few automobile owners will tolerate. Even when warm an air-cooled engine is still more noisy than a water-cooled engine. The cooling fins vibrate and give off a "ringing" sound. This can be dampened with rubber inserts between fins as is frequently done on motorcycle engines. If a blower and shrouding are used, these also add noise. The inside of the shroud may be coated to also dampen noise and drumming. This adds cost and, while effective, is still not equal to a liquid cooled engine.

THE TWO-CYCLE ENGINE

This basically simple low-cost engine is now used primarily for chain-saws, snowmobiles, some motorcycles, and nearly all mopeds. It is an excellent racing engine as has been proven many times. Tuned exhaust systems boost power somewhat like a supercharger. Reference 35 can be helpful, as can reference 36.

The small two-stroke engine can be quite efficient as a constant speed engine, and might serve to power a generator or alternator for a hybrid electric.

Environmentally it is not fully acceptable, for reasons of noise and exhaust emissions. Some progress has been made in these areas with fuel injection and other improvements, but the advantages of simplicity and low cost suffer. When displacement is kept below 50 cc, as in a moped, noise and exhaust emissions can be kept within the law with little difficulty. In fact, the exhaust emissions regulations exempt engines below 50 cc.

Mixing oil and gasoline for lubrication is a bother to many users. There are designs to eliminate it by metering oil at the carburetor. If a premix of oil and gasoline is used, it should not be stored because the oil tends to "sour" the gasoline, causing rapid deterioration, resulting in loss of power and hard starting.

THE ROTARY COMBUSTION ENGINE (WANKEL PATENTS)

This fine engine got off to a good start in the early 1970's but became unpopular in this nation because the early designs used more fuel, had some difficulty in meeting later emissions requirements while reducing fuel consumption, and long life of the rotor seals was questionable. Research and development has continued in Japan and Europe as well as in this country to the point where these problems have been largely solved. Over a million vehicles have been built using this type engine, and it is now certainly a mature design deserving serious consideration.

Its advantages include smooth power, it is 50% smaller, 30% lighter, has fewer moving parts than a piston engine, and it out-performs a comparably sized piston engine. Some twenty companies are now working on rotaries to power motorcycles, boats, chain saws, automobiles, and for industrial applications.

It is understood the Wankel patents expire in 1982. This may provide additional stimulus to rotary development.

Present use of the rotary combustion engine in this country in a motor vehicle is limited to imports. One of the lesser-known applications is in some of the scaled-down Grand Prix racers rented by the lap at mini-tracks in different parts of the country. They use a single rotor Sacks engine of 294 cc's, developing 28 hp at 6,000 rpm. The engine was originally designed for motorcycles and snowmobiles, and uses a Salsbury torque converter variable ratio automatic V-belt transmission.

Reference 37 will provide additional information on this engine as well as their 8 to 10 hp smaller engine. Both engines are air-cooled and lightweight.

The next larger rotary is the SUZUKI production engine, model RE-5, for motorcycles. This engine produces 56 hp at 6,500 rpm, with a displacement of 497 cc and has a water-cooled housing with oil cooling to the single rotor. Its power goes to a 5-speed constant-mesh transmission. For a more technical description the reader is directed to reference 38.

The MAZDA RX-7 sports car represents one of the latest developments, and uses its two-rotor engine of 1146 cc to develop 100 hp at 6,000 rpm. This car and its engine was full described in various magazines, such as Road and Track, May 1978.

For further information on the continuing progress reference 39 is suggested.

Control of exhaust emissions is discussed in connection with the MAZDA article, but not with the smaller engines.

THE DIESEL ENGINE

The CAFE requirements have recently made the diesel engine popular in this country as well as in other parts of the world. Ultimately it is expected that diesels will account for 25% of the engines in passenger cars. Over the highway tractor-trailer units have been nearly 100% diesel powered for many years because of their fuel economy and long life.

Recent automotive diesel engines are too numerous to be described here. The reader can learn about them from trade and technical magazines from month to month. There are some very interesting developments that are worth repeating herein in anticipation of their effect on vehicle design. These will apply to truck diesels first because of the greater savings possible for the truck owner. Later it is quite possible these same features, or at lease some of the features, will find their way into passenger car diesels.

TURBO-CHARGING AND TURBO-COMPOUNDING (GASOLINE OR DIESEL)

Turbo-charged engines are not new. They consist essentially of using the exhaust gas to drive a small centrifugal turbine/blower to force more air into the cylinders. Turbo-compounding is the next step. Its history goes back years to use on aircraft engines. The exhaust gas is used to drive the turbine as in turbo-charging, but rather than have the turbine connected to a blower, it is connected through a two-stage set of gearing to the crankshaft to provide additional power from the exhaust. One of the problems is that there must also be some sort of torsional isolator to permit a smooth flow of power from the turbine to the pulsating crankshaft.

THE TOTALLY INSULATED ADIABATIC ENGINE – THE ULTIMATE DIESEL

In a normal diesel engine about one-third of the energy goes to the cooling water and its large radiator, cne-third to the crankshaft for usable power, and one-third to the exhaust. By turbo-charging and turbo-compounding we can extract considerable energy from the exhaust. The next problem is "How to utilize the heat energy that goes to the coolant?" Adiabatic simply means the change in volume or pressure without the loss or gain of heat. It is easier said than done.

Diesel researchers are proposing that the engine be completely insulated *with the elimination of all coolant, radiator and water pump.* All high-temperature internal surfaces would have to be insulated to prevent the heat of combustion from traveling outward. This would be accomplished by applying a ceramic coating to the cylinder bores, cylinder head, valve faces and guides, piston crown and exhaust ports. Then practically all the heat and expansion of combustion must drive the piston downward for direct usable power, and on the upward stroke pass out the exhaust to then drive the turbo-compound turbine which is geared to the crankshaft. Very little heat is lost through the walls of the engine and the exhaust gas energy is harnessed also, leaving the tailpipe relatively tame.

How much has efficiency improved? Diesels are normally considered about 25% more efficient than gasoline engines, that is, 25% better fuel mileage. Fuel consumption traditionally has averaged about 0.4 lb/hp/hr. Recent improvements cut this about 10% to 0.36 lb/hp/hr. The ultimate engine is expected to achieve an order of 0.28 lb/hp/hr. Of course, additional refinements in electronically controlled fuel injection are included along with improvements in bearings, materials, and optimizing of all details of design. Such an engine is expected in the 1980's for heavy trucking and related applications. It will indeed be a hard-working engine which in turn will affect vehicle design as we shall see.

AERODYNAMIC BODY DESIGN AS AFFECTED BY THE ADIABATIC DIESEL AND THE 1982 NOISE EMISSIONS STANDARDS

Truck operators are always interested in fuel savings because fuel is a major item of operating costs. Wind resistance is a major reason for high fuel consumption. That large box of a body pulled (usually) by a cab-over-engine (COE) tractor is anything but aerodynamic. This will begin to change in the next few years. Since the adiabatic engine won't need a radiator up front, the tractor nose can become well rounded and internal airflow is nearly eliminated except for the engine inlet air-cleaner, which is often outside anyway.

Truck diesel engines are noisy and the 1982 noise emissions standards are stringent – 80 dBA. Lots of noise escapes downward, under the engine. The hard-working adiabatic engine may be more noisy. Present thinking dictates a belly pan to fully enclose the powerplant. Now if a belly pan is required anyway, why not make it aerodynamic in design to reduce wind resistance on the underside.

Truck and automobile companies have resisted belly pans ever since they were suggested. Now it looks like the truck manufacturers will have to use them to meet 1982 noise standards, so will probably do the job properly and include aerodynamic design. When this happens it will also be relatively easy to include the remaining underside portion back under the trailer. Efficiency goes up again. The improvement by eliminating internal flow can be of the order of 10% or more, then a well rounded nose will improve air flow at least 15 to 20%, and a well designed underside can add another 15% or so. At highway speeds the improved aerodynamics may improve fuel consumption as much as the adiabatic engine.

Bus diesels will be affected in like manner.

Who knows, the automobile diesel may also inadvertently be the one to first have an aerodynamic belly pan on cars after the trucks show the way.

ENGINE MODIFICATIONS FOR RACING

Since racing is exempt from emissions regulations, engines may be modified as permitted by race sponsoring bodies. Few engine changes are permitted for so-called "stock" classes. However, "modified" and "formula" classes allow extensive changes. These include head porting, compression increases, hotter cams, tuned intake and exhaust systems, carburetor changes, ignition changes, and even dry sump lubrication. In fact, the list is nearly endless and the subject of many excellent books on the subject. Engine design as mentioned earlier, is not within the scope of this book. Engine design and the competition engine in particular, is a specialized field unto itself. The vehicle designer, in working out the installation, must deal with the differences due to the tuned exhaust system, which must be carefully designed for agreed-upon peak rpm and desirable maximum torque range for the engine under consideration. To better acquaint the reader in this field reference 40 is recommended. More recent books on the subject have been written and are advertised in weekly and monthly magazines on racing.

THE COMPLETE POWER TRAIN INSTALLATION IN TODAY'S AUTOMOBILES

Power train staff engineers, research and development engineers, and laboratory staff technicians do a tremendous job of bringing a new engine and all its related accessories up to certification. The vehicle designer must work closely with these people to be certain that every detail of his installation in no way conflicts or negates any part of the operational aspect of the engine as developed for certification of emissions control, performance, fuel economy, reliability, and safety.

Small companies lack the facilities, funds, and experienced personnel to do complete development and certification. They must therefore be most careful when using the complete power train as developed for a different motor vehicle.

EMISSIONS FROM GAS CAP, THROUGH THE ENGINE, AND OUT THE TAILPIPE

Starting at the gas cap realize that if the engine is to use only unleaded fuel, the fuel tank inlet pipe must include a restricted opening so that the service station pump nozzle of standard (larger) diameter cannot be inserted.

Next, the fuel tank itself must have a vent provision to conduct fumes to a separate canister of charcoal so that fumes do not escape to the outside atmosphere. Now we are ready to discuss the engine itself.

If the engine is small, possibly about 1600 cc or 100 cubic inches displacement or less, the emissions problem is a little less difficult to deal with. For example, the SUBARU four-cylinder horizontally opposed boxer engine with its lean-burn SEEC-T (Subaru Exhaust Emissions Control Technique) runs without a catalytic converter or air pump, and uses regular gasoline. The California version incorporates exhaust recirculation. The HONDA CVCC engine also does not need a catalytic converter and uses regular fuel. Stratified charge is achieved with a third valve (two inlet valves). The Mitsubishi also uses two inlet valves and calls its design MCA-jet. These engines come in several sizes and are installed in DODGE COLT and CHALLENGER and PLYMOUTH ARROW and SAPPORO. The TOYOTA lean-burn engine is another approach and is described in reference 41.

As engine displacement goes up it becomes more difficult to achieve the same low level of emissions per mile of driving. American engines, traditionally larger, have more sophisticated controls. Precise controls of every aspect of combustion at any altitude and any temperature as well as idle, full load and deceleration have been developed. This led to the introduction of electronic controls with every aspect coordinated. Sensors are in the exhaust system, the intake manifold, the transmission, the crankshaft position, the throttle, coolant temperature, and sense the EGR position, spark advance position, and barometric pressure. All of this information is constantly fed into the micro-processor which, for every possible combination of inputs, then continuously readjusts the settings on the electronic actuators at each component. All this is done nearly at the speed of light. The HMOS micro-processor performs 16 bit operations using a clock time of 200 nanosec. A nanosec – 10^{-9} second – is about one light foot! The progress in these innovations, together with new design and manufacturing techniques in semiconductor technology, tends to double every two years and has been for the last fifteen years. References 42 and 43 provide more information on this interesting subject. A little more about electronics will follow in the next chapter on electric vehicles.

VIBRATION CONTROL – INTERNALLY AND EXTERNALLY

Over the years the V-8 engine with its hydraulic torque converter transmission has set the standard for smoothness in motor vehicles. Admittedly the 12-cylinder engine is even smoother, although less readily available. Now with downsizing, the V-8 is losing popularity to the V-6 and the ubiquitous "four banger". As a result, vibration control has become more of a problem. Each power stroke affects vibration at least three ways: 1) Torsional pulses in the crankshaft and all rotating parts through to the tires, as well as the reaction couple laterally at engine mounts, 2) Vertical bouncing of the engine mass as with the recoil action of a gun when fired, 3) Rocking couples fore and aft, due to each cylinder being some distance from the engine center of gravity. Additionally, if the engine firing order results in unevenly spaced pulses, this introduces out-of-phase vibration. Then we have the problem of frequencies, induced and natural. Every mass, such as the frame/body mass, has a natural frequency. This means that when excited it will vibrate like a tuning fork at its own natural rate or frequency. Then, if the engine is running at an rpm that coincides with or is a multiple of this natural frequency, the engine then excites the frame/body mass resulting in more vibration and even in some cases a "drumming" effect at a critical speed. Wind buffeting and road shocks transmitted through the suspension have their influences, too (as exciting forces).

Regularly spaced undulations on the roadway can add "heterodyning", or amplification at critical frequencies. Each wheel assembly has its own natural frequency and the springs have a natural rate. The problem is to move the rates or frequencies of these different masses upward or downward so that they do not combine. Since the engine is an exciting force, care should be taken to have its vibration out of phase with the others, and/or to reduce the exciting forces (engine vibration) as much as possible. This can be done internally with counterweights on the crankshaft and with changes in internal stiffness. Externally, damping at the motor mounts and isolation of torsional vibration along the shaft is used.

In the case of four-cylinder engines the addition of two smaller counterweighted shafts running in opposite directions and properly located, dampen vibration internally quite effectively. The Mitsubishi "silent shaft" design is a good example. Reference 44 describes this in detail.

It is clear the problems of vibration, natural and exciting frequencies is quite involved. Complete analysis with the aid of a computer is most helpful. Lacking this, experience, patience, and lots of testing with resulting changes is the next route. Experienced designers seek rigidity to limit deflections in certain structures, determine natural frequencies and often change them by moving masses or changing natural frequencies to higher or lower positions to avoid critical points, and influence the exciting forces by damping or isolation. In any event, the problems cannot be ignored if smoothness and reduced fatigue failures are to be obtained throughout the operating range.

GEAR RATIOS

It is rarely practical to use the same gear ratio as used in the original power train for another installation. Efficiency, gross weight, and aerodynamic drag are seldom identical. These must be calculated for the new application and compared with the power available at the wheels. Remember too, it is virtually impossible to obtain full terminal velocity (top speed) because of slight headwinds, temperature and roadway variations. It is best to figure top speed about 10% less to assure full power at rated rpm, will be reached in normal, average conditions. Then the engine will overspeed slightly under slightly better than normal conditions. Most first calculations are usually a bit optimistic.

If the power train is used essentially intact, and the ring and pinion gears are in the transaxle along with the transmission as in a front drive unit, other optional ratios may not be available from the manufacturer. Then the additional labor cost of tear-down and reassembly must be added to the cost of a new ring and pinion of the desired ratio. This can become expensive.

If the desired change in ratio is small, possibly it can be obtained by a change in tire and wheel size. This is not difficult if the change is made at the rear tires, but is seldom desirable if for the front driven wheels, because it may affect steering and/or suspension clearances.

Sometimes a different transaxle with the desired ratio can be fitted with an adapter plate and other modifications, as was done with the VIKING rear engine design described in the last chapter of this book.

If a front engine rear drive is planned, a different rear axle with the desired ratio becomes less of a problem. This is covered in more detail under the next sub-title.

GENERAL ARRANGEMENT AND RELATED PROBLEMS

Figure 33 depicts a "conventional" front engine rear drive in simplified form with the usual live solid axle and also with the alternative – an independent rear suspension (IRS). Most American and many foreign cars have such an arrangement. The CORVETTE uses an IRS rear axle as does JAGUAR and MERCEDES-BENZ, to name a few. A bit smaller and lighter rear axle is available in the model 28 SPICER DANA, reference 45. This axle is available as live solid axle or as an IRS unit with a load capacity of 2,000 lb. and a maximum torque output of 1,000 lb/ft. It uses automotive-type hypoid gears in four different ratios: 5.17, 6.83, 7.83, and 12.25. This axle is also used for electric vehicles. A search of available axles of different sizes, capacities, and ratios in this country and foreign sources provides a nearly endless variety to suit nearly any application.

"CONVENTIONAL" FRONT ENGINE - REAR DRIVE
– FIGURE 33 –

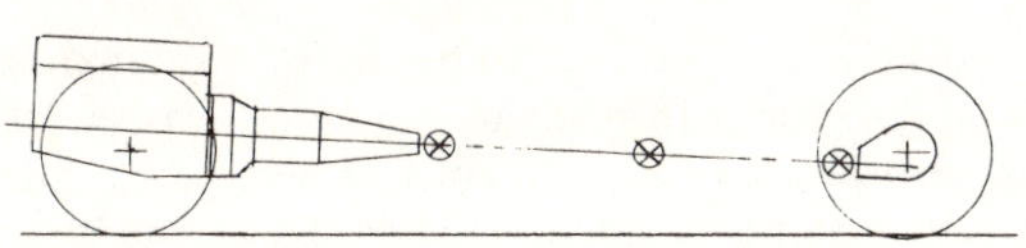

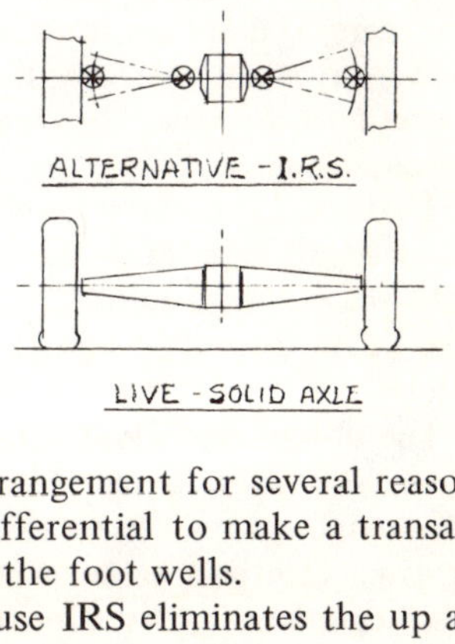

Figure 34 is generally considered a more desirable arrangement for several reasons.

1– Putting the transmission aft and integral with the differential to make a transaxle with independent rear suspension allows more room in the foot wells.

2– The drive shaft tunnel is lower or non-existent because IRS eliminates the up and down movement of the transaxle.

3– Superior weight distribution and balance are achieved. Better ride and improved handling result.

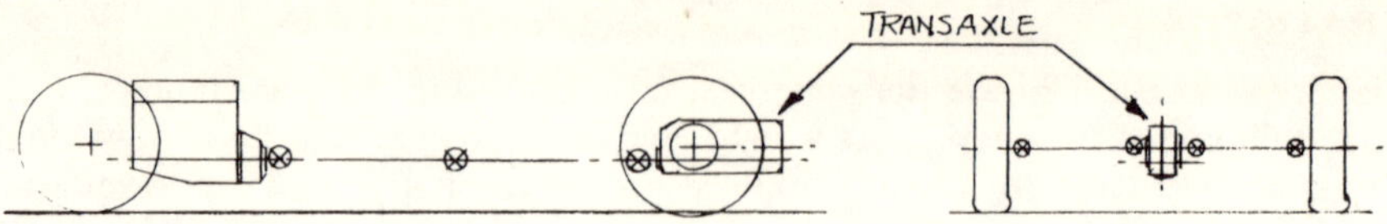

FRONT ENGINE AFT OF AXLE – MORE WEIGHT AFT
– FIGURE 34 –

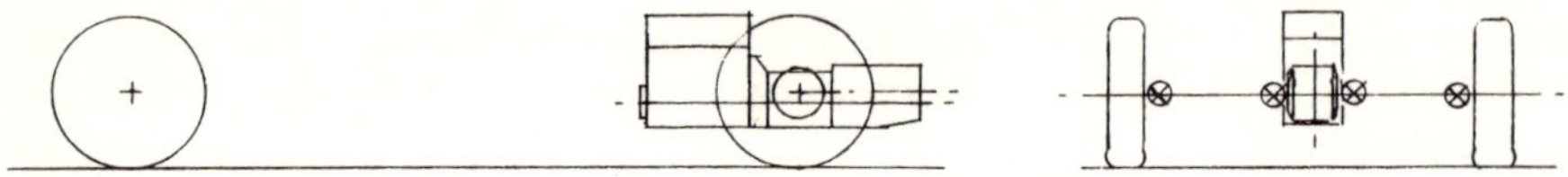

"MID-ENGINE" WITH TRANSAXLE BEHIND WHEELS
– FIGURE 35 –

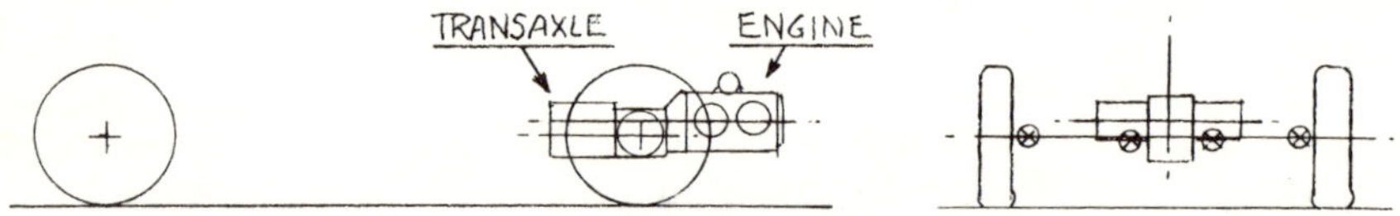

REAR ENGINE – TRANSAXLE AHEAD OF WHEELS
– FIGURE 36 –

The PORSCHE 924 and 928 are examples. Individual transaxles are not readily available, but they can be obtained by modifying units attached directly to the engine as in the case of the old VW "Beetle." The transaxle portion can be separated from the engine and an adapter plate and bearing added, along with shaft modifications to add a universal joint for driveshaft attachment. This type of modification has been used with other engines as in the VIKING described in the last chapter. This has also been done many times with electric vehicles. These changes are by no means limited to the VW transaxle. This can be done with a number of the newer units as they become available.

The so-called "mid-engine" is depicted in figure 35, in which the engine is just ahead of the rear axle. This is almost universally used for racing as in Formula cars, Indianapolis type racers, and for some sports cars. Now that front drive power trains are available in great variety, many will also be used for sports cars and certain racers by moving the complete unit to the rear. The steering link is replaced with a suspension arm and the exhaust is re-routed.

The mid-engine arrangement allows the nose of the car to be more aerodynamic with a low wedge shape. More weight is moved to the rear over the rear axle for high acceleration without wheel spin, although larger tires are needed to control oversteer tendencies. Noise to the passenger compartment is more of a problem as is access to the engine for servicing.

The ubiquitous VW "Beetle" and the ill-fated CORVAIR located the engine *behind* the rear axle as in figure 36. This provides more room for the rear seat and makes servicing the engine a bit easier. However, it also increases the polar moment of inertia to further aggravate the oversteering tendency. Larger tires and suspension changes help, but the arrangement is now considered obsolete from a design standpoint for passenger cars, but acceptable for buses with dual rear tires.

Front wheel drive is now "in" in this country, although it has long been a favorite in Europe for certain small cars. Although a few people remember the CORD FWD, most think of G.M.'s TORONADO and ELDORADO as our first front wheel drive cars. They both used an arrangement as shown in figure 37 for their large V-8's.

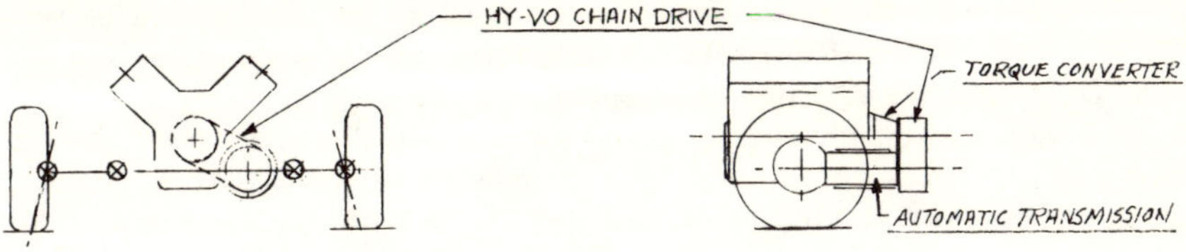

LARGE V-8 FRONT WHEEL DRIVE
– FIGURE 37 –

SMALL V-4 FRONT WHEEL DRIVE
– FIGURE 38 –

Figure 38 shows what can be done with a small V-4 as in the case of the old FORD TAUNUS 12M of Germany. This is an interesting case because this power train, as pictured in figure 39, was moved to the rear for the original experimental MUSTANG sports car of 1963. The suspension was changed, as can be noted in the MUSTANG story, as was the radiator location (see chapter 22).

The general arrangement of engine, transmission, and differential seems almost endless, as may be noted for the front wheel drives shown additionally in figures 40, 41 and 42. Figure 40 is similar to the early French DYNA-PANHARD. Figure 41 is the way the modern Renault LE CAR R-5 does it. The new wave of front drives have the engine mounted transversely as in figure 42. This was done originally in the Austin MINI of England in 1961, followed by HONDA of Japan, and others since. References 46, 47, and 48 provide more information on several of these power trains.

Another new version is provided by TOYOTA in its TERCEL/CORSA FWD, as depicted in figure 43, with the engine mounted above the differential in a fore and aft arrangement.

Scale drawings such as the TOYOTA FWD arrangement are convenient for making feasibility studies. Such drawings may be copied from company literature, magazine articles, and/or often from company authored S.A.E. papers or other media data. If and when a project gets past the feasibility stage, direct contact with the manufacturer will yield installation drawings – full size and in complete detail if serious vehicle manufacturing intent can be shown.

SMALL URBAN CAR POWER – THE MOTORCYCLE POWER TRAIN

Small cars for low-speed urban uses need small lightweight engines of low power. Modified motorcycle engines and similar size engines have been used in many such applications. Even small racers have been built with these engines. After WW II Formula III racers, later called Formula IV, had these power trains. Figure 44 depicts the general arrangement.

Figure 45 shows how it was done with the small BMW-ISETTA of Germany shortly after WW II. The rear tread was quite narrow, only 20.4 inches, and utilized a simple chain-drive to a live axle, with a single suspension system and no differential for both rear wheels.

Three-wheelers with the single wheel at the rear use an arrangement as shown in figure 46. The little MESSERSCHMITT KR-200 was an early example. The author's MINI-K and DUA-DELTA used this arrangement.

Over the years motorcycle engines have been developed tremendously, just as have automobile engines. Today, production bike engines are available that are water-cooled as well as air-cooled. Any number of cylinders up to six with displacements in excess of 1,000 cc are common. Extra smoothness with countershaft balancing is available. Automatic transmissions come on several models, and shaft drive is becoming more popular. With shaft drive it becomes easy to locate the motorcycle engine forward and with an extended shaft drive the rear wheel or wheels for three- or four-wheel vehicles.

CHAIN-TO-SHAFT DRIVE – ENGINE TO WHEELS

It is possible to adapt a chain drive design to shaft drive as shown in figure 47. The motorcycle engine is turned 90 degrees in planview and the small sprocket is used to attach the "U" joint as shown. If a major drive train design is contemplated, it is recommended the designer acquaint himself with complete information on universal joints and drive shaft configurations. There are about 19 varieties of universal joints, both constant velocity and non-constant velocity, such as the hooke joint. Each of these have certain strengths and weaknesses that should be understood. Drive shafts, based on

FIGURE 39

— FRONT WHEEL DRIVES —

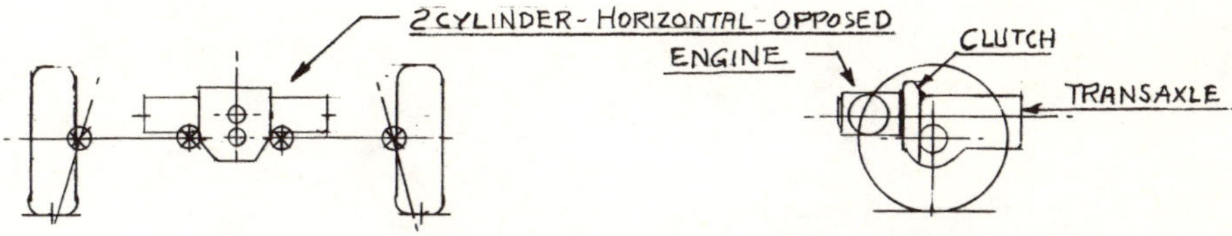

SMALL "BOXER" ENGINE WITH TRANSAXLE AFT
– FIGURE 40 –

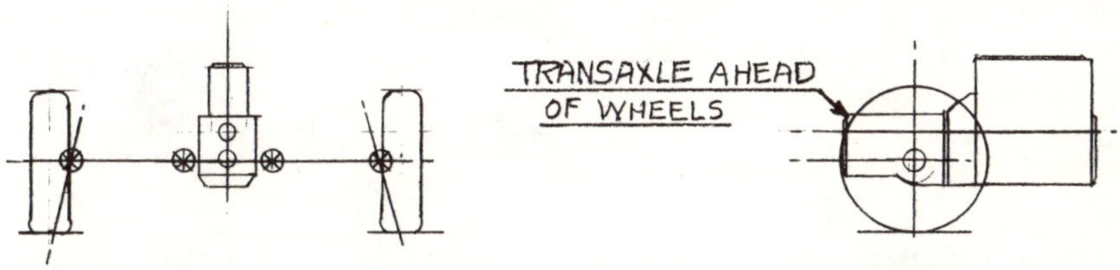

4-CYLINDER VERTICAL ENGINE AFT
– FIGURE 41 –

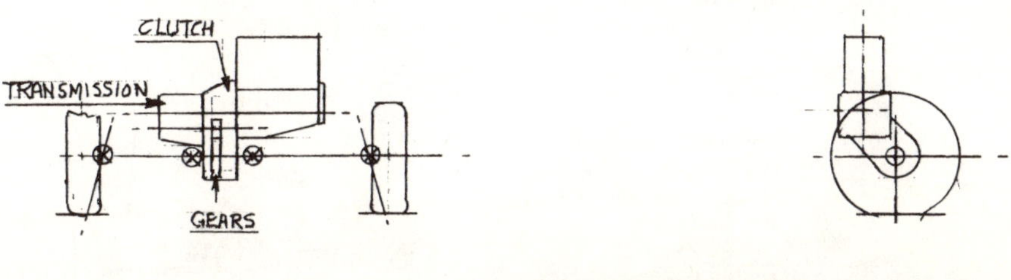

4-CYLINDER TRANSVERSE ENGINE AHEAD OF WHEELS
– FIGURE 42 –

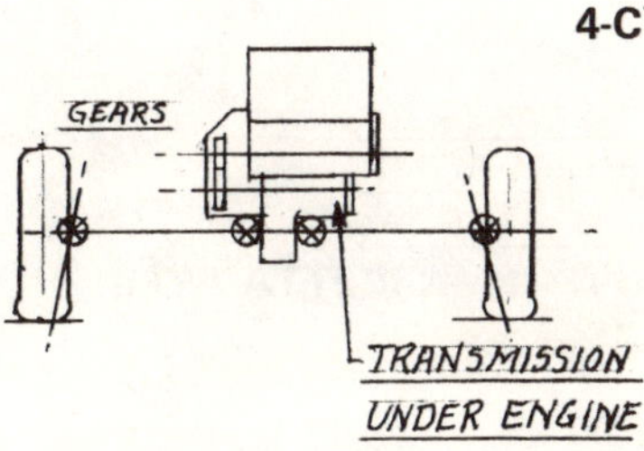

ALTERNATE ARRANGEMENT

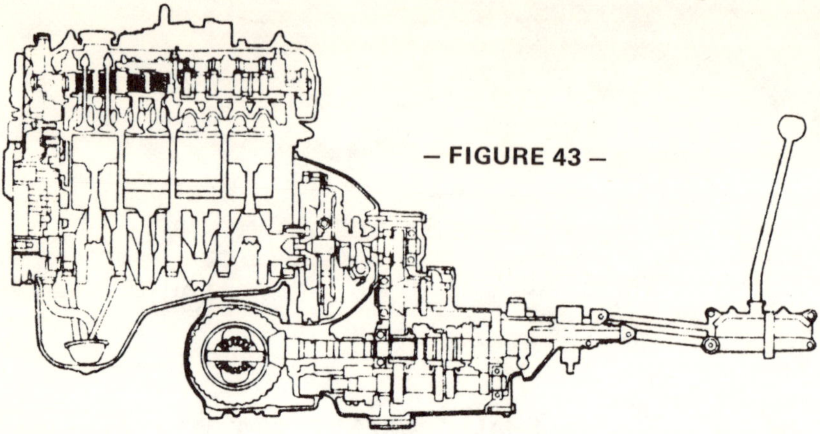

— FIGURE 43 —

Heart of the new Toyotas is this engine, mounted longitudinally over the transaxle.

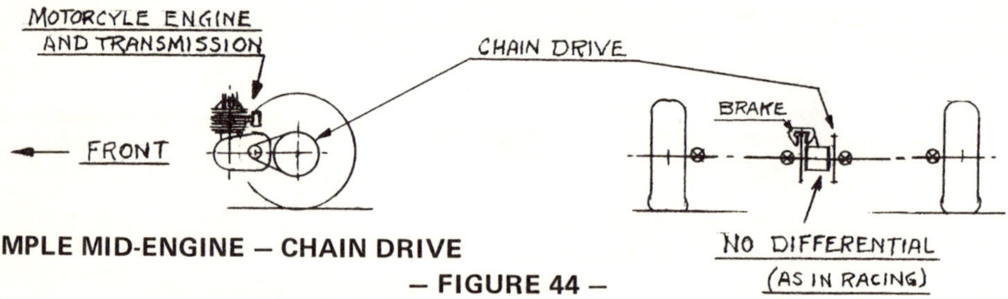

SIMPLE MID-ENGINE — CHAIN DRIVE

— FIGURE 44 —

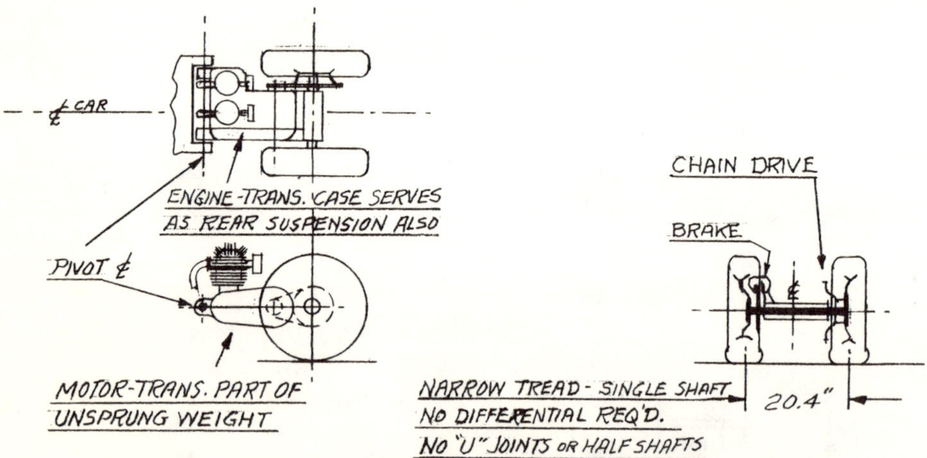

SIMPLE NARROW TREAD DESIGN - BMW-ISETTA TYPE

— FIGURE 45 —

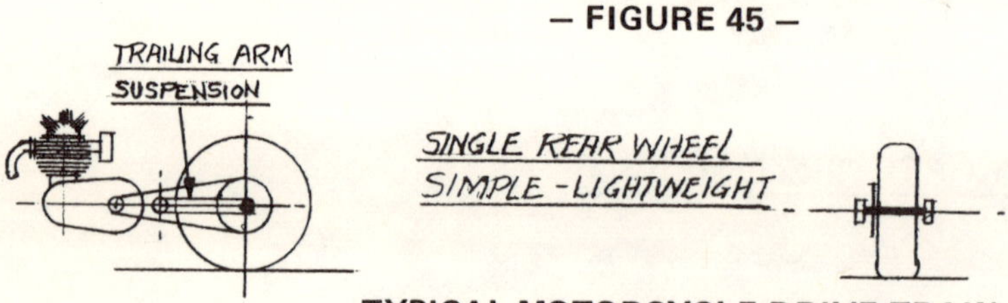

TYPICAL MOTORCYCLE DRIVE-TRAIN

— FIGURE 46 —

length and diameter, have critical speeds and require special attention to avoid problems. The library can provide excellent books on the subject. If you prefer the latest information all in one place, I recommend reference 49. This is a complete manual of more than 450 pages.

SYNCHRONOUS (TIMING) BELT DRIVE

It is not too practical to use chain and sprockets all the way back from a front engine to the rear, because of the problems of length adjustment and idlers, as well as lubrication in some designs. An excellent solution is the synchronous or timing belt.

The first timing belts had trapezoidal shaped teeth. These are still available and are rated for loads from sub-fractional to over 150 hp. The latest designs used a curved tooth and carry from 120 to 200 percent higher horsepower ratings than the trapezoidal tooth design.

A dragster motorcycle with two four-cylinder engines (reference 50) uses a curved tooth 8 MM pitch synchronous belt that transmitted nearly 600 hp at speeds to 10,000 rpm.

Figure 48 depicts how the belt might be used. The sketch is not to scale, and the toothed sprockets are shown larger than need be for clarity. Accurate alignment must be maintained at all times, but length can remain fixed and no lubrication is needed. The belt and sprockets should be covered for safety.

Complete engineering design data can be obtained from manufacturers who advertise in trade magazines and have distributors in all major cities.

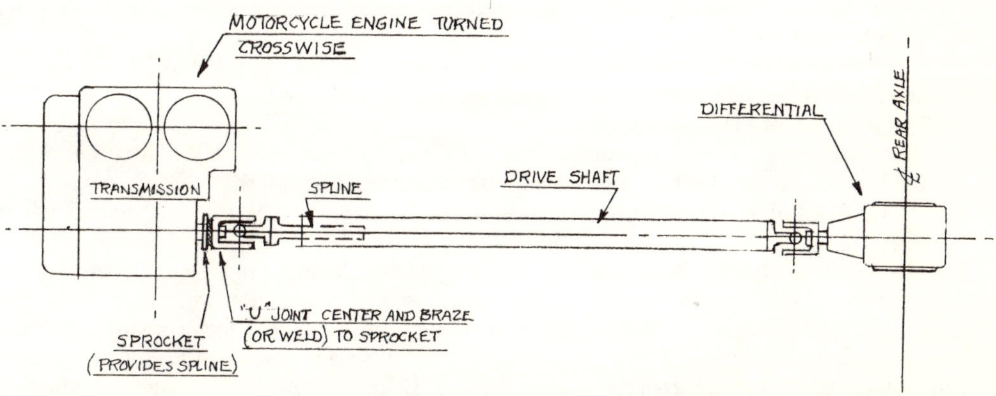

PLANVIEW – MOTORCYCLE ENGINE DRIVE TRAIN ADAPTATION
– FIGURE 47 –

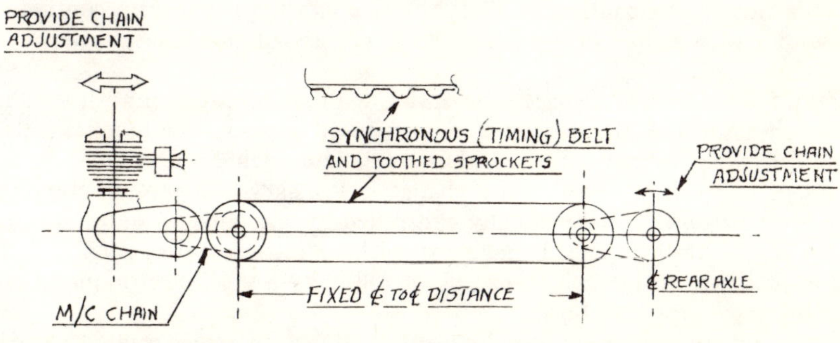

SIDEVIEW – MOTORCYCLE DRIVE BELT DESIGN
– FIGURE 48 –

EMISSIONS CONTROL FOR THREE-WHEELERS

Starting in 1979 motorcycles, including three-wheelers, require emissions control, although, due to small displacement, the equipment need not be as sophisticated and expensive as for cars. This makes it convenient to use a modern motorcycle power train for a small urban three-wheeler. Major savings in development costs are therefore possible, and this may justify renewed interest in three-wheelers. Emissions requirements, both exhaust and noise, are discussed in chapter IX.

THE V-BELT AUTOMATIC TORQUE CONVERTER TRANSMISSION

The V-belt automatic is used on many mopeds, a few motorcycles, and almost exclusively on snowmobiles and small all-terrain vehicles. Its use on small automobiles was pioneered by the production DAF of Holland. Originally, two V-belts were used, one for each rear wheel, to eliminate the differential and "U" joints for the independent rear suspensions. This is shown in figure 49, and in greater detail in figure 50. This drive principle was also used for a small Formula racer as shown in figures 51 and 52. The drive utilizes cams and counterweights for centrifugal force on the driver sheave, and a spring on the driven sheave for belt tension. An override vacuum control on the driver sheave connects to the engine to delay or to speed up shifting, depending on engine speed and throttle setting. Reverse is obtained by shifting bevel gears laterally on the centerline pinion gear from the driveshaft. The bevel gears are on the shaft that supports and drives the forward or driver sheaves as may be seen in figure 50.

There is some question of precisely equal rpm for both rear wheels if the two V-belts do not wear equally.

DAF was taken over by VOLVO and a later model, the 70 hp VOLVO 343, overcomes this problem. The two V-belts now drive a differential with conventional short half shafts and constant velocity joints.

In this country racing snowmobiles of 80 to 100 hp use but one V-belt successfully, although wear, as with tires on a car, when used hard must be faster.

Engineering data for various applications can be furnished by manufacturers such as Salsbury, reference 51. Drives from 2.5 hp up are available. Engagement speed and shift patterns may be changed to suit nearly any engine by changing flyweights, ramp angles, and springs. V-belt torque converter transmissions are simple, lightweight, and not too expensive. Their use is increasing and improvements in belts and mechanisms continue.

One of the problems is that spring tension must be adequate to prevent slippage at full power, so is therefore more than adequate at low power. This causes considerable frictional drag and a loss of efficiency at low power conditions. Anyone who has ever tried to push a vehicle, even a motorcycle, with this type of torque converter, quickly understands this. When continuous high power conditions, or nearly continuous, exist such as with a snowmobile or in racing, this is not as serious because drive efficiency is excellent then.

A recent development replaces the spring with air pressure on both the driver and the driven sheaves. The air pressure is then reduced when torque requirements are low and increased when torque increases, thus minimizing belt friction drag. A further development uses the air pressure to also change ratios, as well as provide grip on the belt. Then the shift pattern can be manually controlled by the operator or be automatic with a hook-up to the throttle and demand load. This air control uses small volume and a reasonable pressure change, which can be provided by a small electric pump as used to supply compressed air for air horns.

Another interesting recent development used a lead screw rotated electrically to shift the driver sheave with a simple spring for grip on the driven sheave. Operation is controlled electronically and may be "tailored" to any application or shifted manually. Reference 52 provides the manufacturer.

– FIGURE 49 –

The writer has used V-belt drives on several small vehicles with complete satisfaction. The DUA-DELTA, described later in this book, uses a 750 cc snowmobile engine of 45 hp and a tuned exhaust system. The V-belt torque converter was modified slightly with different weights, ramps, and springs to match the engine nicely for street use. These modifications were available from the transmission manufacturer.

– FIGURE 50 –

REVERSE DRIVE WITH V-BELT TRANSMISSION OR MOTORCYCLE GEAR BOX

When the driver sheave of a V-belt transmission is mounted directly on the engine crankshaft, the reverse must be incorporated behind the transmission. Motorcycle gear boxes normally lack a reverse, although modifications for side-car use sometimes replace one speed, usually low or first, with a reverse gear. This isn't always convenient to accomplish, though.

One of the simplest designs is used on the DAF as shown in figure 50. The pinion and bevel gears may be located after the transmission, instead of ahead of it, providing the gears are larger to accept the higher torque load.

– FIGURE 51 –

– FIGURE 52 –

Another simple design for reverse uses an electric starter motor and a fractional hp V-belt somewhat as shown in figure 53. The narrow V-belt is required to get around the very small pulley sheave on the starter. Of course, the starter pinion gear and spring are replaced with the small pulley sheave. Tightening the belt provides smooth engagement in reverse. When disengaged, the belt lies against the guides without rubbing in the V grooves of the sheaves. The schematic sketch shows the design as used on the author's DUO-DELTA. Note that the large sheeve is mounted on the driven sheeve of the torque converter. It can be on either end of the jack shaft. Note also that this provides double

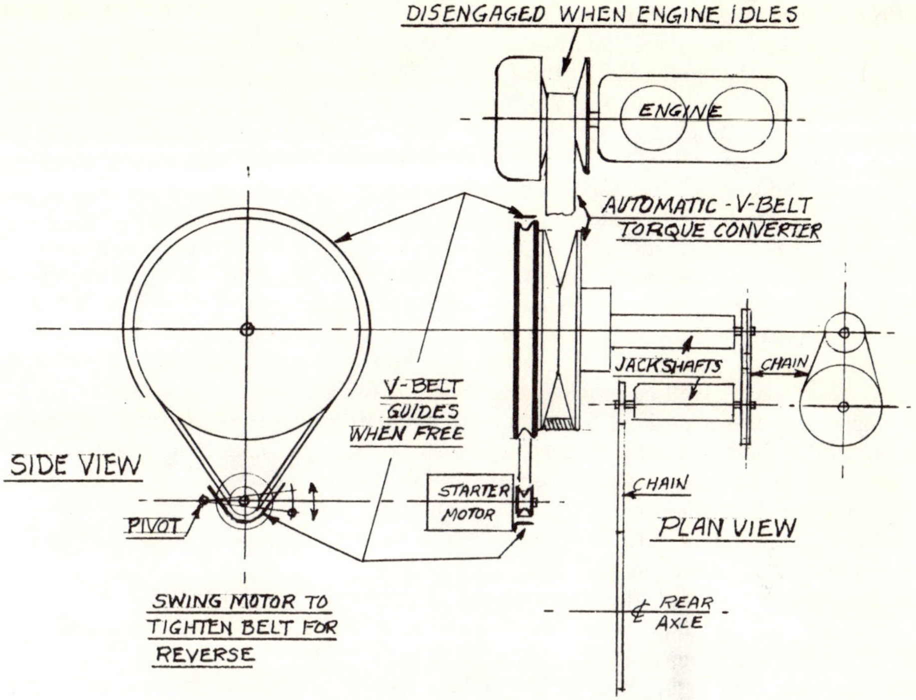

REVERSE DESIGN WITH STARTER MOTOR AND SMALL V-BELT
– FIGURE 53 –

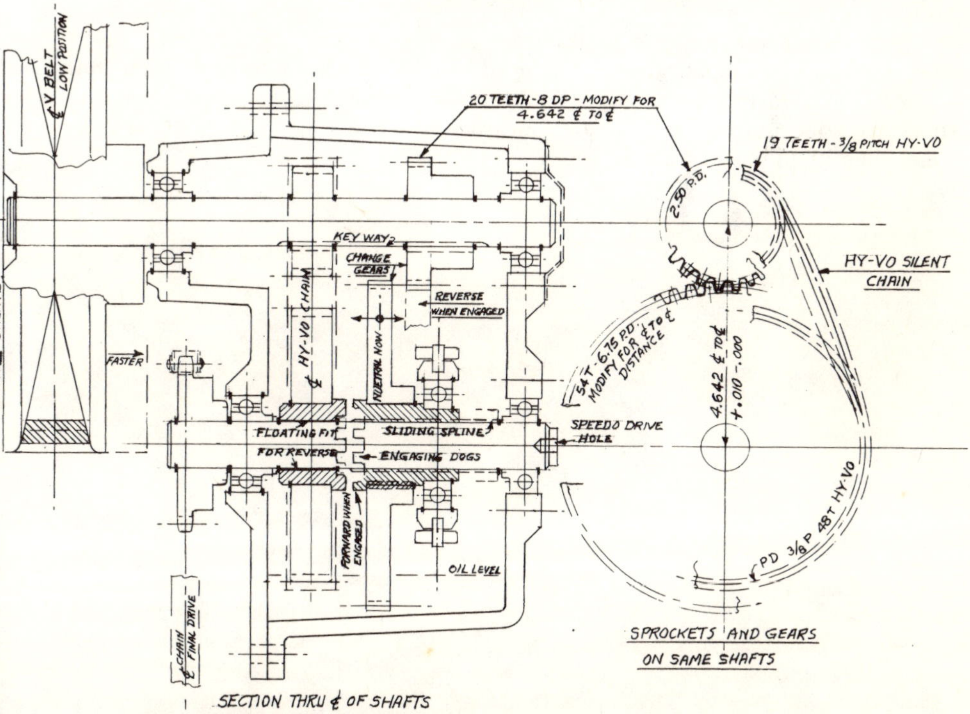

HIGH SPEED TRANSMISSION – Forward - Neutral - Reverse
– FIGURE 54 –

reduction of rpm to the rear wheel. This is important if sufficient torque is to be available at efficient electric motor rpm. Reverse speed is limited to about 8 or 10 mph, while reverse drives directly connected to the final drive chain seldom work adequately as they lack power. Another requirement is sufficient battery capacity. A large 12-volt automobile battery is needed. In fact, two batteries may be needed in some applications. This reverse is limited to parking, backing up short distances, and to small grades in reverse. It is, however, quite satisfactory when properly designed.

Figure 54 provides the details of another design which is more powerful, uses engine power, and is lightweight, although more costly. In using V-belt torque converters, the high speed position is essentially an overdrive. This then required a large reduction between torque converter and rear wheel(s), often best obtained by double reduction through an additional jackshaft. This second jackshaft is incorporated in the reverse mechanism shown in figure 54. Therefore, the unit is compact and replaces one stage of speed reduction, as well as provides the reverse. Forward speed is through the HY-VO silent chain. This also provides 2 to 1 reduction and determines the exact center-to-center distance for the shafts. It is then necessary to choose reverse gears that, when matched, will have this exact same center-to-center distance. Gear catalogs of suppliers provide this information. Even then it isn't always possible to obtain exact center distances that match the chain requirements. Gears may be cut for the application with slight modifications to achieve the center-to-center distance ordered.

In this design note that the chain always remains engaged and the gears shift in and out. These are called change gears, designed for easy engagement. The lower shaft has a spline on which the large gear and shifter unit slide. The large sprocket for the chain "floats" on the same shaft but on the spline. The unit is shown in "neutral". When pushed toward the sprocket the dogs disengage and the gears engage for reverse. Power should not be applied except when fully engaged in either forward or reverse. The shifter lever must be positive in action and have an over center hold or other means of lock in position.

This forward-neutral-reverse transmission design may require some development effort for different applications, but need not be expensive in production quantities.

Mechanical drives with sources of supply are the sole subject of MACHINE DESIGN magazine, reference issue, June of each year (reference 53). Material such as this is most helpful in power train design.

Chapter XI

POWER TRAINS – ELECTRIC AND HYBRID

Electric vehicles have improved considerably in the last few years, due to the progress in electronic controls, motors, and batteries. Since the energy problems of the mid 1970's Congress passed Public Law 94-413 to provide funds for "THE ELECTRIC AND HYBRID RESEARCH, DEVELOPMENT, AND DEMONSTRATION ACT" of 1976. This is providing additional stimulus to take electrics into the marketplace in production quantities in the 80's. To fully understand the progress made up to 1978, the reader is referred to "STATE OF THE ART ASSESSMENT OF ELECTRIC AND HYBRID VEHICLES" prepared by NASA as authorized by this act. (reference 54)

If your background and training does not include electricity and electronics, and you want to learn about them, the library in most cities can provide excellent books on the subject. References 55 and 56 are recommended in addition to whatever else is found of interest.

To know what is going on the the industry you may want "ELECTRIC VEHICLE NEWS" (reference 57). The February issue of each year includes a directory of sources of supply in the industry.

If you prefer a club, the ELECTRIC AUTO ASSOCIATION (EAA) has members in nearly every state, and many chapters with monthly meetings (reference 58). EAA NEWS is published monthly.

Electric vehicle design, although similar in principle to internal combustion engine (ICE) powered motor vehicles, requires a bit different approach due to the electric power train. Batteries are heavy, bulky, and store far less energy than a small gasoline tank on an automobile. I hasten to agree with those readers who point with pride to the hundreds of ICE vehicles that have been converted to electrics. Although successful in their time, such vehicle conversions are stop-gap designs and do not represent the true potential of the electric vehicle. Batteries stored in the trunk compartment with a few more under the hood, the extra weight carried on heavier springs and larger tires without stronger frames, larger brakes, spindles, and bearings do not represent good design. An aerodynamically inefficient body makes matters worse, although its use eliminates a great deal of design and construction effort.

ELECTRIC VEHICLE DESIGN REQUIREMENTS

Electric vehicle design represents a real challenge to the designer to achieve acceptable performance with limited power and at reasonable cost. Minimum weight, proper balance, ample strength, maximum efficiency in every detail including aerodynamics, and best utilization of limited space, are absolute requirements.

While a golf car may be completely satisfactory with a top speed of 16 mph, a small urban car would be lacking at double that speed, 32 mph. A freeway electric that barely reaches 55 mph would also not have sufficient top speed. As can be seen, power requirements go up fast. A small urban car should reach a top speed of 50 mph on the level to be able to accelerate and move with traffic on boulevards with some grades. Any freeway car must be capable of 70 mph on the level to move at 55 mph on a slight grade and with some headwind. If overloaded, or if the batteries are not up to charge, this could be difficult indeed. These numbers are a bit arbitrary and need not apply for all communities or vehicle uses, but they are realistic for most situations. The Federal government established minimum performance and design requirements as outlined in chapter IX.

The electric power train consists essentially of the electric motor, its controls, the power transmission system to the wheel(s) and the battery. These will be discussed in that order.

THE ELECTRIC MOTOR – AC or DC

Electric vehicles have been built with AC as well as DC power. Generally, large vehicles requiring more power, such as buses, vans, and trucks, often, but not always, use alternating current motors. They offer higher hp per pound of weight and eliminate the brush wear problem of DC commutator motors, but require a complex and expensive inverter system for control.

By far the great majority of electric vehicles are not large and use direct current motors. Our discussion will be limited to direct current power trains. The advantages of direct current motors are:

- They don't need sophisticated controls
- They have high starting torque
- They operate directly from battery power
- They are reliable and simple
- Motor types provide a wide range of characteristics for any application
- There are many sources of DC motors

MOTOR TYPES CATEGORIZED BY COOLING METHOD AND FIELD EXCITATION

Continuous rated power is limited by the motor's ability to absorb or dissipate heat. Present-day insulation on windings is far superior to that of early motors, so that today's motor may run hotter longer without burnout and thus be rated higher for the same size. An enclosed frame motor is protected from dust, dirt, and water, but the enclosure limits heat extraction and therefore has a lower hp rating.

If an open frame design is used, air cooling may be drawn through the motor by a fan. Under such conditions the continuous horsepower rating may be 70% or more higher than an enclosed motor. Where this puts power output on the efficiency curve is another matter, and so weight and efficiency as well as cost must be considered as "trade-offs" in the final analysis.

Field excitation methods must also be considered. Essentially this includes series wound, shunt wound, compound wound, and permanent magnet motors. In this later category we will also discuss the "pancake motor" and a recent brushless development of it.

SERIES WOUND MOTOR

This motor has the field windings wired in series with the armature as the name implies. The series wound motor has been most popular for small, slow speed, electric vehicles such as lift trucks, golf cars, and certain commercial vehicles, primarily because it offers nearly constant horsepower throughout its rpm range. Under heavy loads, the torque per ampere ratio of the series motor is higher than that or any other type, thus reducing battery drain (relatively) during acceleration, or while negotiating grades. Damage from overheating is possible, as during a long, step climb, or against a headwind. Protection must also be provided against the possibility of overspeeding under no load. Usually this consists of altering the field windings to make it a compound motor, or to include a "panic switch" to stop the current.

SHUNT WOUND MOTORS

Field windings are in parallel with the armature in constrast to wiring in series with the previous motor. Another difference is in the size of the wire. The series field consists of a few turns of large cross section conductors in series with the armature. The shunt motor field consists of many turns of smaller wire which are connected to a separate controller. The advantages of the shunt motor over the series motor are:

- Reduced pulse discharging of the battery
- Field control is smaller, less expensive

– Regenerative braking may be incorporated a bit easier

To incorporate regenerative braking in a series motor the connectors to the series field must be reversed. For a shunt motor operating above base speed, it is accomplished by increasing the shunt field current. Since the shunt field current is only a fraction of the armature current, it is much easier to control. Field weakening of a series motor to obtain higher speed is cumbersome because of the high currents that must be handled.

COMPOUND MOTOR

This motor is essentially a shunt motor with the addition of some field windings in series with the armature to provide higher torque at low rpm as with the series motor. The low current separately excited shunt field still provides control. Compounding can be added so that it limits the speed of the motor at low torque outputs, but doesn't appreciably alter series performance.

PERMANENT MAGNET MOTOR

These motors eliminate field windings by the use of permanent magnets, thus achieving a constant field – similar to a shunt wound motor. Constant field strength means it will have constant speed for any given voltage to the armature, within close limits. The P.M. motor is slightly smaller, lighter, and less costly by the elimination of field windings. It is popular for lawn mowers, sweepers, low-speed vehicles including some small motorcycles. It is less affected by overheating, but lacks low-speed high torque, as in a series motor, by about half, although efficiency is generally higher. It is less affected by temperature variations than a wire wound field type motor.

RARE EARTH PERMANENT MAGNET MOTOR

High cost has limited the use of rare earth magnets in motors until recently. Samarium-cobalt magnets offer up to six times as much energy as alnico. New designs have four times the horsepower, 66% higher speed, and much smaller size. Mischmetal, a mixture of cerium, lanthanum, neodyminum, praseodymium and other rare earths is in greater supply and less expensive than samarium, so will be more popular for many new motors. A magnet of mischmetal with about 15% samarium has an energy product rating of 75% that of samarium-cobalt magnets at much lower cost. Without doubt rare earth permanent magnet motors will eventually find their way into certain electric vehicles as developments continue.

THE PANCAKE MOTOR

In this motor the armature is the "pancake". The field is on each side of the armature (making a sandwich), rather than surrounding the outside diameter. At this writing horsepower is about 4 hp, but due to the narrow width (and large diameter) the motors may be "stacked" or assembled on the same shaft in modules as needed to achieve higher horsepower. A joint effort between motor manufacturer and vehicle designer is needed to obtain a desirable unit in production quantities at reasonable cost.

ELECTRONIC COMMUTATION IN DC MOTORS – A NATURAL FOR ELECTRIC VEHICLES

Conventional DC motors use brushes for commutation and this has disadvantages. Brushes wear and arcing puts an upper limit on motor rpm under load. Interpoles may be added to achieve higher rpm, but these increase weight and cost.

Electronic commutation was introduced on fractional horsepower motors first several years ago.

More powerful semiconductors have increased the horsepower limit greatly. Motors to 100 hp and more can now use electronic commutation economically, thanks to the rapid progress in electronics. Any kind of synchronous motor can be made with electronic commutation. If the motor has permanent magnets it becomes a brushless DC motor. Slip rings are needed for a wound rotor. Brushes on slip rings are no problem, since they do not switch current – arcing is slight.

An electronic commutator has a number of power transistors to replace the conventional mechanical commutator. These transistors are controlled by shaft position sensors on the motor. The motor is started, stopped, and speed and torque are regulated by control circuits between the sensors and power switches. Efficiency is improved because commutation is exact and power is only on to the extent needed. Reference 59 provides an excellent description in greater detail.

The added cost of electronic commutation is offset by the simplicity of adding motor controls. This is what makes it a natural for electric vehicles. An example of this is an electric bicycle motor developed in prototype in England (reference 60) and described next.

LOW LOSS DC SYNCHRONOUS MULTI-POLE TRACTION MOTOR

This interesting design is an example of a motor designed for a specific application resulting in improved efficiency and less weight, as well as having integrated features. This motor was produced at Patscentre International in England in cooperation with Sedgeminster Technical Developments. Although conceived for an electric bicycle, the design can be used for other electric vehicles by scaling up and/or by paralleling several of the pancake units. They also have a small industrial version. The article describes the motor essentially as follows.

The motor has a *stationary* circular electromagnetic coil sandwiched between two rotor discs faced with strontium ferrite permanent magnets cemented in place. The discs enclose the drum-like assembly and a planetary gear train is inside the coil. These gears drive a hub ring carrying the spokes of the wheel.

The rotor discs have the PM magnets so arranged that the N and S poles are alternating. The stator ring has 18 magnetic poles formed by blocks of "U" shaped laminations packed with two circular coils wound concentrically together. Current is switched between the two windings to reverse the effective polarity, and thus electromagnetic flux every 20 degrees or 1/18th of a turn.

The electronic commutator is a simple light emitting diode (LED) illuminating a photo-transistor through a rotating disc perforated with 18 holes. The light-chopper interrupts the beam, and the solid-state circuit is arranged so that two back-to-back switching transistors feeding the coils are alternately conducting.

The same transistors are used for speed control by shortening their "on" times relative to the "off" periods to lower the effective power fed to the coils. This pulse system is regulated by the twist grip throttle which turns a potentiometer in the timing circuits.

A separate box holds the external electronics including a circuit for regenerative braking to recharge the battery by converting the motor to a generator during braking. This is activated by twisting the throttle past "closed", thus feeding current back to the two small 12-volt lead-acid batteries. Normal range is 40 miles and top speed is 15 mph.

Reduced losses inherent in this design account for a claimed efficiency of about 85%, which is much higher than for a conventional motor.

A larger pancake motor similar to this with rare-earth PM magnets and more powerful electronic commutation could provide efficient power at reasonable weight and cost for electric passenger cars.

MOTOR EFFICIENCY AND OPERATING RANGE (RPM)

Undersized motors must work harder, tend to overheat on hills, and are not a efficient as they might be, although some designers are tempted to use them in expectation of weight and cost savings.

A few general rules are:

1 – Maximum motor efficiency occurs when average or normal power output is held at approximately *half* its maximum power output. This is because a larger motor will have *lower armature resistance* at the given power.

2 – It is also desirable to keep motor rpm in the upper half of its range. Since field control can be used (as explained later) it is also desirable to keep motor rpm up *to avoid overexcitation* and resulting *overheating* at low rpm.

3 – Excess motor rpm is also to be avoided with brush commutation. Commutation *sparking* can occur at high rpm due to the high motor current possible at high speeds with field control. The addition of commutating poles carrying armature current to avoid this requires motor redesign if a suitable motor is not available. This can be costly, both for a prototype and for later production models.

4 – Motor operating range (rpm) might ideally be a single chosen rpm – then putting full responsibility for vehicle speed change on the transmission. This would also simplify motor controls. However, an operating range of 3-to-1 or even 4-to-1 is possible with brush commutation without difficulties at either end of the range.

5 – An operating rpm range of 4-to-1 may be satisfactory for a slow-speed vehicle, such as a golf car or in-plant machine, but is not sufficient for street or highway vehicles. Even 5-to-1 is, at best, only good for a low speed of say 8 mph, and a top speed of about 40 mph. If the vehicle is to be able to creep along at about 5 to 7 mph in slow traffic and still have a top speed of 70 mph on the highway, a 10-to-1 ratio at the drive wheels is needed. Although this can be done, motor efficiency suffers unless part of the speed change is by the transmission.

MOTOR SPEED CONTROLS – GENERAL PRINCIPLES

The factors which determine the speed of a motor are the voltage across its armature and the strength of the field. If voltage is reduced to half, speed is halved. Likewise, if voltage to the armature is doubled, speed is doubled providing field strength is unchanged as in a permanent magnet motor.

If the magnetic field is *weakened*, motor speed *increases.* If the field is *strengthened,* motor speed *decreases.* By this method the speed of a *shunt* motor may be varied from *normal to about four times normal* in most cases. If more speed range is required it is usual to alter voltage to the armature – reducing voltage from normal to further lower speed.

Shunt motor speed control by means of a variation in field resistance is most satisfactory and efficient. The voltage loss (I^2R) in the field circuit resistance is small, since he field current is small compared to the armature current.

This method of speed control is also used for *compound motors,* although for any speed setting there is some variation in the speed with load, in contrast to the shunt motor in which speed is but slightly affected by load. This variation is desirable for motor ehicles in that drivers then have some feedback as to the load imposed on the motor.

Series motors sometimes have part of their current shunted around their field coils o weaken the field and thus increase motor speed.

TEPPED CONTROLS – RESISTANCE COILS

This is one of the oldest forms of speed control. Small low-performance electrics such as golf cars and some in-plant vehicles usually have resistance coils graduated into or 5 steps on a switch to provide easy, relatively jerk-free starts. This is inefficient because full strength current from the battery is always used. It is simple and less costly. Range may be reduced as much as one-third in stop-and-go driving, due to inefficiency. Foot throttle control to the switch may be by levers with a strong spring return, or a small rotary switch at the throttle may operate contactors at the resistance coils.

STEPPED CONTROLS – VOLTAGE SWITCHING (MANUALLY OR BY CONTACTORS)

Voltage switching is more efficient than resistive control. In voltage switching battery wiring routes are switched in different combinations of parallel and series selectively to achieve progressively increasing voltages in a manner similar to shifting gears if done manually. Power should be "off" during shifting to avoid arcing.

Heavy duty contactors may also be used to switch wiring routes and thus voltages. In either case the system gets quite complicated if more than 3 or 4 steps are provided. Many more contactors are needed just to add another step. Resistors are still needed for parking and low jerk starts. As vehicle performance improves, this method of voltage control becomes less satisfactory because of the need for more steps.

BATTERY SCANNING

In this system all batteries are not used all the time. Voltage is increased by adding batteries to the wiring circuit either manually or by contactors. It is a little less complex than voltage switching. It also still requires resistors for starting. While satisfactory, it is not used a great deal since it offers few advantages.

ELECTRONIC STEPLESS CONTROLS

Electronics provide a giant step forward in controls for multi-stop road going electric vehicles. The electronic power semi-conductor is the invention that makes it all possible, appearing in the early 1960's. It is used essentially as a switch, pulsing full power to the motor, then cutting it off repeatedly far faster than mechanical switches can operate. It is very efficient – about 95 to 97%.

Electronics is a whole new ballgame and has a language all its own. You must learn the symbols to understand what is going on. Again, the library will provide knowledge needed both in textbooks and in recent electronics magazines. Reference 61 lists recent S.A.E. books that deal specifically with automotive and electric vehicle electronics.

STEPLESS CONTROLS – THYRISTERS (SCR'S)

Silicon Controlled Rectifiers (SCR'S) or thyristers as they are often called, were the first solid-state electronic controls and were favored over the transistors of that time, because they could accept higher loads, had more overload capacity and were usually less expensive. The electrics built by General Motors and described in "SPECIAL PURPOSE URBAN CARS" (reference 62 and in chapter XXII) used SCR'S for stepless control. The control unit, along with its smoothing inductor and dual battery charger, was a bit bulky and weighed 63 pounds. Another example is "SUNDANCER: A TEST BED ELECTRIC VEHICLE" (reference 63). In both examples, SCR control is compared with stepped controls previously used. Wiring circuits and features are described in considerable detail in the references.

STEPLESS CONTROLS – TRANSISTORS

One of the earliest uses of a transistor control system is described in reference 64. In this example a small switching transistor as used on electric powered wheelchairs is utilized for *field current* control.

Transistors are much faster than SCR's in switching. For example, SCR's rarely switch faster than 1 kHz while transistors switch at rates from 1 to 20 kHz. This gets the "shudder" out of starts.

Now that very high current switching transistors are available and the price is dropping with increased demand, it is reasonable to expect transistors to take over the stepless control job of electrics. Complete units are available from sources listed in the February issue of reference 57 each year.

As developments continue, electronic commutation and speed control will be combined within the DC motor itself as mentioned earlier in the section on motors. This will be a great help in getting costs down as well as offering simplicity and compactness. The know-how is available, it's just a matter of production in components of a size useful for electric vehicles.

CONTROLS OTHER

BRAKING – DYNAMIC AND REGENERATIVE

The motor field can be reversed to cause the motor to become a generator and to provide some retardation. This electrical energy may be dissipated as heat through resistance coils in which case the method is referred to as *dynamic braking.* On a long, steep decline a great deal of heat is created and may require a ducted fan to carry it away. Although this method provides assistance to normal brakes at the wheels, and is relatively simple, it is inefficient.

A better way is to direct the electrical energy back into the batteries, in which case it is called *regenerative braking.* This, however, must be controlled carefully. If the field current rises too fast, the wheels will lock. Or the batteries may be recharged too quickly under some circumstances, thus damaging the batteries. Further, braking effort is not even. It is quite strong when the motor, as a generator, is spinning at high rpm, and non-effective at low or idle rpm. The return current must overcome the normal battery current. For example, a 12-volt battery, to be recharged, requires an alternator current of about 13.7 volts. Again, based on experience, a 24-volt battery should be recharged at 27.4 volts. The voltage regulator controls this. Battery systems of 36, 48, 72, or 96 volts would require increased return voltages proportionately.

Regenerative braking and dynamic braking offer better "feel" for the operator. Instead of "free wheeling" or coasting when he lets up on the throttle, the reaction is somewhat like "engine compression braking" if designed correctly.

Regenerative braking can be more effective if motor rpm is kept reasonably constant with an automatic transmission to compensate for the changes in vehicle speed. Then recharging occurs down to lower vehicle speed and this gives the driver the "feel" he expects.

When an electric vehicle is operated over considerable distances without stops, as on a highway, regenerative braking does little for the batteries. In stop-and-go city driving, especially if the streets are hilly, the effectiveness is much higher. Overall results have been reported from less than 10% to over 25% recharging, based on driving conditions and design efficiency. Possibly an average of around 15% would be reasonable to expect.

Another helpful condition is that the batteries do not suffer as much from "layering" of the charge in the electrolyte. Frequent short-time recharging helps the battery, just as does a waiting period between uses. The battery tends to stabilize a bit better.

To design regenerative braking into an electrical system with proper safeguards for a vehicle operating under all kinds of conditions, becomes a little more complicated and does add cost. In some cases it may not be justified. In the electrics of the future it will be a "must". Electronic controls, along with a proper transmission, makes the job a little easier, less costly, and worthwhile for increased range, as well as better driver reaction. Reference 65 describes an automatic field control (A.F.C.) in use with regenerative braking in considerable detail.

STORED ENERGY FLYWHEEL

Storing energy in a flywheel is, of course, a very old idea and has been used in many ways in various applications, such as on buses and trains. AIResearch, a division of Garrett Corporation in Los Angeles, under contract to the Dept. of Energy and the U.S. Postal Service, designed and built a flywheel power system for use in a JEEP type electric postal vehicle with results as shown in the performance comparison above figure 55.

Work in this field is continuing and eventually components may become available in production quantities.

BATTERY RECHARGING

Vehicle accessories are usually served by a separate 12-volt battery with a ground return as in the typical ICE vehicle. The main traction batteries operate at much higher voltages in ungrounded systems. Thus, an electric vehicle normally requires two separate charging systems. The 12-volt system may be charged by an independent 12-volt charger, or by a DC to DC converter operating from the main traction batteries. If an independent charger is used, it must have similar controls and protective devices just as used for the main traction batteries.

Custom built battery chargers usually consist of a transformer, a rectifying circuit and control devices. It is kept in the garage for overnight recharging. The designer need only provide the vehicle with a "plug-in" socket and appropriate wiring. There are many excellent sources for such chargers.

On-board chargers are provided in some vehicle designs and these may be included in the electronic speed controls as a complete unit.

TRANSMISSION AND TRANSAXLES

As the desire for improved performance continues, the advantages of a variable ratio transmission become more evident. Acceptable acceleration in traffic, with reasonable current draw on the batteries, is of prime concern. Hill climbing ability is also a must in many areas. Any electric vehicle with a top speed above 40 to 45 mph, can be improved considerably by the addition of a variable ratio transmission of some type.

MANUAL

Early designs were helped with a two-speed gear box – low range for hills and around town in stop-and-go driving – high range for faster streets and sustained cruising.

Any standard gear box can be used. The clutch is not needed if the motor controls provide jerk-free starts, or if preferred, the clutch may be retained. In any event, an adapter plate is needed along with a bearing and shaft extension. Chain or V-belt drive from the motor to the shaft will provide the proper ratio between motor and transmission.

The old VW "Beetle" transaxle has been used by many early builders. In fact, many VW's have been converted. Again, an adapter, bearing, and shaft extension is required. The transaxle arrangement is ideal in that it includes differential and half-shafts to the wheels in a single unit. Modern transaxles are available as ICE front-wheel drive, as well as rear-wheel drive. In fact, a desirable FWD transaxle may be used at the rear as easily as at the front if it suits your purpose.

AUTOMATICS

Automatics, when used with an electric motor, seldom include the torque converter. It is not needed if controls provide jerk-free starts, and it is not efficient as well as adding weight and space. Modifications in addition to the usual adapter plate, bearing, and shaft extension include the small oil pump drive for pressure to the automatic shifting mechanism. Many excellent small front-wheel drive automatics are now available from ICE powered vehicles. These can be used either at the front or rear, and modification need

PERFORMANCE COMPARISON

	PRESENT BATTERY JEEP	BATTERY/ FLYWHEEL PROPULSION JEEP	PERFORMANCE IMPROVEMENT BY UTILIZING GARRETT BATTERY/FLYWHEEL PROPULSION SYSTEM
ACCELERATION 0-30 MPH	25 SEC	12.5 SEC	12.5 SEC
TOP SPEED	33 MPH	40 MPH	7 MPH
DAILY START-STOP CYCLES	350	500	150

The system is depicted below:

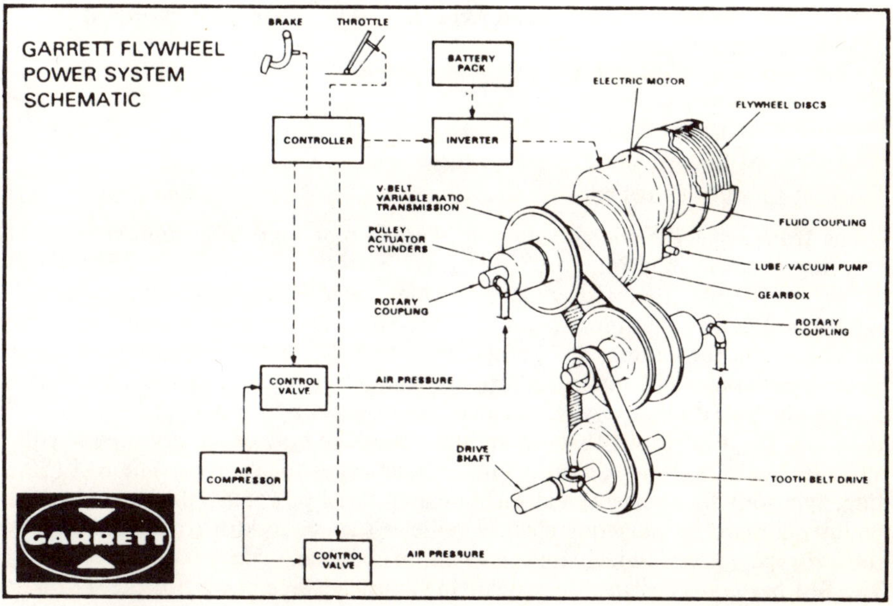

FIGURE 55

Principle flywheel characteristics are:

Rotor Diameter	=	7.5 inches
Rotor Material	=	4340 Steel
Number of Discs	=	6
Flywheel Weight	=	33.2 pounds
Speed Range	=	36,000 RPM to 28,000 RPM
Energy Stored	=	0.134 KWH

In-service results by the Postal Service are described in reference 66.

not be too difficult if the transaxle is chosen carefully.

V-belt automatic torque converters are also available and may be used quite successfully with proper adaptation. This type of transmission was discussed more fully in the previous chapter on POWER TRAINS – (ICE).

BATTERIES – LEAD-ACID

This old standby has been with us for a very long time and probably will be used for nearly all electrics through most of the 1980's. For this reason, most of our discussion will be limited to it. It is the only battery available in mass production at this time, and at reasonable cost.

The late Frederick J. Port, president and chief executive officer of ESB Inc., delivered an excellent address to the Electric Vehicle Symposium IV in 1976, titled "ENERGY SOURCES". It was reproduced in reference 67 and provides a reasonable assessment of the battery problem.

Among many other things Mr. Port pointed out, satisfactory and useful electric vehicles can be built today with existing lead-acid batteries by utilizing modern technology in other aspects of vehicle design – it isn't necessary to wait for further improvements in batteries. His supporting data is quite convincing.

The lead-acid battery of 1976 was used as a basis of comparison as to the amount of improvement that could be expected as follows:

	Lead-Acid (EV-106)	**Lead-Acid improved**
Specific Energy (Watt/hrs. per Kilo) at 1 hr rate	24	40
Cycle life	400	800
Initial Cost/kwh	$40	$45
Lifetime Cost/kwh	10 cents	5.6 cents

Going from a specific energy value of 24 to 40 represents 67% improvement. This is quite an improvement, and Mr. Port felt that the battery industry has an 85% probability of reaching it. EV-106, the battery identification, means Electric Vehicle, and 106 means the battery will deliver 75 amperes continuously for 106 minutes. A year later batteries, in production, had the minutes at 75 amperes up to 135 – a 27% improvement. Now that the Federal government is funding additional research, it is quite possible Mr. Port's predictions for the lead-acid battery will be reached, possibly by the early 1980's.

It should be pointed out that we are discussing the type of battery used in golf cars. Basically, lead-acid batteries are available in four types for vehicles. The SLI (Starting, Lighting, Ignition) for conventional gasoline and diesel powered vehicles; industrial and semi-industrial batteries powering electric vehicles such as forklift trucks and large buses; and golf car type batteries which are somewhat in between.

The SLI battery is designed to provide high peak specific power for the starter at low initial cost. The industrial battery is designed for long life and high energy, but is heavier and more expensive than golf car batteries. The semi-industrial battery falls between the industrial and golf car battery. The golf car battery is designed for relatively low initial cost, high power, and high specific energy.

The SLI is not used for electric vehicles because it is not designed for repeated deep discharging and recycling. It is kept a near full charge by the alternator on ICE cars and is rarely deep discharged. This type battery is fine for those hybrid electrics wherein the batteries are constantly charged by an auxiliary generator and never allowed deep discharge.

The industrial battery is bulky and quite heavy, but provides two- to three-times the cycle life of the golf car battery. Its bulk and weight are not primary concerns for forklifts and in-plant trucks and large buses and trucks.

Golf car batteries are used for smaller electric passenger cars because they offer a good balance in weight, cost, energy, and cycle life. These are being constantly improved and now, with the help of government funds for research and development, we should see the pace quicken.

There is still a great deal the designer can do to get the most out of his battery pack. The rate at which a battery is discharged determines the amount of battery energy that may be utilized from charge to discharge. Heavy surges of current for acceleration and hill climbing discharge the battery faster than optimum. A full range of ratios or infinitely variable ratios is even better.

There are also advantages in going to higher voltage, lower current systems to increase operating time between charges on a battery. Heavier wire sizes, shorter lengths, and good, tight connections, also reduce losses.

BATTERIES – OTHER TYPES

The electric vehicle's future looks bright indeed, when the possibilities of other types of batteries are included. While not available in production quantities, and at competitive prices at this time, it is reasonable to think that the immense amount of effort being expended on these alternative batteries will eventually pay off.

ZINC-NICKEL OXIDE BATTERY

This battery looks like the one that will be available next. General Motors has made some progress with it. Its specific energy is expected to reach 60 watt hours per kilo, at the 1 hour rate. This is 50% higher than the "improved" lead-acid mentioned earlier, but it may also cost about twice as much. Its cycle life is expected to reach 1,000 eventually, with an initial cost of $100 per KWH and provide a lifetime cost per KWH of 10 cents. Cycle life is the difficult part of the problem, and nickel is expensive, although good, and may be recovered from old batteries.

ZINC-CHLORINE BATTERY

This battery is expected to have a specific energy of 110, nearly double that of the Zinc-Nickel Oxide battery. Cycle life is down to 350 and even that may be doubtful in the minds of some engineers. Plumbing and auxiliary refrigeration are required which makes it more of a problem for vehicles. Research continues and breakthroughs may improve its chances.

SODIUM-SULFUR BATTERY

Specific energy of 125 watt hours per kilo at the 1 hour rate and a cycle life of 500 with a lifetime cost per KWH of 10 cents projected for this battery, makes it look good indeed. It is a high temperature battery with operational temperatures above 300 degrees C. Insulation is needed as well as start-up heaters. Many problems still exist, but research continues.

BATTERIES OF THE MORE DISTANT FUTURE

Zinc-Bromine battery development is underway by EXXON. LITHIUM DISULFIDE batteries are being worked on by BELL LABS, as well as EXXON. GENERAL MOTORS scientists are working on LITHIUM-IRON DISULFIDE batteries. A lifetime LITHUM-MOLYBDENUM battery system is being developed by a University of British Columbia professor. It is next to impossible to keep track of all the battery developments under way throughout the world.

In summing up, it might be fair to say that many companies now feel the lead-acid is good enough to use to enter the electric vehicle market with at this time and that, as other batteries become available, the market for electrics will enlarge and be just that much better.

HYBRID ELECTRIC VEHICLES

A hybrid electric vehicle utilizes two sources of energy, one of which is electricity. Usually the other source is an internal combustion engine powered by gasoline or diesel fuel.

Early hybrids aimed at reducing emissions by using the electric mode in the cities and the petroleum powered engine for highways. Essentially, a hybrid also has the potential of reducing petroleum usage and offers better performance than a straight electric (with lead-acid batteries) particularly more range. Public law 94-413 includes hybrids along with electrics primarily for this reason. Depending on the viewpoint taken, the hybrid can be seen as (1) more complicated and costly than a straight electric vehicle, or (2) the answer, at least in the near term, to providing a practical electric vehicle until better batteries are available. Neither viewpoint is 100% correct. Hybrids can be built that are relatively simple in design at reasonable cost, and straight electrics no longer need wait for better batteries. There seems to be an honest requirement for hybrids as well as for electrics. The hybrid's two energy sources may be arranged in either a series or parallel configuration as described next.

SERIES CONFIGURATION

Figure 56 depicts a series system. Note that the internal combustion engine drives an AC alternator-rectifier or a DC generator which is used, through the controls, to provide power to the electric motor or to charge the batteries. At times a combination of generator and battery current goes to the motor.

References 54 and 58 (Oct. '78) describe a generator hybrid vehicle designed and built by Dr. Karl V. Kordesch. Details are as follows:

VEHICLE – Austin A-40 two door sedan – curb weight 860 kg.
BATTERY – 96-Volt, 100 Amp/hr. Globe-Union – 180 kg.
ENGINE-GENERATOR – 16 hp, 7 kw, 110-Volt (Sears Industrial Alternator)
ELECTRIC MOTOR – BAKER 96-Volt DC, 20 hp peak, 10 hp continuous
SPEED CONTROL – 4-speed manual transmission plus electric circuit

The car can be driven at 60 to 70 km/hr (about 44 mph) unlimited distances based on gasoline tank capacity and can attain a maximum speed of 100 km/hr. (62 mph). SLI automobile type batteries are used since they are needed primarily for peak power and usually not deep discharged.

The engine generator delivers up to 70 amperes at 90 volts and can be considered as a continuous primary supplier of power. The battery then serves as a load-peaking device to produce additional power for acceleration or hill climbing. "Load sharing" between generator and battery occurs automatically. Under fully charged condition (120 volts) the battery receives only a charge current of 2 to 3 amps, but at 90 volts the full power of the engine-generator is available for driving or charging. The charge current tapers between 90 and 120 volts. During city driving the battery frequently supplies half the current and the generator the other half.

Dr. Kordesch has developed an excellent performing hybrid using available components. He is to be congratulated for his design.

PARALLEL CONFIGURATION

Figure 57 depicts a parallel hybrid system. In this case the I.C. engine and the motor may both be connected to the transmission and differential at the same time. The vehicle may also be driven separately by the I.C. engine or by the electric motor. In some designs the motor always spins with the engine (like a flywheel), even though it may not contribute any power. Or a clutch between engine and motor may disconnect the engine for electric-only operation. The parallel hybrid is normally a bit more complicated than the

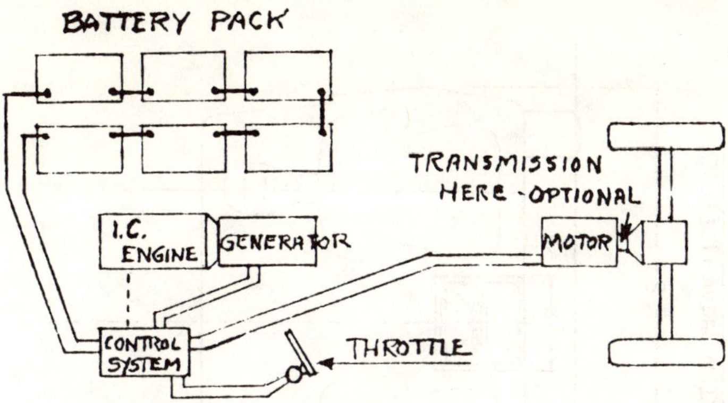

SERIES – HYBRID CONFIGURATION
– FIGURE 56 –

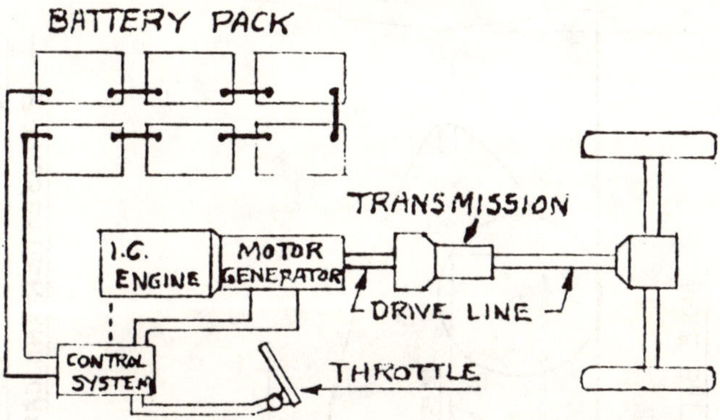

PARALLEL–HYBRID CONFIGURATION
– FIGURE 57 –

series hybrid, but it permits operation in either the I.C. engine mode only, or the electric mode only, or a combination of both.

The General Motors model 512 hybrid electric urban car described in the last chapter of this book is another version of this type of arrangement.

Reference 54 lists eighteen hybrid electric vehicles of different sizes and types throughout the world. These include passenger cars, trucks, buses, a van, and a taxi. Many have been built by large companies and some are in daily use. Without doubt there is a great deal of interest in hybrid electrics and they are finding their proper uses in the marketplace.

GENERAL ARRANGEMENT OF COMPONENTS

At first thought it appears that a great deal of latitude is possible in the arrangement of the electric components – more so than in the case of a conventional internal combustion engine design. Then it dawns on the designer that batteries take up a great deal of room and their weight affects handling, ride, and chassis design far more than realized. Other things like controls and the motor may be moved around a bit, but in the final analysis there are more components and considerable wiring circuits, including large cables, that must be provided for in working out a good all-around satisfactory vehicle design.

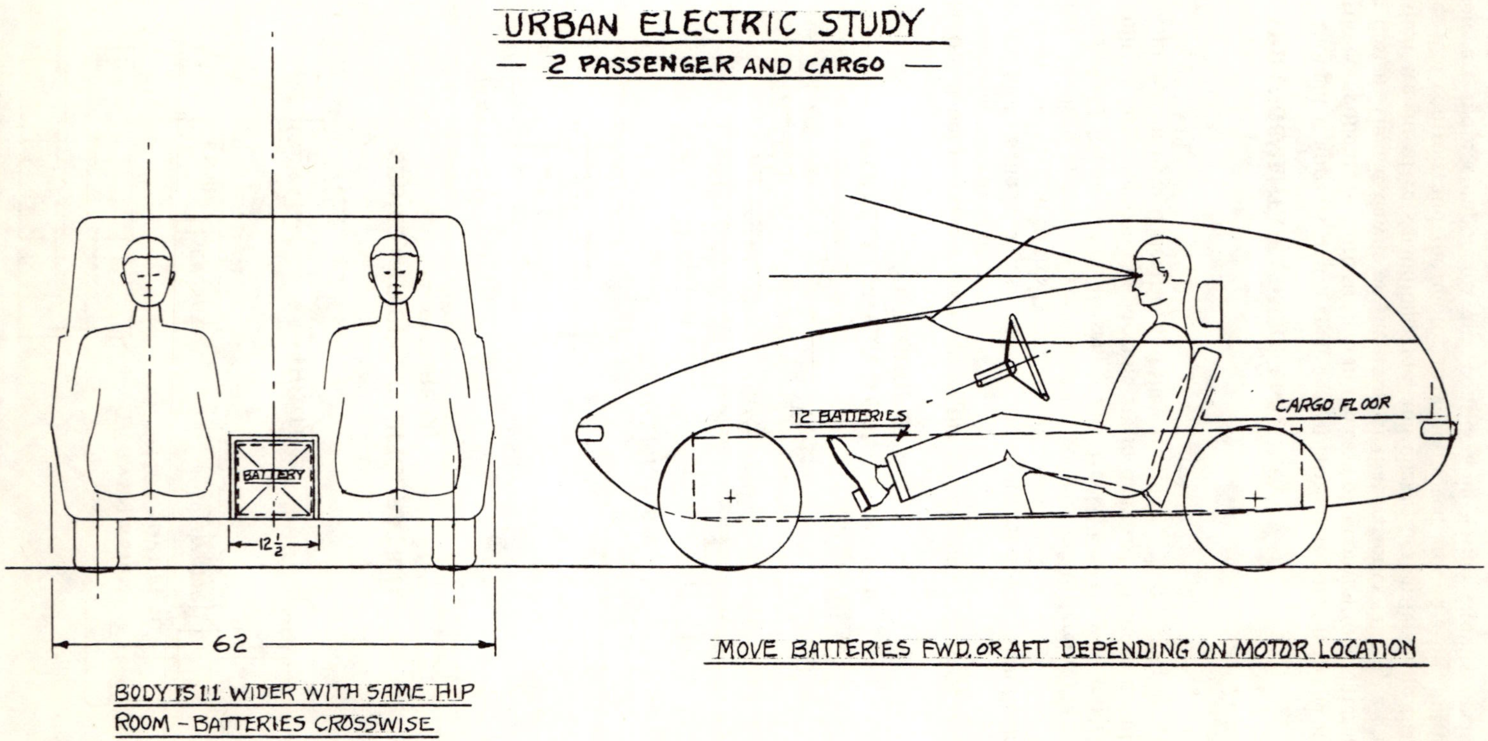
URBAN ELECTRIC STUDY
— 2 PASSENGER AND CARGO —
BATTERY
12 1/2
62
BODY IS 1.1 WIDER WITH SAME HIP
ROOM – BATTERIES CROSSWISE
12 BATTERIES
CARGO FLOOR
MOVE BATTERIES FWD. OR AFT DEPENDING ON MOTOR LOCATION

– FIGURE 58 –

BATTERIES UNDER THE SEAT

Golf cars almost always locate the batteries under the seat. This helps create a short vehicle but the arrangement is not ideal for faster vehicles. The seats must be high enough to clear the batteries and this makes them higher than needed for comfort, as may be noted in the section on HUMAN FACTORS. Additionally, the seats must fold for access to the batteries. Further, with this much weight concentrated near the center of gravity, the ride becomes "choppy" due to a low polar moment and short wheelbase. This arrangement is satisfactory for in-plant vehicles, both 2-passenger and 4-passenger back-to-back when slow speed and easy entry and exit are factors.

BATTERIES BETWEEN THE SEATS – ALONG THE VEHICLE CENTERLINE

This arrangement has been used many times in recent designs. Locating the batteries between the seats within a box frame is ideal structurally and it does distribute the weight along the wheelbase nicely for balance and ride. The disadvantages are that the batteries must be in a rack that moves on rollers, either out the front or out the back, every time the batteries are to be watered or for any other service check. A special dolly must be provided for them to be rolled onto, and private garages seldom have sufficient room for such equipment.

The intrusion within the passenger space is considerable and a wider body is required. This can be observed by comparing figure 58 with figure 29 in the chapter on HUMAN FACTORS (see page 66). The extra width is undesirable aerodynamically because of the increase in frontal area. Narrow widths are also desirable for parking clearance.

BATTERIES FORE AND AFT

If batteries are to be located out of the passenger compartment, they must be either in front of or behind it or divided between the two positions. For safety they must also be low and their racks must be well anchored to the frame. This is a good location for service access, ride and low frontal area.

In a hybrid design it might be desirable to keep the batteries under the hood only, since fewer batteries are needed. The engine-generator then could occupy the space aft, near the motor in a rear-wheel drive arrangement.

Electronic controls, an auxiliary battery for lights, etc., and other items are not too bulky and so may be positioned for balance and service access as the design details are developed.

Chapter XII

TIRES, BRAKES, STEERING AND SUSPENSION SYSTEMS

Before dealing with the problems of ride, handling, and cornering in the next chapter, we should start with the choice of tires and brakes and the geometry of steering and suspension. These things must be understood thoroughly to determine their inter-relationships and to see how they affect the dynamics of ride, handling and cornering.

The possible combinations are nearly endless. We will attempt to show the effect of each by itself in the order the designer most likely would consider them. After the prototype is completed, testing will serve to indicate further refinements or modifications.

TIRES

The choice of tires begins with the tire manufacturer's specifications. The designer knows the approximate load to be applied to front and rear tires and can find design load and maximum permissible load ratings for indicated tire pressures in the specifications. This can be determined for several tire and wheel combinations, to allow for design changes in brakes, steering, or suspension. Usually the smallest size is desired to keep unsprung weight down and to reduce cost (except for racers).

The tire manufacturer's specifications contain a great deal more than size, pressures, and design loads. Tire construction, materials, height-to-width ratio, and basic differences in design must be understood if the vehicle designer is to make an intelligent choice. Let's start with tire design. *The way the plys run is the greatest single factor in determining how a tire will behave.*

THE BIAS TIRE

Bias-ply tires have the plys laid diagonally in opposite directions at an angle of from 25- to 40-degrees of the tire centerline. They overlap each other successively to achieve the number of plys desired. This has been the standard tire construction for many years. The bias tire deflects primarily in the tread and this causes tread "squirm" as the tire rolls under load. This tread distortion is a sidewise pinching and scrubbing motion which causes more tread wear. It also reduces traction, particularly on wet pavement. At high speed a "standing wave" of distortion occurs just behind the contact patch because the tread cannot return to its normal shape quickly enough. All this flexing of the heavy tread creates heat and eats up energy.

THE BELTED-BIAS TIRE

This tire is like the bias ply with the addition of a circumferential belt under the tread. The angle of the belt cords is about 30 degrees from the tire's centerline. Belted-bias tires were introduced in 1967 in the U.S. The belt adds some stability to the tread and this tire is an intermediate step to the radial. It is measurably better than the bias-ply but not as good as the radial.

THE STEEL BELTED RADIAL TIRE

The steel belted radial tire is beyond all doubt the most popular type of tire at this time. It was pioneered by Michelin, who is also the leader with its new "TRX" which will be described later. In the radial tire the body cords extend from bead to bead in "hoop" fashion, at 90-degrees to the tire's centerline. The circumferential belts under the tread are at from 10- to 30-degrees to the tire centerline. The steel belts restrict the radial cords as to overall diameter, but the side walls flex to provide the cushion. Since the tread does not "squirm" and remains in good contact with the road surface, traction is better on both dry and wet surfaces. Cornering ability as well as traction is better. Radial tires also do not wear out as fast but cost more. The added cost is balanced by the added mileage obtained. Since radials deflect in the sidewalls rather than the tread, they also have less rolling resistance and therefore offer an improvement in fuel economy.

The disadvantage of radials is in low speed "harshness". Below about 40-mph the radial does not ride as smoothly as the bias-ply or the belted-bias with equal pressure; usually about 24 psi. Above 40-mph the radial ride improves and is superior to the other tires.

Increasing tire pressure reduces rolling resistance in any tire but also means a stiffer ride. Tire pressures must go up to reduce rolling resistance and thus improve fuel economy for the Federally mandated fuel mileage standards.

SECOND GENERATION RADIALS

The need for increased pressure and better ride led to the development of the second generation radials exemplified by Michelin TRX, Goodyear elliptic, Firestone low-flange metric (LFM), and others. These are shown in comparison with the conventional radial in figure 59. The sidewall is made more flexible by providing greater bulge or more angle as shown. Since the belt at the tread maintains the same circumference regardless of tire pressure, the flexibility of the sidewall is not changed much and a soft ride is maintained with higher pressure and thus lower rolling resistance. Additionally, the larger diameter special rim provides more space for brake and suspension while maintaining essentially the same rolling radius.

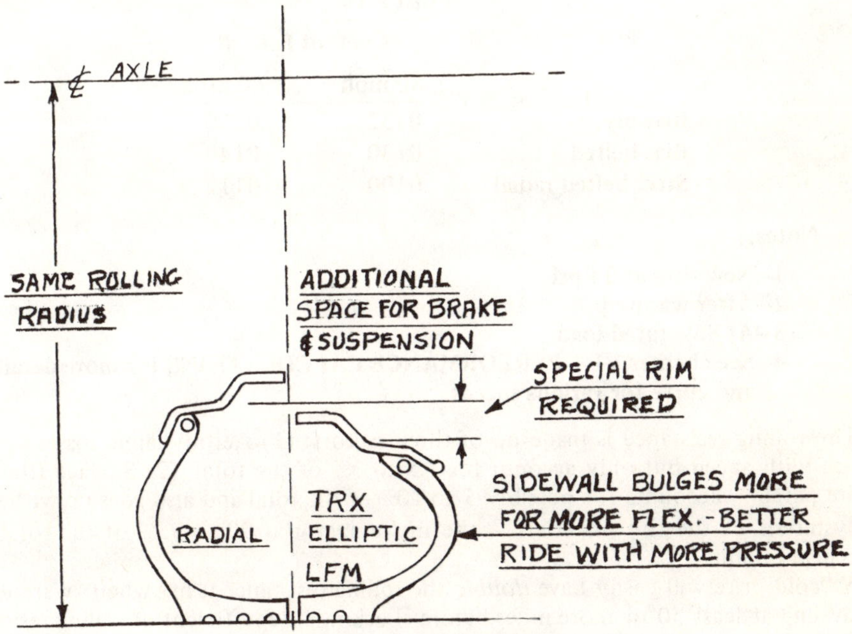

TIRE CROSS SECTIONS COMPARED
– FIGURE 59 –

TIRE DESIGNATION

The new tire designations tell more than size. They indicate type, width, height-to-width ratio, construction, and rim size. For example: P215/75R15. The "P" means passenger car. The "215" means the tire is 215 millimeters across at the widest point. The 75 means tire height (rim to tread) is 75% of its width. High performance tires have a lower number meaning they are wider. Tire construction is indicated by the "R" (Radial). If the letter were "B" it would be bias-belted, if "D" bias-ply. The last two numbers indicate the wheel diameter in inches. Tread, traction, and temperature ratings will be added but are not available as this is being written.

Racing tires are in a class by themselves with special material formulas. These are changing with every racing season. Essentially, racing tires are more "sticky" for better traction, but as a result, the tires wear much faster. A harder rubber increases mileage, as with a truck tire, but the ride suffers. Sports car tires and other high performance tires tend to favor greater widths and better traction at the expense of mileage.

ROLLING RESISTANCE

With the Federally mandated fuel economy standards, tire rolling resistance has become important and the subject of much study. Electric vehicles with limited power must also have tires of low rolling resistance.

Older books on automotive design include a formula for figuring rolling resistance, but this does not include all factors and results may be a bit misleading. We have decided to tabulate data by tire type under typical conditions and list possible variations due to wear, pressure, and percentage load as well as percentage "warm up" of the tires. Table IV lists the coef. of rolling resistance at 30 and 60 mph on a smooth surface under the conditions listed. It should be noted the variation with speed is not a straight line variation – the curve starts out flat up to about 30 mph and then slowly becomes steeper at higher speeds.

TABLE IV

	Coef. of R.R. at	
	30 mph	**60 mph**
Bias-ply	**.0132**	**.0150**
Bias-belted	**.0130**	**.0140**
Steel belted radial	**.0100**	**.0112**

Notes:

1–New tires at 24 psi
2–After warm-up
3–At 85% rated load
4–See chapter VI – PERFORMANCE CALCULATIONS for more details and curve for various speeds.

Tire rolling resistance is made up of three factors: (1) Aerodynamic losses – which increase with speed but only account for 1.5 to 3% of the total; (2) Surface friction at the tire patch – accounts for possibly 5 to 10% of the total and also goes up with speed; (3) Hysteresis of the tire materials – the major factor at 90 to 95% of the total. This causes heat buildup.

A "cold" tire will easily have *double* the rolling resistance it has when "warmed up" by traveling at least 50 or more miles before checking its coefficient of rolling resistance. The coef. drops rapidly and most of the excess resistance is overcome in the first fifteen or so miles of "warm-up". The flexing under load and rotation with resulting heat build-up reaches equilibrium by relaxing the internal stresses in the distance stated. This is another reason why fuel economy suffers in short distance city driving. Electric vehicles, of course, have the same problem.

Inflation pressure has a direct effect also. If a tire is rated at 28 psi and pressure is altered 4 psi up or down, the rollowing resistance will change about 10% also. Increased pressure yields lower resistance.

Percentage of rated load carried is also a factor; nearly a straight line variation. A tire that is "underloaded" by 10% will also have about 10% lower rolling resistance.

New and worn tires also differ in rolling resistance. A worn tire may have only two-thirds as much rolling resistance as the same tire when new. Obviously, there is less material to flex in the worn tire.

From the above it is clear that rolling resistance can vary considerably under different conditions. If the reader is interested in a more detailed study reference 68 is recommended.

TIRE ADHESION – CORNERING AND TRACTION FORCES

Tire material, height-to-width ratio, and tire construction are the major factors here. Radials are, of course, superior.

"Stickiness" of the tire's material, as with a dragster tire, yields the most adhesion. A wider tire patch on the road surface also helps. Race cars with an aft CG location and higher power require very high adhesion on the rear tires for traction under acceleration and for high speed cornering ability. Therefore, their rear tires are made of soft, sticky material and are quite wide on wide rims, to achieve high cornering forces as well as acceleration without wheel spin.

Adhesion may be used for traction under power, or for cornering to the limit, but you cannot have both at the limit at the same time. If the tire's adhesion is fully used for acceleration in a straight line, there is precious little remaining adhesion for side forces in cornering. Conversely, if the vehicle is cornering at the limit of adhesion and power is increased to the rear wheels, "oversteer" results and eventually a spinout occurs. This is why the tire cannot do it all itself – suspension geometry and other factors mentioned in the next chapter come into play, and may be used to modify the forces to achieve the most from the tires and create a superior vehicle.

Tire adhesion is quite evident when we compare the lateral acceleration, in "g" units, of different type cars. "g" stands for gravity; 1 equals a side force equal to the weight of the car. Modern racing cars reach a value of 1.15 g (dragsters even higher in traction). Modern sports cars achieve values of .80 to .85 g. Family sedans come out at from .65 to .72 g. Remember this is the "g" force for the complete vehicle. If the front tires are not working at their limit the rear tire values must be a bit higher to average out for the vehicle. Road tests in motoring magazines often provide the cornering ability of vehicles in "g" units for comparison. High performance vehicles frequently have larger or wider tires on the rear to increase adhesion there for greater traction and higher cornering forces. This goes with an aft CG location.

USE AND ABUSE OF TIRES TO IMPROVE HANDLING

Handling, as we shall learn later, is often a compromise in a high performance car. The weight bias is to the rear for high acceleration without wheelspin. The rear weight bias causes "oversteer" and adversely affects handling. It is not unusual to use larger and/or wider tires at the rear to reduce the "oversteer". Ideally, neutral or very slight understeer makes a vehicle easier to control during cornering. If the tires are kept the same size front and rear, a change in air pressure will help some. Reduced pressure in the front tires and/or increased pressure in the rear tires will reduce "oversteer" somewhat.

If air pressure and/or tire size differences do not improve handling sufficiently, the next step is to alter roll stiffness, front to rear, or to change suspension geometry, usually at the front. This causes the outside front tire (in a turn) to either be overloaded or to not roll properly for the direction in which it is aimed. In either case, adhesion suffers and tire wear increases. This is an abuse of the tire to achieve improved handling. Roll resistance also increases excessively while in the turn. Suspension geometry to accomplish this is quite common in many early cars and in some contemporary cars, but is not recommended. This will be discussed in greater detail later in the description of various suspension designs.

SUSPENSION SYSTEMS – SEQUENCE OF DESIGN

After choosing the tire size and type, the designer can start to work inside it. The wheel, brake, hub, bearings, spindle, hanger, and steering come next. If a front drive is used, the wheel should be "deep dish" to allow space for the steering kingpin and universal joint elements. Usually all these parts and components are manufactured by large suppliers or other automotive manufacturers, so the designer need only review what is available and make a choice. The choice of wheels is almost endless in steel, cast and forged aluminum alloys, and even in magnesium. If the design is for racing or a sports car, often some modifications for reducing the weight of parts and special suspension units are preferred. If a small industrial vehicle is being designed, usually front and rear axles, brakes, spindles, and other somewhat standardized parts can be chosen directly from supplier catalogs.

If the suspension system is to be unique or designed completely for other reasons, it is necessary to start with a cross-sectional drawing of the wheel, brake drum or disc and calipher, bearings, hub, spindle, and hanger to establish the kingpin angle and suspension attachment points along the kingpin axis. This would appear about as shown in figure 60. Although this indicates a front wheel with independent suspension, a rear wheel assembly with independent suspension could be similar.

CROSS SECTION OF WHEEL, BEARINGS, BRAKE, AND KINGPIN AXIS

Figure 60 depicts a recent bearing assembly unit development worthy of consideration. Flanged bearings have been around for a very long time and most bearing manufacturers create special designs on request when quantities justify the development costs. This particular design was created by SKF for the new front-wheel drives and is now in its third generation of development. NEW DEPARTURE also has a similar bearing. The inner raceway includes a large flange to accept the brake and wheel. This eliminates the need for a separate hub for the wheel. The outer raceway includes a hefty flange for the hanger, either for front or rear wheel applications. Weight and space are conserved as well as cost. The unit also includes seals. The inner diameter may be splined to accept the drive-shaft, again for either front or rear drive. Hanger design is also simplified in that it need only be designed to accept the bearing flange, not the bearing diameter along with seals and retaining means. In fact, if independent rear suspension is planned, the same hanger may be designed for both front and rear applications in many cases.

Next in sequence, after bearing choice, the brake assembly is designed. Disc brakes are used nearly 100% at the front because of their non-fade characteristic. They also can reduce weight. Excellent brakes are available in various sizes and capacities. Each supplier provides literature and the formula for choosing adequate units within different wheel sizes and for different *gross* vehicle weights. The amount of offset required at the disc is determined by the clearance needed within the wheel for the calipher. If a brake cylinder is used on one side only, an anvil with its pad is on the opposite side of the disc and the brake unit must "float" to maintain equal pressure on both sides of the disc. This "floating" feature must be designed into the hanger. The anvil is smaller than the brake cylinder so is located between disc and wheel to reduce the amount of offset for the disc.

Brake units, not shown in figure 60, are normally located horizontally, either at the front side or rear, or both, to avoid interference with the kingpin axis and suspension members. Bleed outlets also favor front or rear location for the brake units. Recent Federal brake regulations should also be reviewed for supplier compliance. Brake pad wear indicators are not always a feature on some units but should be.

Proper, approved, brake lines must be located to accept steering and suspension motion with minimum bending. It is also desirable to keep brake lines high to avoid fouling on road obstructions.

Master cylinders and reserve fluid capacity, as well as approved fluid and compatible

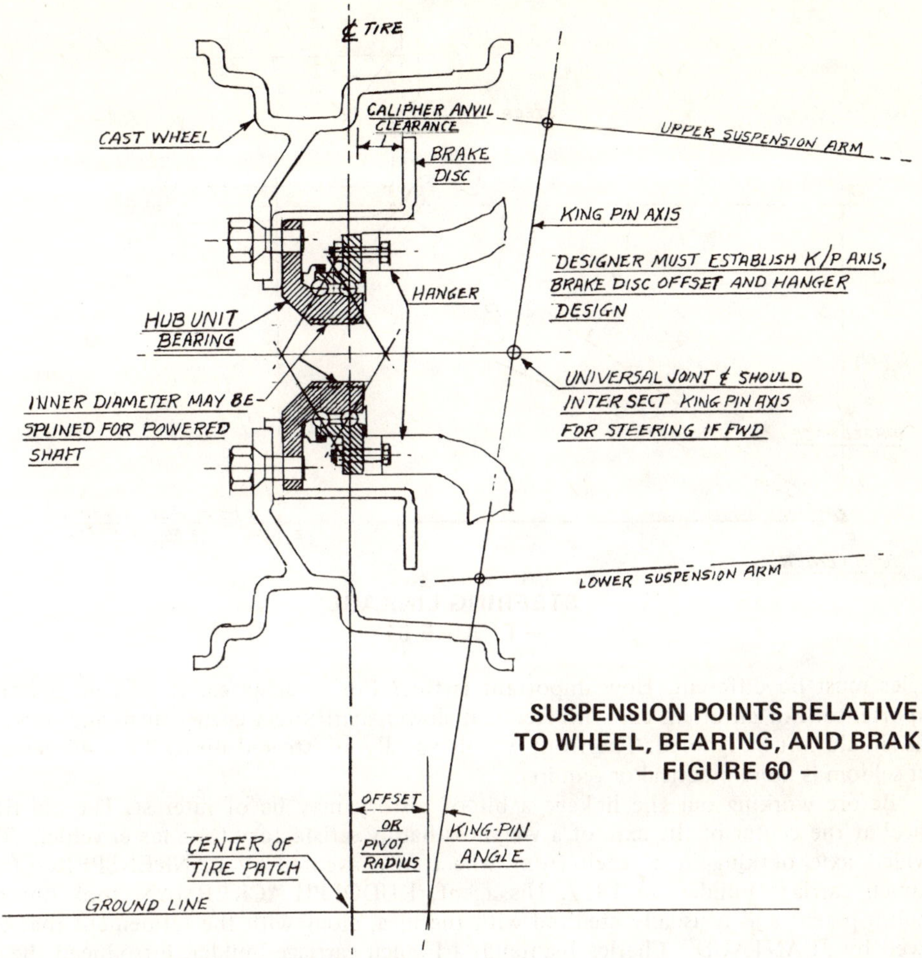

SUSPENSION POINTS RELATIVE TO WHEEL, BEARING, AND BRAKE – FIGURE 60 –

seals throughout the brake system must be determined and used.

The parking brake is an independent system for safety, but may use the same discs or drums as the service brake system. In any event is must be designed into the space provided, usually on the rear wheels, before the designer can proceed with the suspension system geometry. Again, certain suppliers have mechanical parking brake units that may be used with their hydraulic service brake units or mounted nearby.

STEERING LINKAGE

In addition to its many other functions the tire is turned relative to the vehicle's direction of travel for purposes of steering. If the vehicle is moving rapidly, and depending on its over- or under-steer characteristics, the front wheels may be turned out or into the turn during a steady state turn. The relative cornering power of the front to rear tires determines the direction and amount of steering angle required. This is a separate subject and will be discussed later.

If the vehicle is moving very slowly around a sharp turn or is being maneuvered into a parking spot, the steering wheels will go along the path in which they are aimed, and the rear wheels will follow. The action is not necessarily the same as with high speed steering during cornering. At this time we will concern ourselves only with steering at slow speeds. Figure 61 depicts the problem. Note that both front wheels have the same turn center along the rear wheel(s) axis but, due to tread width and wheelbase length, the steering

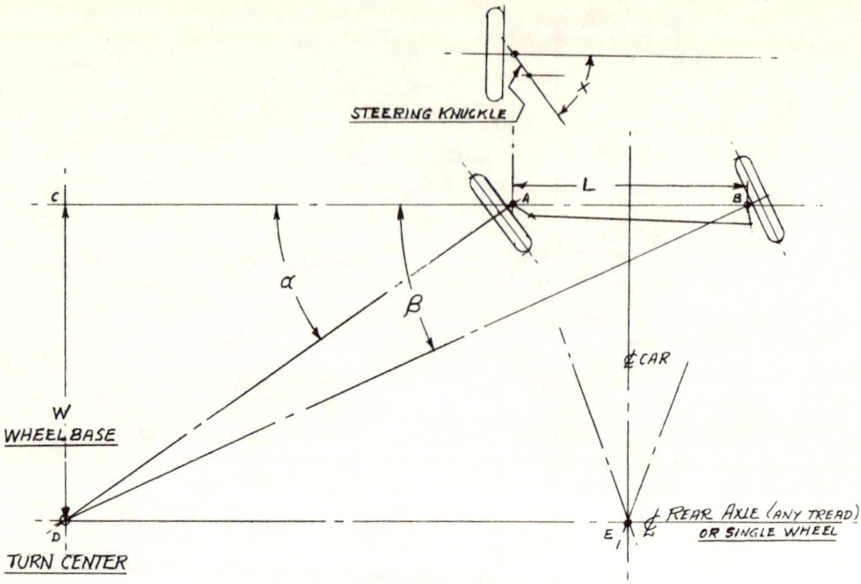

STEERING LINKAGE
– FIGURE 61 –

angles must be different. How important is this? For a racing car it is of no practical importance. For an urban car with frequent slower sharp street corner turns and parking, this should not be ignored. The front wheels usually are steered up to 35 or 40 degrees, but seldom is more practical or required.

Before working out the linkage a bit of history may be of interest. The old fifth wheel at the center of the axle of a wagon is hardly satisfactory for a faster vehicle. The divided axle or kingpin at each front wheel was invented by LANKENSPERGER, a Munich carriage builder, in 1817. His agent, RUDOLPH ACKERMAN, took out the English patent and is usually credited with the idea, along with the refinement that followed by JEANTAUD. Charles Jeantaud, a French carriage builder, introduced the refinement in 1878 known as the *Jeantaud diagram,* which we will now deal with. Older books on chassis design go into the subject in considerable detail and may be found in most libraries. Examples are references 69 and 70.

JEANTAUD DIAGRAM

Jeantaud decided that he could approximate correct steering if he aimed the steering knuckles at the intersection of the rear axle and the centerline of the car when the front wheels are straight ahead. Later it was found this was true only if the wheelbase was short or the tread width at front quite wide. Today, the intersection on the centerline would be ahead of the rear axle by a considerable amount for most cars.

Figure 61 and its identified points and angles can be used for an analytical approach as follows:

Note that $\frac{A\ C}{C\ D} = \text{Cot. a}$ and $\frac{B\ C}{C\ D} = \text{Cot. B}$

Therefore $\frac{B\ C - A\ C}{C\ D} \left(\frac{L}{W}\right) = \text{Cot. B} - \text{Cot. a}$

This equation makes it easy to plot different values of B and a for any ratio of L/W.

Unfortunately, the equation only tells us what the angles should be, but not the method of obtaining the correct knuckle design. The older books mentioned earlier and others, provide pages of data, both graphical and analytical, to determine the steering knuckle angle when the knuckle is to the rear of the wheel kingpin axis and also to the front. If the designer is using available spindles and knuckle and keeps his L/W ratio close to that of the original design, this need not concern him. If, however, the spindles and steering linkage must be designed completely, the problem must be dealt with. Many years ago I worked up TABLE V as shown herein to save myself a lot of time and work. It is reproduced for the reader's use also.

TABLE V

W	L		W	L	
Wheelbase Inches	Tread Inches	Angle X	Wheelbase Inches	Tread Inches	Angle X
100	42½	72°	100	60	66°
90	38		90	54	
80	34		80	48	
70	30		70	42	
100	45	71°	100	62½	65°
90	40½		90	56	
80	36		80	50	
70	31½		70	44	
100	48	70°	100	64	64°
90	43		90	57½	
80	38½		80	51	
70	33½		70	45	
100	51	69°	100	65½	63°
90	46		90	58	
80	41		80	52½	
70	36		70	46	
100	54	68°	100	68	62°
90	48½		90	61	
80	43		80	54	
70	38		70	48	
100	57	67°	100	70	61°
90	51½		90	63	
80	45½		80	56	
70	40		70	49	

Angle X is measured as shown in figure 61 with the front wheel straight ahead. Different L/W ratios may be extrapolated from the data provided.

Figure 61 depicts a continuous tie rod between knuckles and assumes a continuous one-piece axle. Independent suspension introduces another problem. The tie rod must then be divided to match any foreshortening due to suspension travel. For this reason it is well to locate the steering knuckle in the plane of either the upper or lower suspension member and have the tie rod length the same as its nearby suspension member to achieve a parrallelogram in planview. If the tie rod is not essentially at 90 degrees to the vehicle centerline, the relationship to the steering knuckle angle is affected. If so, a graphical study is needed to check out all conditions during suspension travel, to determine changes and how this can be corrected.

KINGPIN AXIS AND INTER-RELATIONSHIPS WITH TIRE CENTERLINE, CASTER, CAMBER, AND OFFSET (PIVOT RADIUS)

The kingpin axis must be established before we can proceed with the suspension design. The kingpin axis must be depicted in front view and side view. Heavy vehicles

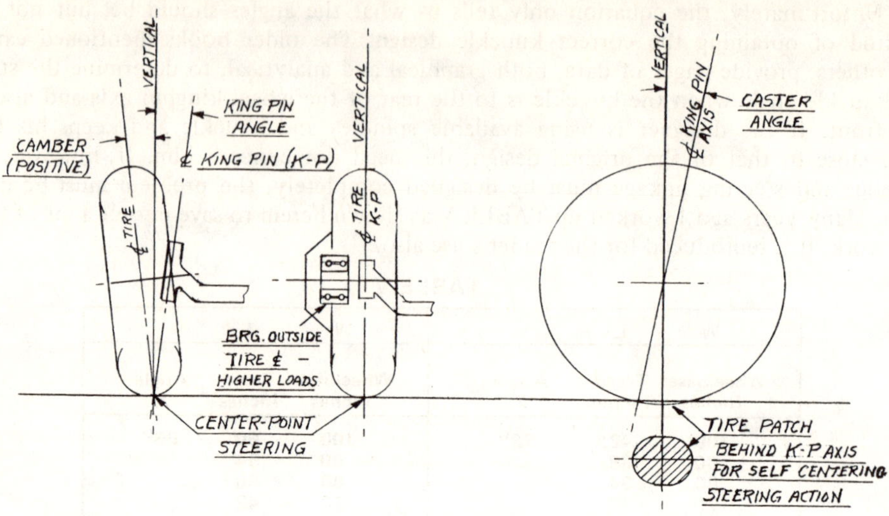

– FIGURE 62 – – FIGURE 63 – – FIGURE 64 –

tend to have the K-P axis intersect the ground line very near the centerline of the tire in both views (unless power steering is used) to keep steering forces within acceptable limits. It is also advisable to use an anti-friction bearing in the kingpin to take the high loads, especially in heavy trucks.

Figure 62 depicts centerpoint steering and camber of the front wheel. This was the method used in early designs, especially with wood-spoke wheels. The spokes were "dished" with the rim outward and the lower spokes were normally in a vertical line under load during straight-ahead motion. The outward tilt of the tire at the top, it was later learned, reduced tire cornering power. The tendancy of the tire to want to follow a circular path away from the vehicle is shown in figure 66. Another method of achieving centerpoint steering is shown in figure 63. In this case the kingpin axis is vertical along the centerline of the tire (as with a motorcycle or bicycle). This has the disadvantage of then moving the wheel bearings further outboard, resulting in higher loads on the bearings. This approach also requires a very deep dish wheel.

Going back to figure 60, we see the kingpin axis in front view with some offset from the tire centerline at the goundline intersection. This angle is typical of most cars with the power to the rear wheels.

Caster angle is measured from the vertical as depicted in figure 64. This, of course, causes the wheel to "trail" because the tire patch is behind the point where the K-P axis meets the ground. This is a strong self-aligning force tending to cause the wheels to return to straight-ahead on level ground, or to steer downhill when the car is tilted sidewise on a slope. Caster angle is in the 3- to 5-degree range for high performance cars. Heavier vehicles usually have less or none, depending on the stabilizing gyroscopic force of the spinning wheels to keep the car aimed straight ahead.

Now visualize the wheel being turned (steered) around the kingpin axis up to estreme angles to understand the inter-relationship of caster, camber, and K-P axis. For example, a vertical tire centerline in figure 60 will take on positive camber when steered in one direction. A large caster angle will cause even more positive camber if the wheel is on the inside of a turn and less if the wheel is on the outside of a turn. It is advisable to avoid large angles of the kingpin axis in front and side views to keep the self-aligning feature from becoming excessive and to avoid tire angle changes and its effect on cornering.

OFFSET – (PIVOT RADIUS)

To avoid an excessive angle of the kingpin axis in the front view it is often offset from the tire centerline as depicted in figure 60. This is sometimes called the "pivot radius". In some cars the amount of offset comes to from 1 inch up to about 1.5 inches. Since the K-P axis does not intersect the tire centerline, road irregularities cause feedback into the steering system. Unequal brake forces due to different tire pressures will cause the vehicle to pull to the low tire. A flat tire or blow-out will cause very high turning forces if the car is quite heavy.

NEGATIVE STEERING PIVOT RADIUS

The disadvantages of the offset, especially with a flat tire or other unequal retarding force, such as brake failure on one side or a different friction coefficient along the side of the roadway, caused engineers to consider moving the kingpin axis past the center of the tire to the outside slightly. This is referred to as a negative pivot radius. The AUDI-80 was probably first with the idea in 1972. Instead of a strong pull to one side, the vehicle continues straight ahead because the negative pivot radius causes a small corrective force or turn in the opposite direction. The amount of negative offset should be small. The AUDI is reported to use 0.70 inches (18 mm); the larger OLDS TORONADO 0.49 inches (12.5 mm).

In order to achieve zero or even a negative pivot radius it is necessary to get the lower suspension joint inside the brake. To do this it is often moved up near the axle centerline as shown in figure 65. This then requires the upper suspension member to be moved up also to keep loads within reason. CITROEN and others use zero pivot radius instead of either negative or positive.

Computer analysis can be used to determine the desired amount of offset (if any). Or an adjustable upper joint may be moved in or out to change the angle in the testing stage of a prototype vehicle.

VARYING CAMBER ANGLE AFFECTS CORNERING AND TIRE WEAR

Figure 62 depicts camber. This angle may vary during suspension travel as well as from steering changes. The effect should be understood before choosing between the

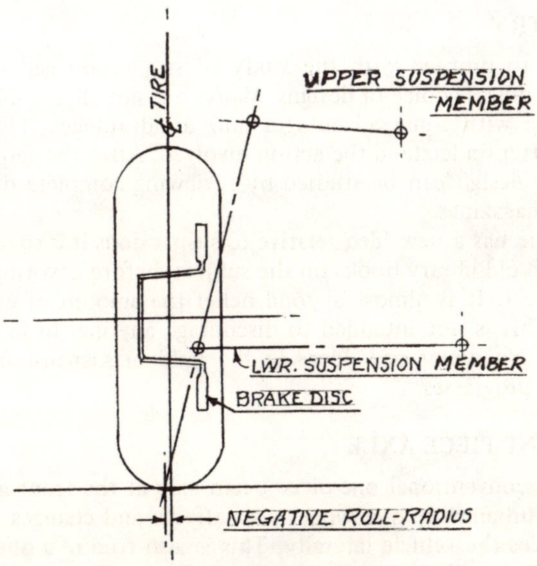

– FIGURE 65 –

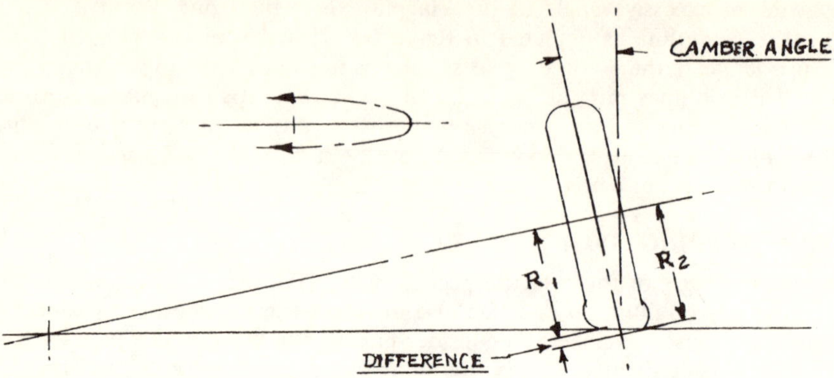

WHEEL TENDS TO STEER IN DIRECTION OF TILT
– FIGURE 66 –

many designs of suspensions. To better understand this refer to figure 66. Note that when a tire is leaning or is tilted the radius R_1 is less than R_2, causing the tire to no longer follow a straight path. If this tilt is opposite to the direction the vehicle is being steered, (because of suspension movement or body lean) the outside front tire loses traction and is being abused as mentioned earlier under tires. The tire patch is being "scrubbed". For many years this has been an accepted method of achieving "understeer" to improve handling in a turn. It results in a loss of cornering power for the wheel/tire. If this occurs on the front it is an under-steering force. If at the rear it causes an over-steering force. While this improves handling if at the front, it does so at the expense of cornering power and tire wear. This was observed quickly on racing cars that did not get around curves as fast as others that avoided the condition. Now that we know better this should be avoided on all vehicles.

SUSPENSION GEOMETRY

We are now ready to proceed with the study of suspension geometry. As can be seen there is an almost endless choice of designs. Many, but not all, are depicted. Features will be indicated, along with some advantages and disadvantages. These sketches are greatly simplified to better understand the action involved. After the geometry is decided on, suspension member design can be studied by reviewing complete drawings and pictures in available trade magazines.

If the reader feels he has a new idea relative to suspensions it is suggested he make a thorough search through old library books on the subject, before devoting too much time and money to the project. It is almost beyond belief the amount of effort already devoted to the subject. This is not intended to discourage anyone. In fact, such a search could be quite helpful. No doubt new ideas in the field of suspensions will always be needed as vehicle design progresses.

(A) – CONTINUOUS ONE-PIECE AXLE

Figure 67 depicts a conventional one-piece beam axle at the front as it rides over a bump. Note that a disturbance at one wheel also affects and changes the angle of the other wheel, and displaces the vehicle laterally. This is also true of a one-piece rear axle. One piece axles originated on horse-drawn wagons and naturally were used on the first automobiles and trucks.

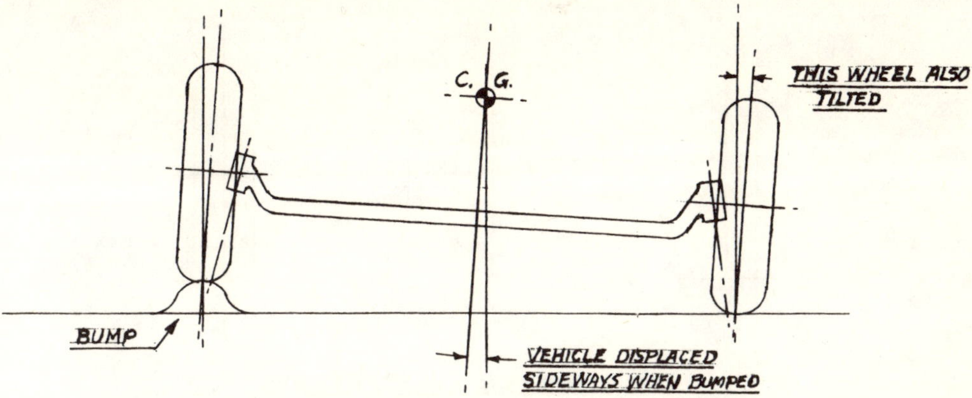

CONTINUOUS ONE-PIECE AXLE – FRONT
– FIGURE 67 –

ADVANTAGES

1 – Rugged simple design. Leaf springs may be used to locate the axle to the frame for further simplicity.
2 – Provides a high roll center – thus reducing body roll in turns. Good for trucks and other high vehicles.
3 – Readily available in supplier catalogs since these axles are used in many slow speed industrial vehicles and almost all large trucks and tractors.

DISADVANTAGES

1 – Gyroscopic forces in *both* wheels cause a precession force in steering initiating "shimmy". Requires a steering damper at road speeds. "Precession" means that when the gyroscopic forces resist the tilt of the wheels, the reaction occurs a quarter turn (90 degrees) *ahead* of the tilt – thus affecting steering. This can be studied by spinning the front wheel of a bicycle, then tilt or lean the wheel and notice the steering reaction at the handlebars. It wants to steer *away* from the lean.
2 – Disturbance at one wheel is conveyed to opposite wheel, thus reducing cornering power.
3 – Lateral body motion with every bump.
4 – Higher unsprung weight – resulting in rougher ride and reduced road holding ability.

Figure 68 depicts a typical "live" or powered rear axle. Since steering is not involved, the gyroscopic forces are not a factor. The other advantages and disadvantages still apply.

(B) – DE DION AXLE

Figure 69 shows the De Dion rear axle in planview. In this case unsprung weight is reduced by mounting the differential on the frame and using half shafts and universal joints to the wheels. It is more complicated and costly, but offers a better ride, although still a one-piece axle connecting the two wheels. The high roll center may be helpful in some cases. This approach was used on early front-wheel drives also, by the addition of steering knuckles at the wheels.

(C) – OVERLAPPING INDEPENDENT SWING AXLES – FRONT

We now begin the study of independent suspensions. Logically, dividing the front beam axle was tried with the pivot at or near the centerline of the car. It was found that

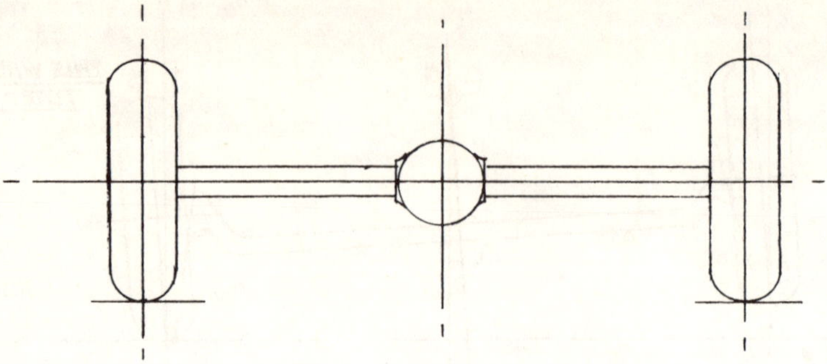

ONE-PIECE LIVE AXLE – REAR
– FIGURE 68 –

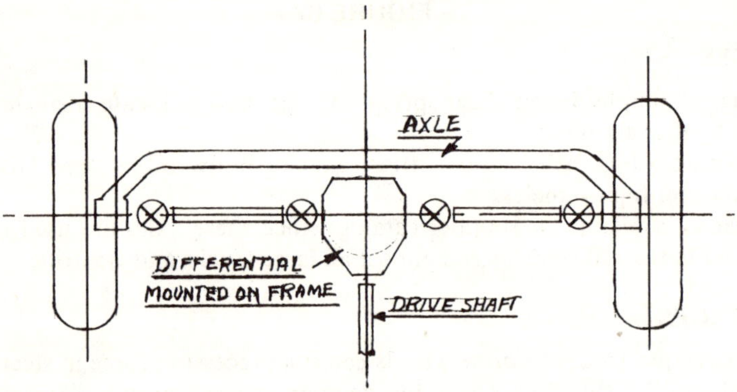

DE DION REAR AXLE – PLANVIEW
– FIGURE 69 –

this caused too much camber change with wheel travel and gyroscopic steering reaction was still quite noticeable. Also, the roll center at the centerline pivot was a bit high and sometimes caused "jacking" which will be described later.

The next step was to move the pivot points into overlapping positions somewhat as shown in figure 70. This provides a longer "swing axle" to reduce camber change and gyroscopic reacitons. The roll center is also lowered considerably, as shown in figure 71, to avoid "jacking" tendancies. This concept is quite an improvement.

ADVANTAGES

1 – Still a rugged relatively simple design.
2 – Better than a one-piece axle for ride and handling.
3 – Minimum or no lateral displacement of body with road irregularities.
4 – Any action at one wheel does not affect the other wheel.

DISADVANTAGES

1 – More costly than a one-piece axle. Radius rods must be added to fully locate the axles fore and aft. This, however, permits the use of coil springs for a softer ride.
2 – Axles are heavier with overlapping feature. Thus, the unsprung weight is still considerable.
3 – "Swing axle" effect may still be greater than preferred for some vehicles.

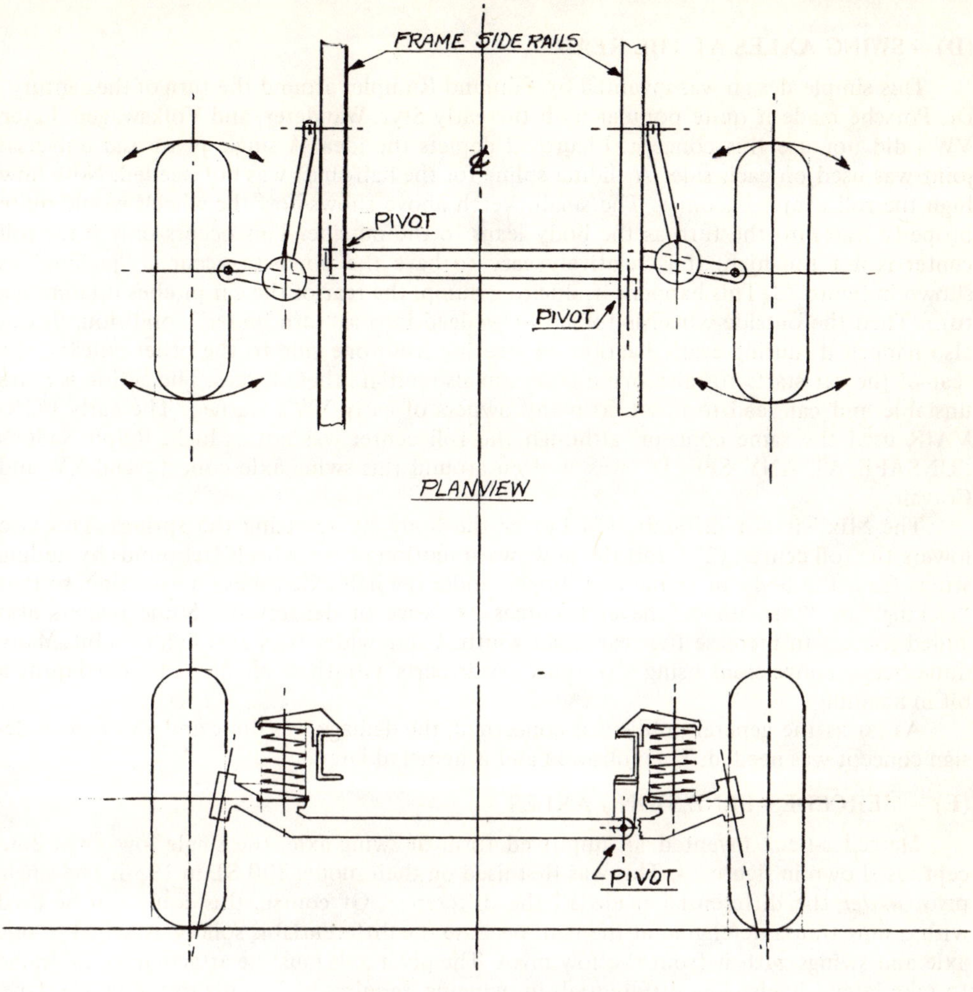

OVERLAPPING INDEPENDENT SWING AXLE SUSPENSION
– FIGURE 70 –

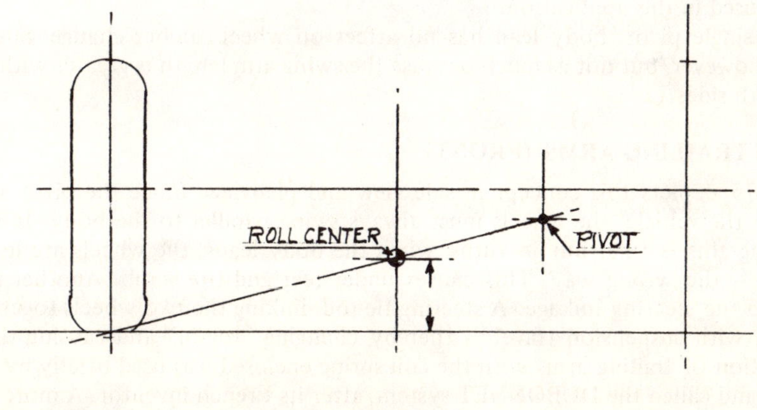

– FIGURE 71 –

(D) – SWING AXLES AT THE REAR

This simple design was invented by Edmund Rumpler around the turn of the century. Dr. Porsche made it quite popular with the early Styr, Wanderer, and Volkswagen. Later VW's did not use this concept. Figure 72 depicts the idea. A single pivot and universal joint was used on each side. A sliding spline for the half-shaft was not needed. Note how high the roll center becomes. The small sketch above shows how the wheels would quite properly lean *into* the turn as the body leans to the outside. This occurs only if the roll center is not too high. It is relatively easy to have the opposite occur – "jacking" as shown in figure 73. This happens if, due to a bump, the rear of the car pitches up during a turn. Then the outside wheel is forced or wedged into a "turn under" condition. It can also happen if, during evasive action, in steering from one side to the other quickly, the rear of the car starts to raise. Wind gusts can also initiate these forces. The action is quite unstable and can lead to upset, as many owners of early VW's learned. The early CORVAIR used the same concept, although the roll center was not as high. Ralph Nader's "UNSAFE AT ANY SPEED" was written around this swing axle concept and VW and Corvair.

The "fix" is not difficult: (1) Lower the body by resetting the springs. This also lowers the roll center. (2) Limit the downward motion of the wheels (rebound) by adding straps from the body or frame and looped under the half-axle tubes on each side, so that "jacking" or "turn under" never becomes excessive or dangerous. Some owners also added spacers to increase the rear tread width. Using wider tires also helped a bit. Many dune buggy conversions using VW and Corvair parts with these changes improved quite a bit in handling.

As far as the general public was concerned, the damage was done and a change in design concept was needed. This followed and is depicted later.

(E) – MERCEDES-BENZ SWING AXLES

Mercedes-Benz invented an improved form of swing axle, the single low pivot concept, as shown in figure 74. This was first used on their model 190 SL in 1956. The single pivot *under* the differential made all the difference. Of course, this could not be used with a unit transaxle/engine at the rear, because the differential is solidly attached to one axle and swings with it from the low pivot. The pivot axis must be attached to the frame to take lateral loads. The differential, in swinging, requires "U" joints to the powerplant.

Since the pivot is not at the half-shaft "U" joint, a sliding spline is needed on the side away from the differential. Under high torque loads from the engine a normal spline would resist sliding. This would cause the suspension to stiffen up or bind, due to the leverage from shaft centerline to pivot. Therefore, a ball bearing anti-friction spline is needed and used in this application.

With a single pivot, body lean has no affect on wheel camber change. Suspension travel does however, but not as much because the swing arm length is greater with a single pivot for both sides.

(F) – TWIN TRAILING ARMS (FRONT)

Figure 75 depicts this concept in side view and planview. Since the pivot axis runs crosswise to the vehicle, the wheels must always swing parallel to the body. In straight-ahead driving this is fine, but in turns, when the body leans, the wheels are leaning to the outside – the wrong way. This causes understeer and tire scrub. Another problem arises due to the steering linkage. A steering tie rod, linking the two wheels together, will fore-shorten with suspension travel – thereby changing "toe-in" and causing tire wear.

A variation of trailing arms with the coil spring enclosed was used briefly by Chevrolet in 1934 and called the DUBONNET system, after its French inventor. A more popular trailing arm design was used for many years by VW. In this case torsion bars running crosswise provided the springing.

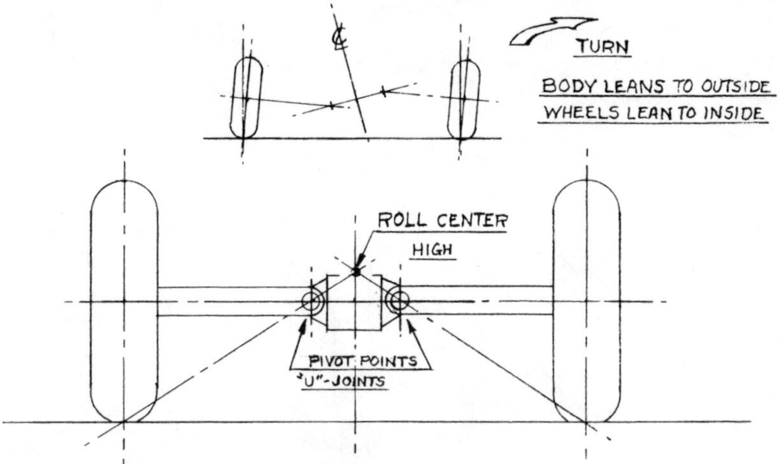

SWING AXLE – REAR
– FIGURE 72 –

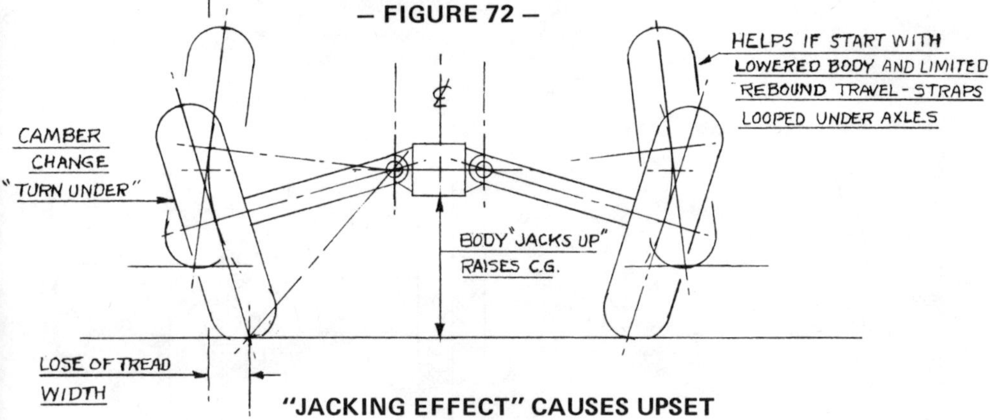

"JACKING EFFECT" CAUSES UPSET
– FIGURE 73 –

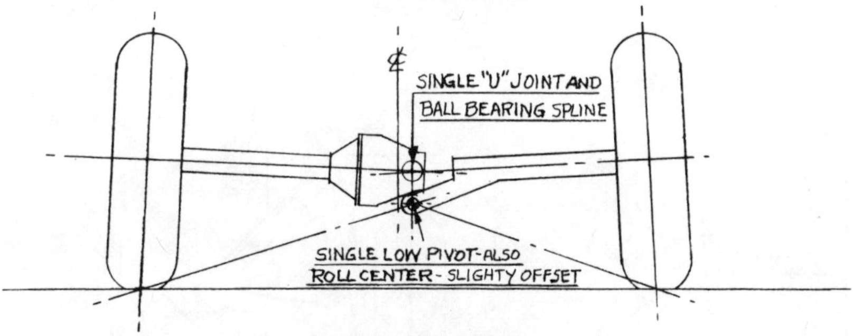

MERCEDES-BENZ SWING AXLE
– FIGURE 74 –

Twin trailing arm suspension is no longer used on cars. It is obsolete in design and inferior in cornering ability.

(G) – TRAILING ARMS (REAR)

This simple idea depicted in figure 76 is used frequently with front-wheel drive vehicles to this day. It provides low unsprung weight and independent rear suspension (IRS) for a better ride. It also conserves space. Loss of cornering power with body lean is still present but is not excessive because power is applied to the front wheels.

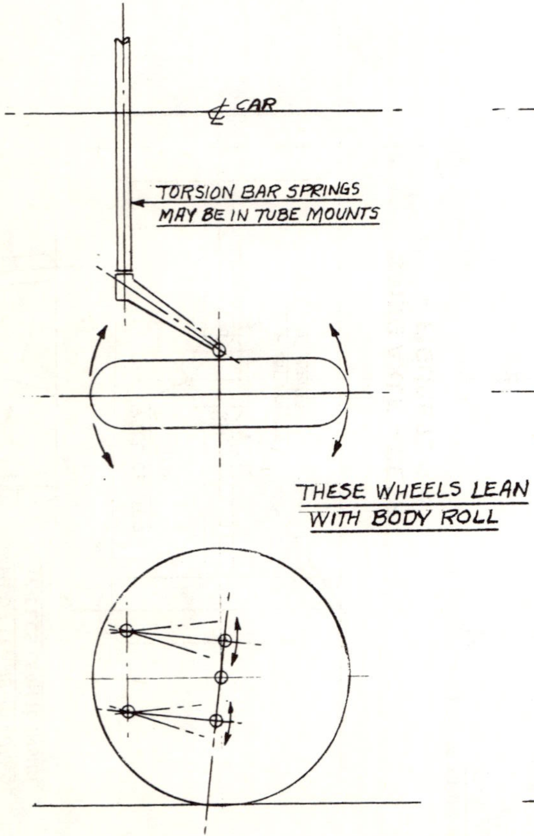

TWIN TRAILING ARMS — FRONT
— FIGURE 75 —

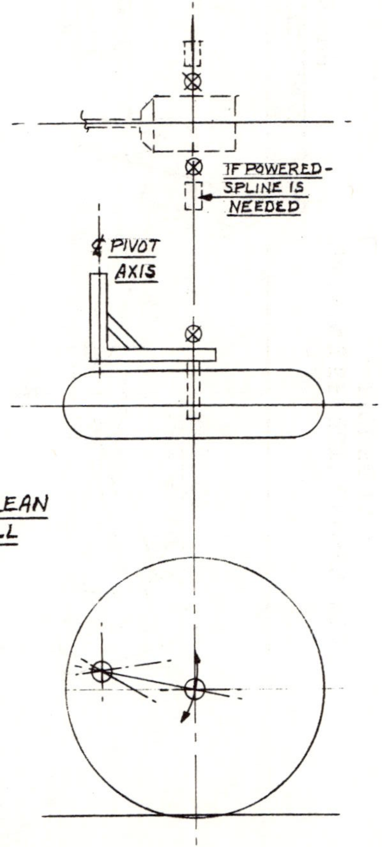

TRAILING ARMS — REAR
— FIGURE 76 —

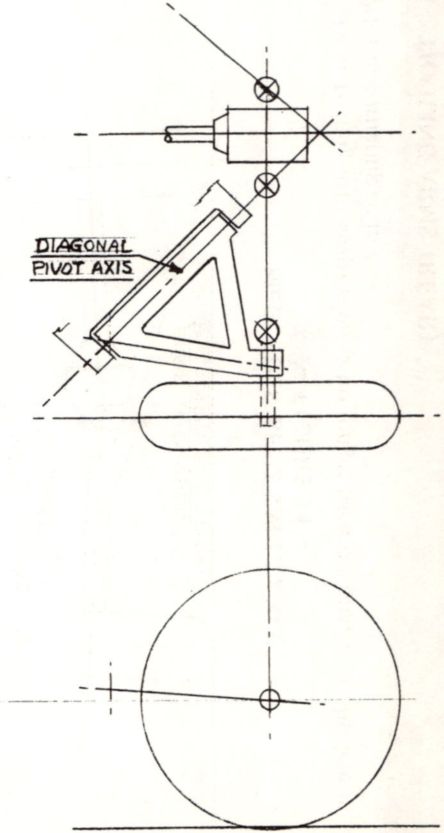

SEMI-TRAILING ARM WITH SOME
SWING AXLE EFFECT — REAR
— FIGURE 77 —

In dash lines a differential and half-shafts are shown to point out that a spline on each side is needed if power is applied to the rear wheels in this type of suspension. If power is applied to the rear wheels it should only be done with a low speed vehicle to avoid loss of cornering power with body lean.

(H) – SEMI-TRAILING ARMS WITH SOME SWING AXLE EFFECT (REAR)

This concept, shown in figure 77, overcomes two of the disadvantages of the pure trailing arm rear suspension system. First, the wheels no longer lean with the body. Since the pivot axis is diagonally located some "swing axle" effect occurs. Second, this idea works nicely with power to the rear wheels because cornering power is not lost when the body leans. If the pivot axis is aimed through the centerline of the inner "U" joint a sliding spline is not required.

Of course, the idea may be used at the rear with power to the front wheels as done by General Motors.

It should be realized there is some slight steering effect with suspension motion up or down. This can be helpful if done right. It can also cause "toe-in" changes between rear wheels and the resulting tire wear if motion is excessive.

Space is conserved and this design is a reasonable compromise between pure trailing arm and swing axle designs.

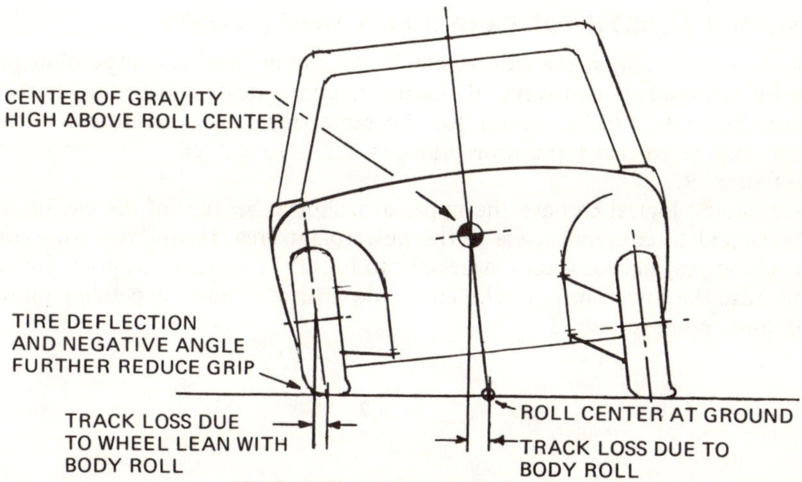

TWO "A" ARMS PARALLEL
– FIGURE 78 –

(I) – TWO A ARMS – PARALLEL (FRONT)

Two A arms for the front suspension is and has been the most common layout for many years. The British refer to them as wishbones. They come in several variations as can be observed by studying different vehicles.

An early version had the two A arms parallel as shown in figure 78. Parallel arms of equal length result in wheel travel parallel to the body, leaning when the body leans, much as with trailing arms. This was an early way to achieve understeer with independent front suspension (IFS). The disadvantages are evident in figure 78. The wheels lean the wrong way for high cornering power and tire scrub is excessive. The roll center is at ground level causing more body lean. The loss of track or tread width is also considerable leading to possible instability with overturn in extreme cases.

The double A arms are a good idea, but paralleling them leaves room for improvement, as we shall see next.

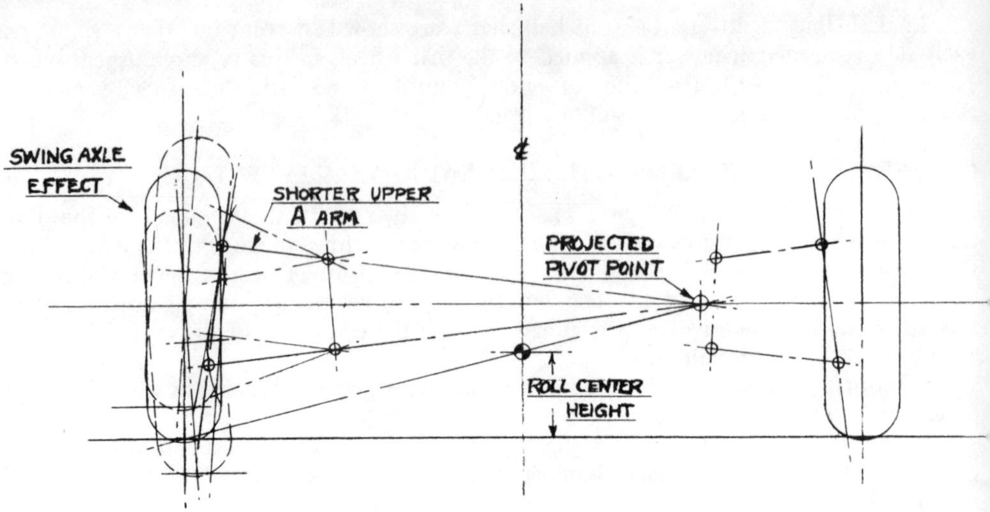

UNEQUAL LENGTH NON-PARALLEL "A" ARMS
– FIGURE 79 –

(J) – UNEQUAL LENGTH NON-PARALLEL A ARMS (FRONT)

The need for some swing axle effect is evident in previous suspension geometry. This can be achieved in two ways. If the upper arm is made shorter in length than the lower arm, its angle will be greater for the same travel and therefore fore-shortened. The other way is to make the arms non-parallel – have them converge somewhat as shown in figure 79.

It is easy and logical to have the upper arm shorter because of the inclination of the kingpin axis, and to conserve space in the powerplant area. Having the arms converge is also easy and logical since it is only necessary to lower the attachment point for the upper arm or to raise the lower arm attachment or do a little of both, depending on where the projected pivot point is desired.

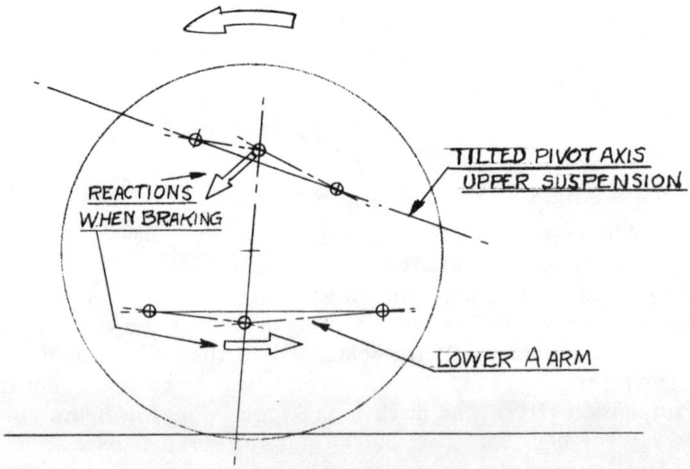

ANTI-DIVE FEATURE – FRONT
– FIGURE 80 –

(K) – ANTI-DIVE FEATURE (FRONT)

The front end of vehicles with soft springing tend to dip or dive when braked hard. This is due to the transfer of weight to the front with braking and the resulting deflection of soft springs. This can be partially alleviated by tilting the upper A arm pivot axis (in side view) somewhat as shown in figure 80. Then, when braking, the twisting couple applied to the kingpin at the upper A arm causes a reaction tending to pull the A arm downward or push the vehicle upward due to the angle of the pivot axis. The braking reaction at the lower A arm is essentially to the rear for stopping because its pivot axis is approximately level.

(L) – ANTI–LIFT FEATURE (REAR)

The same transfer of weight occurs at the rear when braking, except it is away from the wheel, tending to lift the rear of the vehicle. This, too, can be partially overcome by using a trailing arm or semi-trailing arm at the rear. The reaction (wind up) from the brake mounting plate then tends to rotate the trailing arm upward to prevent lift at the rear.

(M) – UNEQUAL LENGTH NON–PARALLEL A ARMS (REAR)

Racing design initiated this development to keep the tires in constant contact with the roadway for improved cornering and to handle the additional power to the wheels without wheelspin. Figure 79 again illustrates the geometry and wheel travel. Half-shafts from differential to wheels must include a low friction spline or provide for endwise mo-:ion in the type of universal joint used either inboard or at the wheels.

This type of rear suspension is usually used with mid-engine and transaxle designs, ɔoth for racing and for sports cars. Although excellent in performance, it takes up too much space for other vehicle types.

N) – TRANSVERSE SPRING REPLACES A ARM

This is an old idea, but still used to simplify or reduce parts in a suspension design. \ good example is the old FORD TAUNUS 12M of Germany as pictured in figure 39 of :hapter X. In this case, the transverse leaf spring replaces the upper A arm. Other de-igns have used a transverse spring at the bottom to replace the lower suspension member. Vhen this is done it is usually necessary to add drag links or rods to the frame to take ıigh braking loads. In a few old examples transverse springs are used top and bottom – naking essentially a very simple parallel arrangement with action somewhat as in figure '8.

O) – HALF-SHAFTS REPLACE A ARMS

The CORVETTE and the JAGUAR E type use their half-shafts unsplined, but "U" ɔinted at both ends as the upper A arms. A single lower arm is also used with trailing rms upper and lower to provide the fore and aft location on each side. Figure 81 de-icts the 1963 CORVETTE rear suspension along with its transverse spring under the dif-erential and the roll center height of 7.56 inches. Later models lowered this height to .71 inches. Eliminating an anti-friction spline on each side does simplify and reduce osts.

If ample space is available the lower suspension member could be removed and an pper suspension member installed above the half-shafts to provide more ground clear-nce.

) – MAC PHERSON STRUT VARIATIONS (FRONT)

Earle S. MacPherson of the Ford Motor Co. invented this strut for front suspensions ı the mid-1940's. It was introduced on the English Fords in 1950, has been used on

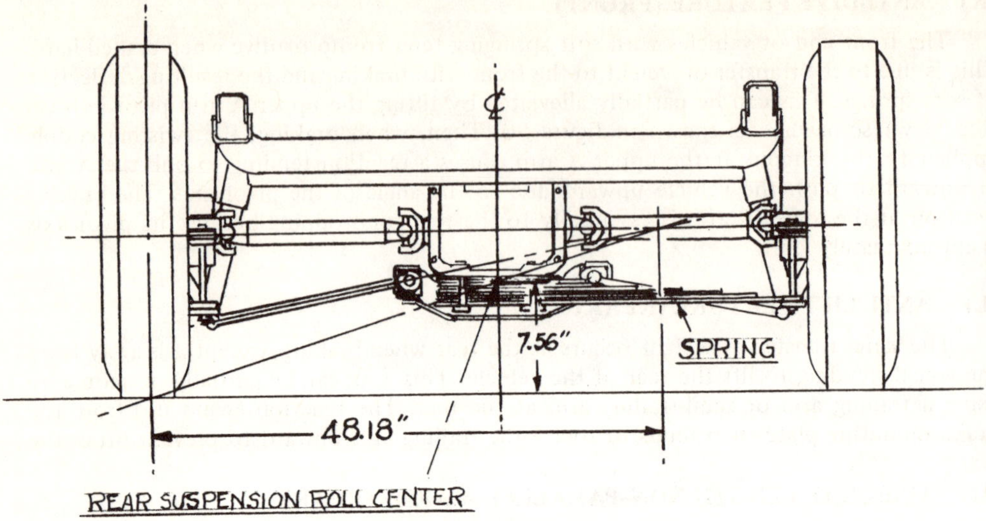

DRIVE SHAFTS REPLACE "A" ARMS
– FIGURE 81 –

several other small cars since, and now is popular again on the 1978 Ford Fairmount and Mercury Zephyr in modified form. Other variations are on the Dodge Omni and Plymouth Horizon front drives as well as on the Ford Fiesta. The principle of this strut is depicted in simplified form in figure 82. Note that the shock absorber is located on the kingpin axis and serves as the steering pivot. The coil spring surrounds the shock absorber also, thereby utilizing the same anchorage. This also conserves space in the engine compartment, simplifies assembly, and reduces the number of parts.

It should be noted that steering will cause wind-up or un-wind of the coil spring, unless a low friction bearing is provided at one end of the coil spring.

This strut, when used at the front wheel of a rear drive car, may be located lower than shown for a front wheel drive wherein the axle shaft and "U" joint must be cleared.

Increasing the offset at the tire patch and tilting the kingpin axis more, provides greater clearance at the top of the tire. These changes and altering the slope of the lower A arm relocates the pivot point and roll center. In any case, much of the same wheel motion results as with unequal A arms.

In some variations the spring is separate from the kingpin axis (inboard) to avoid too much height above the tire.

(Q) – CHAPMAN STRUT (REAR)

Colin Chapman used the MacPherson strut principle at the rear on his ELITE and ELAN models. In this case the spring/shock absorber strut unit may be mounted ahead of, or behind, the drive shaft since steering is not involved, thus keeping overall height within reason. Otherwise the idea remains somewhat as shown in figure 82. Again space is conserved and the design remains simple.

SUSPENSION MEMBER END FITTINGS

Nearly all passenger cars use rubber-like bushings in end fittings to absorb some shock and to isolate noise and vibration. These may be replaced when worn to restore alignment.

Race cars, especially open-wheel designs, prefer more precise geometry, so therefore use solid bushings. This adds harshness and conveys noise and vibration, but is perferred for responsive handling.

An adjustable feature can be obtained by using threaded rod ends but two things should be carefully avoided to prevent failures due to design. First, observe that threaded rod ends are available as either male or female. The female is depicted in figure 83 and should never be used in any structural application in the suspension or steering linkage. The rod end itself will not necessarily fail but the threaded male insert from the suspension arm or drag link will eventually fail due to fatigue stresses in the threads just beyond the rod end tube as shown in the sketch. This can be avoided by using a male rod end and threading it into a female suspension member as shown in figure 84.

The other thing to be avoided involves the quality of the steel and its heat treat in the rod end. Only structurally rated aircraft quality rod ends should be used. Soft non-structural free machining steels will not take the beating imposed on a suspension system, including the steering linkages, so should be avoided. They elongate and break off due to repeated fatigue loads. Through-hardened steel of SAE 4130 alloy or better should be used.

The ball end also deserves attention. The method of retention is important. If retained by limited staking, it may eventually pound out or at least become loose. A careful study of the literature of only the highest quality rod ends will clarify these differences in products available. Additionally, in many applications it is highly desirable to use rod ends that include a lubrication fitting for the ball end.

Since suspension member applications are subject to very severe use, only heavy duty rod ends should be used, well within their design limits. If in doubt choose the next size larger for longer trouble-free service.

In summarizing the portion of this chapter dealing with suspension geometry, the writer wishes to point out that the sketches, figures 67 through 82 are simplified to more clearly deal with geometry only, to show wheel motion, and how the roll center is determined. Detail design of suspension members can be studied by reviewing the literature provided by manufacturers and as depicted in trade magazines.

The addition of springs of different types will be studied next, followed by shock absorbers, bumper stops, and anti-roll bars. These, together with suspension geometry, tires, and other factors, determine vehicle handling.

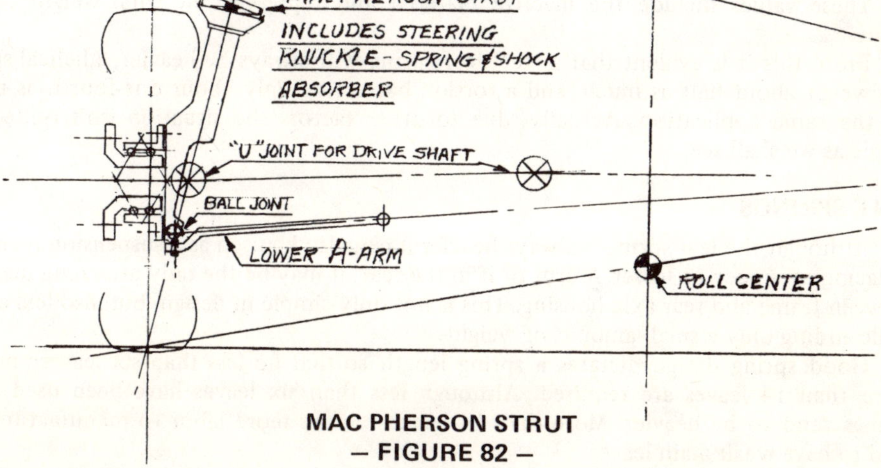

MAC PHERSON STRUT
– FIGURE 82 –

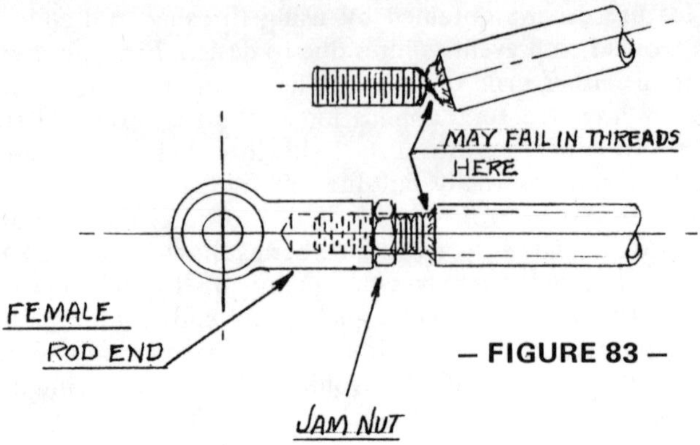

— FIGURE 83 —

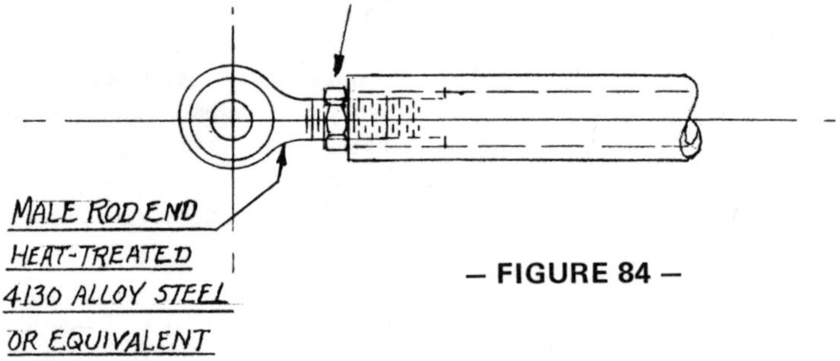

— FIGURE 84 —

SPRINGS

Springs absorb and store energy when deflected, to be returned later, usually a very brief moment later. The stress in the spring builds up during deflection but must not exceed a certain maximum to avoid premature failure. This limits the amount of energy that can be stored. Steel springs predominate in motor vehicles and these safe limits have been followed for many years. More recent developments may raise the values a bit, but relationships remain somewhat as listed below.

- Leaf spring with properly stepped leaves – 300 to 450 inch pounds energy per pound of spring.
- Helical (coil) round wire compression spring – 700 to 1100 inch pounds energy per pound of spring.
- Torsion bar spring – 1000 to 1500 inch pounds per pound of spring.

These values include the inactive material and represent the total weight of the spring.

From this it is evident that a steel leaf spring will always be heavier, a helical spring will weigh about half as much, and a torsion bar spring only about one-fourth as much for the same application. Actually, due to other factors the situation isn't quite that simple as we shall see.

LEAF SPRINGS

Although the leaf spring is always heavier it can also be used as a suspension member, replacing an upper or lower A arm or if in the rear, it may be the only attaching member between frame and rear axle housing. This is not only simple in design, but also less costly while adding only a small amount of weight.

Good spring design dictates a spring length so that no less than six leaves and not more than 14 leaves are required. Although less than six leaves have been used, such springs tend to be heavier. More than 14 leaves require more labor to manufacture and tend to have weak main leaves.

Interleaf friction interferes with a soft ride and is overcome by the use of inserts between leaves.

Overload leaves are added, especially for trucks. These only contact the main spring when it deflects sufficiently, so that the spring is not too stiff when lightly loaded.

Unsymetrical spring design is used to overcome or reduce "wind-up" due to high driveline torque at the differential pinion gear.

All sorts of features may be incorporated and are fully covered in references 71, 72 and 73. These are recommended for anyone about to design a steel leaf spring.

Single leaf springs, tapering in thickness or width or both, have been made in steel and more recently in lighter materials. These are efficient and lightweight.

TORSION BARS

As noted earlier, a torsion bar can store about four times as much energy as an equally stressed leaf spring and nearly double that of a helical coil spring. They may be located either crosswise or along the side of the frame rails, thus conserving space as well as weight. In essence, a torsion bar is a shaft similar to a drive-shaft, with attachment means at each end such as serrations or splines. Usually it fits into a suspension A arm at one end with a crank attachment to the frame at the other end. Road clearance may be altered by adjustment of the crank or by re-indexing the shaft in the spline or serration ends. Present day design practice may be studied by looking at applications on the few cars that use torsion bars.

Silico-manganese steels such as SAE 9260 with a maximum operating torsional shear strength in the range of 100,000 to 150,000 psi are generally used. Surface decarburization should be avoided in hardening. Fatigue resistance is increased by shot-peening and presetting, although presetting only helps in one direction. The bar cannot then be used interchangeably left and right sides.

Diameter of the torsion bar end is usually 25 to 30% larger than the working section to keep stresses in the safe range in the serrations or spline. Many handbooks have excellent design data on torsion bars. Magazines such as MACHINE DESIGN (reference 74) also occasionally include such information.

Costs are usually considerably higher for torsion bars because of the extra machining of the ends and their mating parts.

HELICAL (COIL) SPRINGS

This is by far the most popular spring for independent front suspensions (IFS) and is being used increasingly for rear suspensions. A coiled spring is easy to design, economical to manufacture, has no interleaf friction, and no complex ends as with a torsion bar. Coiled springs of different spring rates may be easily interchanged. Usually a coiled spring application can be as lightweight as a torsion bar design when all other factors are considered.

Race cars frequently use coil springs mounted on shock absorbers as done with motorcycles. Often these may be adjusted for height by turning a threaded cup at one end. The shock absorber end bushings also take the spring loads, thus simplifying design and installation as well as saving weight. Lightweight urban cars could also use coil springs in this manner.

Supplier catalogs often provide standard sizes listed to include wire diameter, free length, solid height, total deflection, carrying capacity compressed solid and load required to deflect 1/8th inch. Some also provide helical spring design tables to assist the designer. References 75, 76 and 77 will also prove helpful.

FLUID SPRINGING AND AUTOMATIC LEVELING

The fluid may be hydraulic or pneumatic or a combination of both. The idea is not new. Air suspension began in this country in 1957 but did not become popular for various reasons. BMC of England later used rubber springs with a hydraulic inter-connection front to rear. Citroen uses air and hydraulics and others around the world are using it, especially with front-drive vehicles. The automatic leveling feature is important because the rear suspension of a front-drive vehicle is very lightly loaded when empty, and may be quite heavily loaded when the trunk is full, along with a full passenger load.

Another version uses a shock absorber with an air spring integral in its design. These may be adjusted for pulling trailers and/or other heavy loads. Usually they are not automatic (but could be so designed) in that the adjustment is made with external air as from a service station. Such air spring shocks are also available for motorcycles, and in some cases these would be fine for small vehicles of three or four wheels.

Reference 78 is useful to the designer as is a study of available literature from suppliers.

SHOCK ABSORBERS – DAMPERS

When a loaded steel spring is bumped its stored energy, plus inertia of the unsprung mass, returns the load beyond its static position, thus initiating a series of oscillations that take some time to be fully dissipated. A motor vehicle under such conditions would pitch and roll increasingly with each bump and quickly become quite difficult to control as well as uncomfortable to ride in. A means of damping the spring action is needed. In this country we started out calling them "shock absorbers" and the name remains to this day. The early scissors-type had friction material at the scissors pivot that could be tightened or loosened for adjusting the resistance. Unfortunately, they also made the ride more harsh because of their resistance in both directions – bump as well as rebound.

Shock absorber design has come a long way since those early efforts. We now have "shocks" in great variety for every purpose and they do more than dampen spring motion. Jumping over all the intermediate developments we now have oil-air types of direct acting piston/cylinder design with end fittings for every application. These usually incorporate a metering valve to allow minimum reaction for slight bumps and much more reaction at the end of the stroke to provide greatly increased snubbing action. Very little resistance is offered in compression (bump) with more control of rebound. Dissipation of energy is by the build-up of heat in the unit from rushing the fluid from one end to the other. This can destroy a cheap unit in a short time, usually 10,000 miles or so under hard driving. So-called heavy duty units do much better.

The means for accomplishing longer life and consistent action, hot or cold is proprietary to the manufacturers and the merits of each should therefore not be discussed here to avoid favoritism. The reader is referred to detailed literature and sketches provided by the manufacturers to arrive at his own answers, including cost vs features. For a complete study, foreign as well as domestic designs, including those for motorcycles as well as for cars of all types, such as street, rally, racing, and off-road, should be analyzed.

The shock absorber must be properly matched to the spring rate, total travel, and vehicle use, to serve its purpose as expected. This always requires some testing and adjustment for best results, especially for competition cars in racing and off-road events. It is quite normal in such cases to use two, three or even four units per wheel and replace them often at great expense.

AUTOMATIC LOAD LEVELING SHOCK ABSORBERS

Lightweight cars, usually front drives, with their disposable load a high percentage of total gross weight and mostly over the rear wheels, can have a great variation in spring height between empty and fully loaded – often too much variation at the rear. This was also discussed under springing and in some cases it is dealt with in spring design. It can also be handled by an added feature in certain shock absorber designs. Electronic sensors in the shock absorber with a delayed action control module (for bumps) directs an air pump to raise or lower the rear shocks. As little as 1/4 inch difference sets it in motion. This feature is probably more easily provided in the shock absorber than in the basic spring itself. Only the difference between empty and full load need be provided for – not the total wieght on each rear wheel.

SUSPENSION TRAVEL STOPS – BUMPERS

Abrupt metal-to-metal contact between frame and axles could impose high shock stresses (even breakage), be bone-shaking to passengers, and could disturb the stability and handling of a fast-moving vehicle.

Rebound beyond design limits could unseat coil springs in some cases, damage shock absorbers not designed for the load, and exceed the angularity allowed in end fittings of the suspension and damage "U" joints whose angularity is exceeded.

Suspension travel in both directions must be kept within design limits by rubber-like bumpers in jounce (bump) and straps or equivalent in rebound. If the vehicle is not too heavy, certain shock absorbers may be used that incorporate limiting travel both ways. A somewhat progressive spring rate can be obtained by using a long cone-like rubber bumper to pick up part of the spring load before it bottoms out.

ROLL STIFFNESS AND ANTI-ROLL BARS

In high speed cornering and sharp maneuvering weight transfer occurs from one side to the other. This causes body lean and affects control and stability, which will be discussed in the next chapter. The roll center may be raised only so far without affecting other features. Therefore, an anti-roll bar may be added either front or rear or both, to asisst in achieving the results desired. An anti-roll bar is essentially a torsion bar cross-wise of the vehicle that connects the suspension on both sides. If both sides go up or down equally the torsion bar arms do not provide any resistance. But if one suspension goes up and the other down, as with body roll, the arms are twisted relative to each other to torsionally resist lean, or body roll. Of course, the anti-roll bar is pivoted in bushings (usually rubber) on each side near the frame rails and at the point of bending to make the arms on each side.

In a few cases the front anti-roll bar is so positioned as to also serve as a drag link in the suspension, thus simplifying the design.

Often anti-roll bars are available in increasing diameters to provide increasing roll stiffness – especially for competition vehicles. This also affects under- or over-steer characteristics depending on which end, front or rear, the anti-roll bar is used.

Chapter XIII

THE DYNAMICS OF RIDE, HANDLING, AND CORNERING

Ride, Handling, and cornering are influenced by tires, suspension geometry, springs and shock absorbers, brakes, steering, sprung-to-unsprung weight ratio, center of gravity location (front to rear and height), polar moment of inertia, frame ridigity, where the power is applied (front or rear wheels), and the inter-relationship of these factors during dynamic maneuvering. Of course, these are not all the factors. Other elements include tire pressure, aerodynamic forces, road surface conditions such as roughness, wet or dry, and banked or level curves, to name a few more. Vehicle size, large or small, is also a factor.

HOW AND WHEN RIDE AND HANDLING WERE IMPROVED

Let's start with the dynamics of ride. After this is understood we'll get into handling and cornering.

Early cars had a choppy ride and the rear seat passengers were easily pitched about on a rough roadway. Independent front suspension (IFS) with coil springs are generally given credit for improving the ride. Actually, the so-called level or flat "boulevard ride" was achieved without IFS and coiled springs, although these came along about the same time. The "AIRFLOW" of Chrysler and DeSoto was the first domestic car to make this big improvement in ride and handling. The year was 1934. The powerplant was moved forward over the front axle instead of behind it as before. Then the passenger compartment could also be moved forward so that the rear passengers were moved 20 inches ahead of the old position over the rear axle. This added luggage space too.

The body was designed to greatly increase the torsional rigidity of the frame to further improve stability. The front axle now carried 55% of the weight to provide understeer, instead of oversteer as before, for better handling. The extra weight up front slowed the ride rate in front for the so-called level or flat ride now universal in passenger cars. All this was accomplished on a front beam one-piece axle and leaf springs.

The Lincoln ZEPHYR V-12 of 1936 and later models had the same features except the leaf springs, front and rear, were located transversely, but the ride was flat and handling was greatly improved.

As cars became lower with the engine in the forward position, independent front suspension became a necessity for reasons of space, although the other advantages were beginning to be appreciated also.

The achievement of acceptable riding comfort was related to the most comfortable walking gait for the human body in early studies. This was fundamental and is still used to this day. It was found that a normal adult walks at a rate (frequency) of 70 to 90 steps per minute, and his torso has an amplitude or up/down motion of about 2 inches. The body seems to be comfortable at this frequency and amplitude in a car as well as during a walk. If the motion slows to a frequency of around 50 cycles per minute it is conducive to car-sickness. If the rate is speeded up to much over 118 cycles per minute, it is into the threshold of discomfort. Of course, an occasional hard bump cannot be avoided and the body will accept it on a random basis.

Generally speaking, a rate of 60 to 80 OPM (oscillations per minute) is considered soft, 80 to 100 is medium, and over 100 is firm. Small sports cars tend to be on the firm side because there is insufficient space for adequate up and down wheel travel to obtain a softer ride. A short wheelbase also somewhat limits the amount of acceptable pitch.

Early studies also dealt with this matter of fore and aft pitch. It was found the front suspension rate should be slower than for the rear suspension, so that when the front wheels hit a bump first the sprung weight (chassis/body) will pitch up slowly and tend to rebound slowly to return to normal position at, or nearly at, the same moment the rear does. This, in effect, tends toward the so-called level or flat ride. The difference in spring rates of the front and rear suspension should be based on wheelbase length, severity of bumps, and the speed of the vehicle. Obviously, these are not the same for all situations, so a level ride is not always completely possible. Shock absorbers have been designed to assist in some ways and other means of transferring loads from front to rear have been devised. One of the earliest was the PACKARD TORSION-LEVEL SUSPENSION of 1955. This will be briefly described later.

TERMINOLOGY AND DESIGN PROCEDURE

There are certain terms used throughout the industry that are useful in studying reports and in communicating with others. The most often used are:

RIDE RATE – The force, in pounds, required at the tire patch to move the wheel upward one inch with the rest of the vehicle held immobile.

STATIC DEFLECTION – The dimension, in inches, from no-load to normal static load measured from the ground to the frame at the centerline of the axle, front or rear.

BUMP TRAVEL (OR JOUNCE) – The *upward* wheel movement, in inches, from the static position to the upper positive stop.

REBOUND TRAVEL – The *downward* wheel movement, in inches, from the static position to the lower positive stop.

WHEEL TRAVEL – The sum of bump and rebound.

RIDE FREQUENCY – The natural frequency, in oscillations-per-minute (OPM) at which one end of the vehicle will tend to vibrate. Shock absorbers are usually disconnected to get the reading.

Specialists in suspension design use many additional terms as may be learned from references 79 and 80 along with other reports on the subject. However, these few terms will serve the designer in preliminary studies and for simple arrangements.

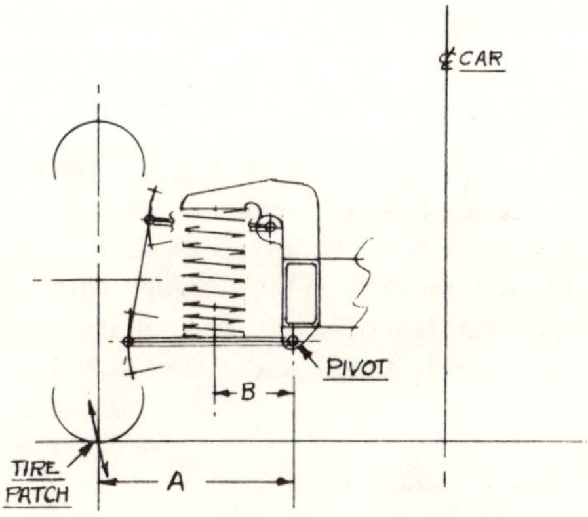

LEVERAGE – WHEEL TO SPRING
– FIGURE 85 –

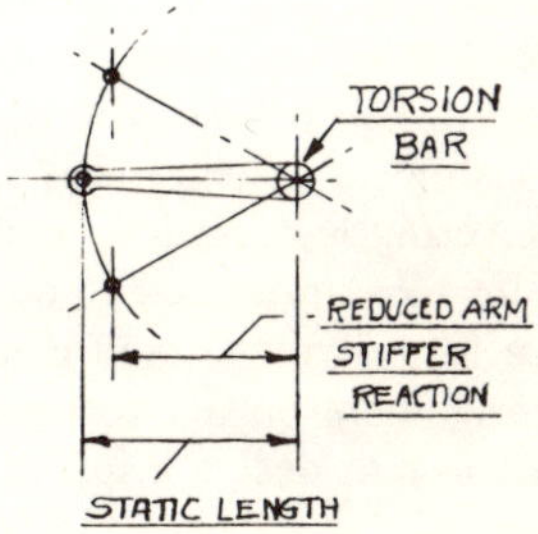

VARIABLE RATE EFFECT
– FIGURE 86 –

Figure 85 depicts an IFS to show that the ride rate is not the same as the spring rate due to the different leverage of the wheel and spring from the suspension pivot point. Not only must the designer determine the lengths of A and B, he must also determine the difference in spring length between static, bump, and rebound positions of the wheel. If the coil spring is angled in toward the top, variable spring action occurs, due to

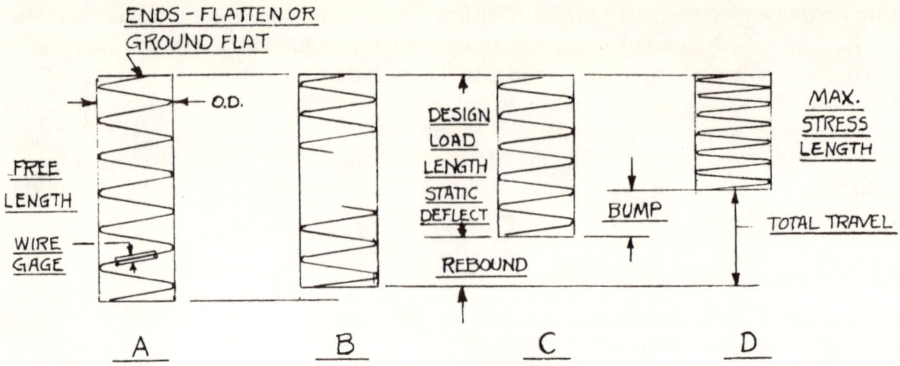

HELICAL SPRING TERMS
– FIGURE 87 –

the change in relationships. This is best determined by a layout showing the wheel an spring travel, along with the change in angularity of the suspension members, especiall the one to which the spring is attached.

VARIABLE RATE SPRING ACTION DESIRABLE

Variable rate spring action, either by design of the spring or by geometry of the sus pension, is often favored. In the mid-point (static position) the spring is "soft" and a both ends of its travel (bump and rebound) the spring *rate* increases or stiffens. Thus small bumps are taken easily with a soft ride while greater bumps and rebounds are re sisted with an *increasing spring rate* before reaching an abrupt stop at the end of travel This is considered more important in bump and may be accomplished by using a lon cone-shaped rubber bumper which takes increasing load before reaching the full sto position. A normal constant rate spring may be used in such a case. A variable rate sprin is designed by having end coils "bottom out" first to progressively increase the sprin rate. While such a coil spring is easy to manufacture, it is not always ideal because o some added noise when coils contact each other. Small rubber bumpers have been adde between end coils to create the same effect. A leaf spring is easily made variable rat by changes in length and arrangement of leaves. A torsion spring naturally achieves variable rate due to changes in effective length of its arm with angularity as shown i figure 86.

Spring design follows the determination of loads and lengths of travel (of the spring in relationship to wheel travel in bump and rebound. The previous chapter listed SA reports covering leaf, helical, and torsion bar designs. Since helical springs are most ofte used, figure 87 is provided to quickly show the relationships in terminology with lengths Reference 81 provides detail design and application data of helical and spiral springs us ful to designers. Spring manufacturers often provide tables and standardized data t quickly choose or design a spring.

STATIC DEFLECTION AND RIDE FREQUENCY – STEEL SPRINGS

Eariler in this chapter OPM (oscillations per minute) was discussed. Static deflectio of a steel spring suspension system and OPM have a relationship - the more deflection th fewer oscillations per minute. For example, a heavier load deflects the suspension mor therefore the OPM is reduced, resulting in a softer ride. Or in terms of the spring itself added weight on a given spring increases the static deflection and therefore lowers the na tural frequency or OPM. Since the spring has farther to travel (at the same rate of trave it takes longer to get there and return. Static deflection is therefore a direct measureme of ride frequency, or OPM. The formula is:

OPM = 188 divided by square root of the static deflection

Example: A static deflection of 4 inches, at a wheel, has a square root of 2. Then 188 divided by 2 = 94. This would yield a medium to firm ride as discussed earlier.

The amount of static deflection at the tire must then be worked back through the suspension design to determine the resulting amount of static deflection in the spring itself. The *sprung load* carried by that spring must also be calculated. Then dividing the load in pounds by the static deflection in inches, we have the pounds per inch of deflection (lb./in.). This can usually be found in standard listings in a spring catalog. If not, the spring must be designed according to standard engineering data provided in the references or by the spring manufacturer.

RUBBER SPRINGS – AIR SPRINGS – FLUID/GAS SPRINGS

RUBBER SPRINGS look like the most promising spring innovation for large off-highway heavy-duty trucks. Some producers feel there is a decent chance they will, by the early 1980's, be more popular than the old standby – leaf springs. The advantages are such that they may be used for many other types of vehicles, including small cars eventually. The primary builder of rubber springs is FIRESTONE INDUSTRIAL PRODUCTS - a division of Firestone Tire and Rubber Company. The springs are solid rubber cores with a hollow center. The core is reported wrapped with bias plys of nylon cord fabric. The height-to-diameter ratio is as high as 3-to-1 in some cases. This, together with changes in durometer of rubber, taper angles, and other material changes, are the secret of the extreme design flexibility offered.

Rubber springs provide variable rate deflection for a nicely controlled ride, whether the vehicle is lightly or fully loaded. They are also much lighter than leaf springs, less expensive, and easy to install.

AIR SPRINGS – Goodyear and Firestone are the major manufacturers of air springs in this country. It appears that about 15% of the class 7 and 8 trucks now being built are equipped with air springs. There seems to be an annual increase in usage or growth rate of 20%. Most air springs are of the rolling lobe type.

Air springs for passenger cars in this country first appeared in production in 1957, pioneered by Cadillac, but they really didn't catch on then as expected. Improvements have since been made, less complicated plumbing is used, higher pressures and more compact designs are appearing. The advantages include improved ride and, of course, the leveling feature for variations in load.

Monroe's load-leveler air spring shock absorbers have been around for some years now, and are used as an assist at the rear for cars pulling trailers or are otherwise tail heavy. These are well known and need not be further described here.

Even motorcycles can now be fitted with air springs at the rear. These units may be of some interest to designers of very small urban type vehicles. These appeared on the market in 1978 and are manufactured by S & W Engineering of Anaheim, CA. They are the invention of Tim Witham, the W in S & W. The unit is essentially a straightforward shock absorber with internal valving and fluid. The air spring fits on the outside, is manufactured by Monroe for S & W. Pressure may be changed at a service station air hose. If a small air pump is provided on board along with a height sensor, automatic leveling is possible.

If air springs are used front and rear on a vehicle and air lines are connected front and rear on each side (separately), the suspensions then become interacting for a flat ride and reduced body roll as described next.

FLUID/GAS SPRINGS – Hydragas units were developed by Dunlop and introduced in 1973 on the AUSTIN ALLEGRO, followed by the MINI and PRINCESS. These units serve as fluid shock-dampers as well as gas springs, and are connected front to rear on each side for a flat ride and less roll. Each unit has a spherical spring chamber with its flexible rolling membrane or diaphram, that separates fluid on the bottom from gas on the top. Below this is a tapered piston connected to the suspension arm and working against a lower diaphram which pushes the fluid through typical shock absorber valving into the upper chamber, to push against the membrane that separates the fluid and gas. The gas, being compressed, is the spring.

Possibly these units, or something similar, will someday be manufactured and made available in this country.

LONGITUDINAL COMPLIANCE – TO OVERCOME HARSHNESS

When a wheel hits a bump at speed, there is an immediate, practically instantaneous, shock driving the wheel upward and rearward. This harsh action is felt by vehicle and occupants. To reduce wear and tear of both, the spring takes the upward motion and the suspension often incorporates a drag link with a rubber bushing to take the rearward motion. Otherwise the tire must do it all. If the tire and wheel are of small diameter the reaction is greater, thus justifying the rubber bushings more for longitudinal compliance. These rubber bushing are used both in the front suspension and in the rear suspension, as can be seen on present-day cars.

INTERCONNECTED SPRINGING – FRONT TO REAR – AND ITS ADVANTAGES

Ideally, when a front wheel hits a bump the reaction should be at the vehicle's center of gravity, not just at the front end, so that pitch is avoided.

This has resulted in intensive study by suspension specialists throughout the world. Some very clever designs have been developed and a few have gone into production. One of the earliest to reach production in this country was the torsion-level suspension of the 1955 PACKARD. This is described fully in reference 82 and later in reference 83. A brief description follows:

a) The front suspension was double A arms IFS and the rear axle was conventional live non-independent. A watts linkage located at the rear axle laterally, and a trailing arm on each side located it fore and aft.

b) Springing was by a single torsion bar running from front to rear on each side for the main load. The torsion bar was supported in needle bearings at each end. At front the arm extended outboard with a link down to the lower suspension member. At the rear the arm went inboard to a stirrup-like link up to the mid-point of the trailing arm. Thus, the load at the front was balanced by the reaction of the load at the rear. The torsion bar and its arms at each end took on the shape of a "Z". This interconnected feature was the key to the design.

c) Changing loads in the rear seat and trunk were handled by a compensator (levelizer) torsion bar added to each side and connected to the rear arm of the main torsion bar. An electric compensator motor with gearing slowly cranked in twist at the forward end which was mid-way of the car. A bi-metal switch provided a 5 to 7 second delay in leveling to avoid constant action with road irregularities. A sensor switch at the rear operated the motor with a variation of but 5/8th inch.

The advantages of this design are remarkable:

1–Individual wheel rates were 66 lb./in. front and 69 lb./in. rear. Ride rates were slightly less than individual wheel rates. The ride is definitely softer than conventional designs and always level.

2–When the front and rear wheels were deflected simultaneously the rates became 100 lb./in. front and 104 lb./in. rear.

3–Roll stability was higher than that of conventionally sprung cars because combined spring rates acted on the body in roll. In hard cornering body tilt was but 3¾ degrees, according to SCIENCE AND MECHANICS magazine, the lowest tilt they ever recorded until that date.

4–The suspension will stay definitely more level than a standard suspension with a minimum of pitching.

5–Whether the car is empty or has six passengers and over 300 pounds of baggage in the trunk, it is always level. Overall height varies by a little over an inch.

6–No rear end squat on acceleration.

7–Less dive on stopping.

8–Reduced torsional stresses in the frame.

The torsion-level system cost Packard $60 to $70 more to manufacture than a conventional suspension. Of this cost the compensating gearbox unit alone cost about $25. Retail cost overall to the buyer was $150 in 1955. Cost today would be several times this, but still acceptable for many buyers.

Packard's torsion-level suspension was the invention of William D. Allison, an automotive suspension specialist who was employed in the Experimental Department of HUDSON MOTOR CAR CO., then as a consultant to PACKARD, and later for FORD and others. His work with interconnected suspensions began in 1937. According to reference 83 he constructed (in his basement workshop) a small car powered by a 2-cylinder motorcycle engine and featured his interconnected suspension. This was in the fall of 1950. It was his demonstrator in 1951 when he contacted PACKARD and others during his two-week vacation from HUDSON. He retired from FORD in 1975.

It is reported that Bill still drives this cycle car around the Grosse Pointe area. Many passengers testify it offers perhaps the ultimate example of how smoothly a small, light vehicle can be made to ride.

CITROEN has, for many years, used an air-oil design in which the height of the vehicle can be changed from the driver's seat as well as offering the many other features of a superior ride.

BRITISH LEYLAND of England has a fluid/gas system that is also interconnected front to rear (the Dunlop described earlier). Both of these designs are described in their motor car literature.

Other designs have appeared from time to time and no doubt the interconnected system will become popular again, in new forms, as efforts are made to improve the ride and handling of smaller, lighter vehicles with considerable variation in loads carried.

HANDLING AND CORNERING

Although a vehicle may provide a satisfactory ride, that does not necessarily indicate it will also handle and corner well. Ride primarily deals with vertical motions at each end of the vehicle. Handling and cornering are a bit more complex. In addition to vertical motion they also deal with horizontal motions related to steering, body roll during turns, weight transfer, tire adhesion and traction, centrifugal force and aerodynamic forces. At the beginning of this chapter a number of design elements were listed that must be dealt with, especially for good handling and cornering. Before getting into these, we should start with a few basics to understand the dynamic nature of the problems.

CENTER OF GRAVITY CONE AND WEIGHT TRANSFER

All forces applied to a vehicle must ultimately be taken by the tires through their tire patches on the roadway. Since the center of gravity is at some distance above the roadway, let's start with figure 88 to see what happens. Vehicle weight is vertical or downward as indicated. When this weight is moved forward (or rearward) under power, there is an accelerating force in the horizontal direction as shown. This could as well be deceleration as in braking, or centrifugal force during a turn. These forces should be drawn to the

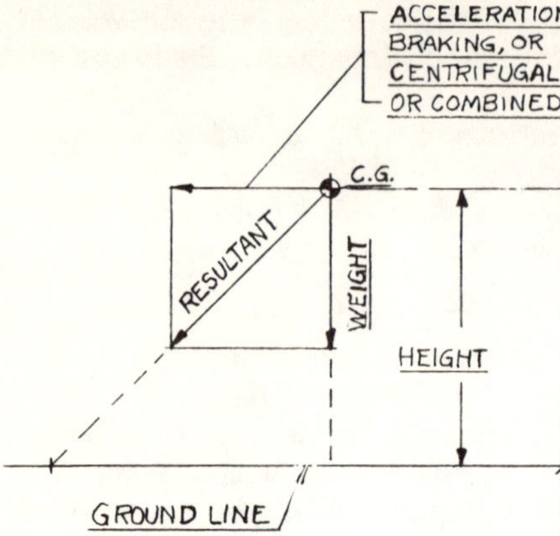

FORCES
–FIGURE 88 –

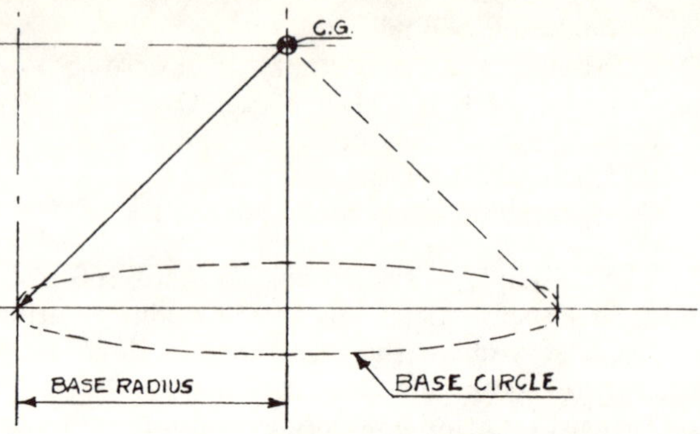

CENTER OF GRAVITY CONE
– FIGURE 89 –

same scale, such as 1-inch equals 1,000 pounds. Then box in as shown and draw the resultant diagonal line. The *length* of the resultant line indicates the force in pounds equal to the combination of the horizontal and vertical forces. The *direction* of the resultant line indicates the direction of the sum of the two forces. If this line is extended as shown to the ground line, we then know at what point the center of gravity is reacting at ground level. The different horizontal forces may occur in any horizontal direction, sideways, forward, rearward, or any combination. Therefore, maximum horizontal forces on the center of gravity, along with vehicle weight could be depicted as a cone in figure 89. The base of the cone, depending on the magnitude of the horizontal forces, may not be a circle. It could be oblong or otherwise. The cone indicates how much the *resultant* from the center of gravity may move at ground level. This is also referred to as weight transfer and is illustrated in figures 90 and 91.

When the base circle of the cone is depicted in planview, along with wheelbase and tread width, it might appear as in figure 92. Obviously, if the base circle is greater than tread width or front or rear wheel centerlines, the vehicle will upset under extreme dynamic handling.

Figure 93, while impractical, shows what occurs if the center of gravity could be at ground level – there would be no cone and no weight transfer in any direction. This also clearly shows that it is desirable to keep the center of gravity as low as possible, to reducε the base circle dimension and minimize weight transfer.

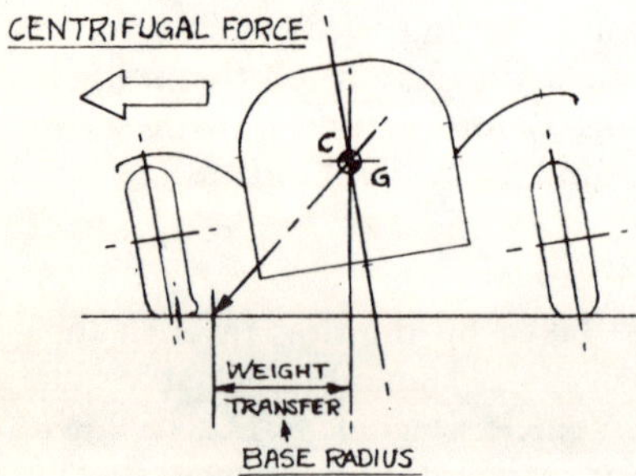

WEIGHT TRANSFER
– FIGURE 90 –

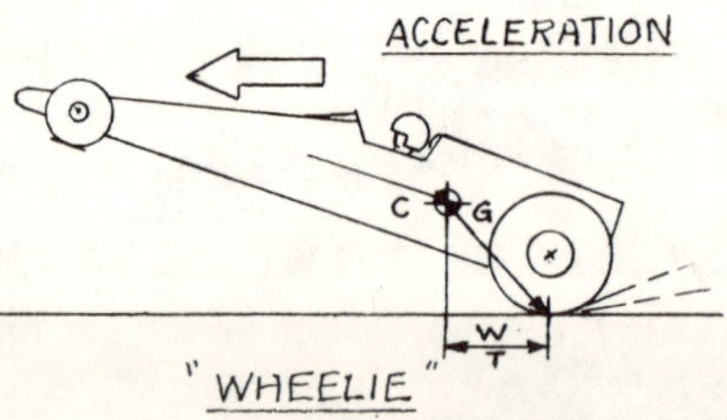

WEIGHT TRANSFER
– FIGURE 91 –

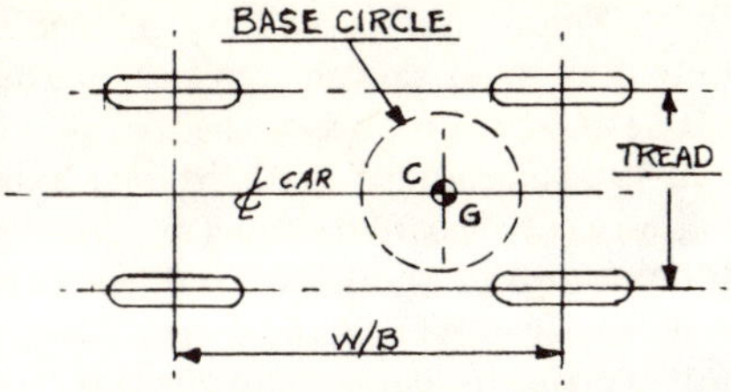

RELATIONSHIP OF BASE CIRCLE TO W/B AND TREAD
– FIGURE 92 –

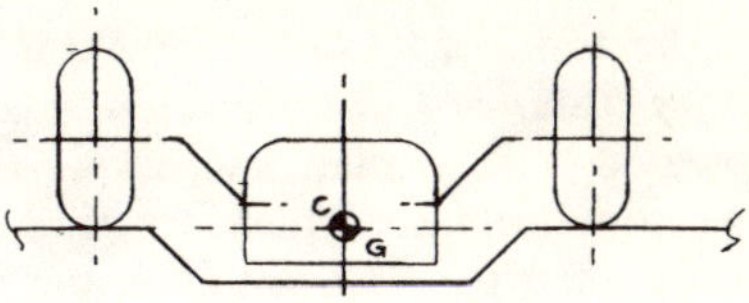

NO C.G. CONE
NO WEIGHT TRANSFER
– FIGURE 93 –

YAW, PITCH, AND ROLL

Yaw, pitch, and roll are illustrated in figure 94. These terms are used with aircraft and water-borne vessels as well as ground surface vehicles. These terms deal with the fact that a vehicle does not always travel in the direction it is aimed. Measurements are in degrees of angle.

In yaw a vehicle (in planview) may be steering into the wind, or against a side force due to a sloping roadway, or from centrifugal force in a turn. The side force will cause the vehicle to "crab" slightly to maintain a straight course, or if in a turn, to maintain the proper curved path.

Pitch, as indicated in the sideview sketch of figure 94, refers to longitudinal disturbing forces such as braking or acceleration, or up or down hilly roadway, or with aerodynamic lift and drag in high-speed vehicles. The addition of forward and/or rearward inverted airfoils as used on racing cars, add strong downward pitching forces.

Roll, as illustrated in the front view of figure 94, refers to body lean as during a high-speed turn, and results from the action of centrifugal force at the center of gravity. Aerodynamic forces such as a strong wind acting at the center of gravity will also cause some body lean.

FORCES MUST BALANCE FOR EQUILIBRIUM AND STABILITY

It is a basic law of physics that forces in opposite directions must balance or cancel each other for equilibrium and stability. This is illustrated in figure 95. The beam scale

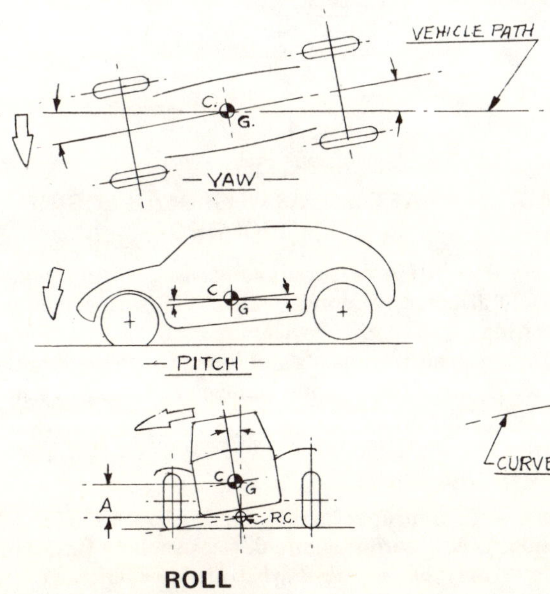

ROLL
– FIGURE 94 –

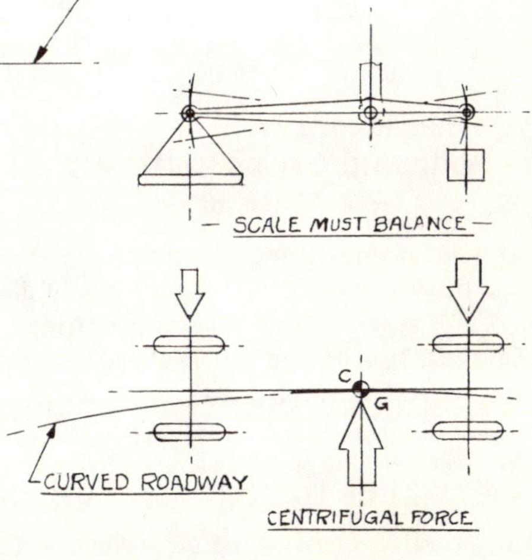

– FIGURE 95 –

at the top is an example. Actually, we are equalizing "moments" which are the *force times distance from the fulcrum or pivot.* The "moments" must cancel each other. The lower sketch, in planview, shows the center of gravity (CG) slightly aft of the mid-point of the wheelbase. Therefore, since the distance from the fulcrum (CG) to the rear axle is *less* than to the front axle, the resisting force at the rear tire patches must be *greater* to create a moment high enough to cancel the distance times force at the front. This requires greater tire pressure or a higher capacity tire at the rear than at the front. The resisting ability of the tires will be discussed as soon as we finish dealing with the disturbing forces.

Aerodynamic force in the form of side winds or just plain center of pressure location must also be considered. Every body shape, streamlined or not, has a center of pressure. It differs with shape and essentially moves forward as speed increases. As a rough start, and without benefit of wind tunnel testing, the center of pressure may be assumed to be at approximately 25% of the wheelbase for a typical body shape. It may be further forward or possibly slightly aft of the 25% point, depending on design. The distance of the center of pressure (CP) from the CG is indicated by "A" in figure 96 for a vehicle with a forward CG such as a front drive. Such vehicles are less disturbed by wind forces because the CP is near the CG. Lots of body overhang at the front and not much to counteract it at the rear could move the CP farther forward and make the matter worse. However, such vehicles are normally less disturbed by wind forces since the CP is near the CG.

Figure 97, typical of many rear engine cars, illustrates why aerodynamic forces are more disturbing with an aft center of gravity. The distance "A" is so much greater. The old VW "Beetle" was an example of this condition.

Figures 96 and 97 show aerodynamic forces acting in a sidewise direction as applied to yaw and roll. It should be noted that pitch is also affected. A race car with an aft CG has been upset, tipping over backward when cresting a small hill, due to aerodynamic forces acting well forward of the CG! The forward inverted airfoil on formula (exposed wheels) racers is needed to keep the front end down and to retain front tire contact for steering and stability.

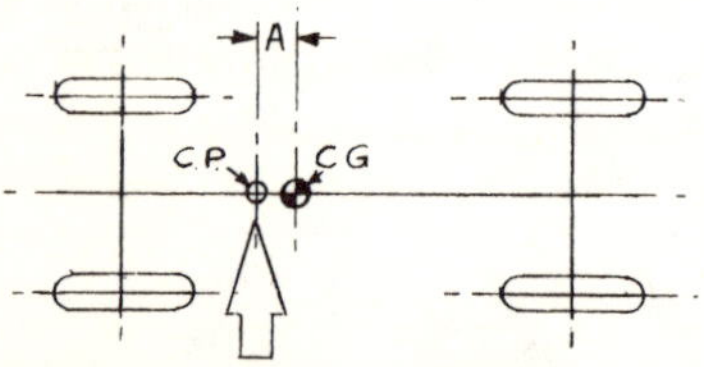

FORWARD C.G. AS WITH FWD
– FIGURE 96 –

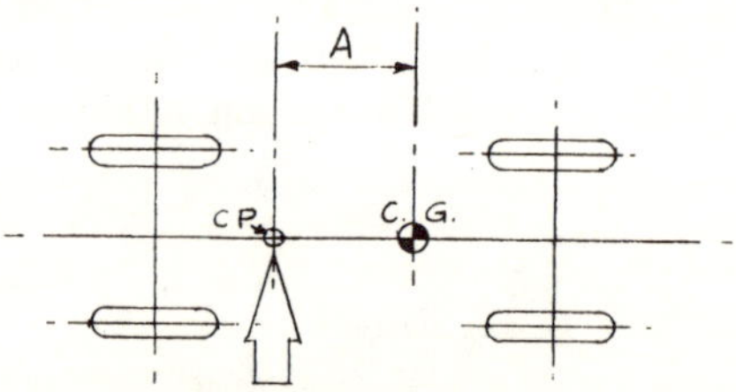

AFT C.G. AS WITH REAR ENGINE
– FIGURE 97 –

If you are designing a fast vehicle do not ignore aerodynamic forces. Any vehicle, even at freeway speed, may be adversely affected by wind forces. This is especially true if the CG is rearward and no compensating aerodynamic devices are used. If the CG is aft, the CP must not be too much ahead of it, even if it means adding fins at the rear and an inverted airfoil at the front to avoid a "squirrelly" handling vehicle at speed. Tires alone cannot handle the problem, especially when airborne.

A SAFETY RULE TO REMEMBER AND USE

RULE – **Balance aerodynamic forces with a proper aerodynamic shape. If this isn't possible with the shape chosen, add aerodynamic devices such as fins, dams, inverted airfoils. Other forces may be balanced with proper choice of tires, tire air pressures, and suspension design.**

POLAR MOMENT OF INERTIA – HIGH OR LOW – AND ITS EFFECTS

Moment of inertia is a measure of a body's resistance to angular acceleration. Polar refers to the masses at opposite ends. To illustrate, consider a ball and a dumbbell of equal weight. Each has its center of gravity at the mid-point. Hold the ball in one hand and the dumbbell in the other and rotate them together with a twisting motion in each hand. Note that the ball turns easily and its direction may be reversed with slight effort. Not so with the dumbbell, even though it weighs the same. The dumbbell has its weight distributed at the ends instead of in the center as with the ball, although both have their weight balanced at the mid-point (CG). The dumbbell requires more effort to initiate the twisting action and more effort to stop it and reverse direction. The dumbbell has a high polar moment of inertia and the ball a low polar moment of inertia. What does this have to do with handling and cornering in a motor vehicle?

A good race car, like the ball, has a low polar moment of inertia. The driver and engine are as near the CG as possible. The wheels are located at the corners with no appreciable weight beyond them. This makes for a very agile handling vehicle. The ride may be a bit choppy, however.

A passenger car has its engine forward in most cases with occupants spread out behind and fuel and luggage all the way back, aft of the rear wheels. The front wheels are under the engine. The major weights, like the dumbbell, are not at or near the CG, but rather spread apart. The supports (the wheels) are under, instead of beyond these masses. This was pioneered by the AIRFLOW, as we learned at the beginning of this chapter, to improve the ride. Large masses at a distance from the CG have a greater polar moment of inertia, so cannot be moved up or down as readily (for a smoother ride). It is also true they cannot be moved directionally (steered) as quickly. This is quite all right for highway (non-race) conditions, because the typical driver has no need or desire to violently maneuver sideways from lane to lane. Other cars also have a high polar moment and speed and safety laws restrict competitive driving.

This is not an "either/or" thing. Designers can compromise at intermediate levels as with a good sports car design that has acceptable ride and better-than-typical agility.

SPRUNG-TO-UNSPRUNG WEIGHT RATIO

The wheel, tire, brake, spindle, half of the suspension, in fact, everything below the spring, is "unsprung" and must somewhat follow the irregularities of the roadway. If these masses had no weight they would have no inertia and would follow readily. But this cannot be. They are exciting forces. The heavier they are, the more difficult it becomes to keep them from bouncing higher than the bump they move over. It therefore behooves the designer to make an extra effort ot save weight on unsprung parts. Not only does this reduce vehicle weight, it also improves the ride and is a major factor for improved handling on uneven road surfaces. Alluminum alloy, or even magnesium alloy wheels are becoming more popular. Eliminating the innertube and its weight some years ago helped. Disc brakes are lighter than drum brakes, and if located inboard as on some race cars, they become "sprung" weight. Spindles, hangers, and suspension members should be kept as light as possible by efficient design without sacrificing strength for safety.

To show what can be accomplished, Colin Campbell, in his excellent book "THE SPORTS CAR", prepared the following table:

TABLE VI

Unsprung Weight of Rear Suspension for 2000-lb. Sports Car

Type of Suspension	Effective Unsprung Weight per Wheel, lb.	Wu/ Wt
Independent with torsion bar or coil spring; inboard brakes	65	7.7
Independent with torsion bar or coil spring; outboard brakes	90	5.6
Swing axle with coil spring	90	5.6
De Dion with torsion bar or coil spring; inboard brakes	75	6.7
De Dion with torsion bar or coil spring; outboard brakes	100	5.0
Live axle with locating links and coil spring or torsion bar	110	4.5
Live axle with leaf springs	130	3.8

Based on equal weight distribution front and rear. Wu/Wt is ratio of total weight (500 lb.) per wheel to unsprung weight for that wheel.
From THE SPORTS CAR by Colin Campbell.

Every major manufacturer makes similar studies and the designer must consider the "unsprung" weight even more now that vehicle size and weight must be reduced in order to keep ride and handling from suffering.

VEHICLE DIMENSIONS REQUIRED FOR CALCULATIONS

Figure 98 is a line drawing of a 1963 CORVETTE as it appeared in reference 84 and on occasion in advertising. This sideview locates the center of gravity (CG) height and distance back from the centerline of the front wheels, the wheelbase, the roll centers front and rear, the roll axis, and the 10-degree anti-dive feature of the front suspension upper arm. The 2-degree caster of the front wheels is also depicted.

Roll axis is the one new term. This is the imaginary line connecting front and rear roll centers. The roll centers are shown in cross-section views of various suspensions discussed in the previous chapter. When the vehicle body leans it pivots along the roll axis. Figure 99 shows the roll axis of later CORVETTES. Note that the rear roll center has been lowered from 7.56 inches to 4.71 inches, to provide a roll axis that is nearly parallel with the ground line.

Open wheel Formula racing cars have a lower center of gravity and a roll axis at approximately ground level to provide minimum tire camber change for best adhesion and higher cornering ability.

Since typical sedans and other passenger cars have higher CG's than shown for the CORVETTE, it follows that their roll axis is normally a bit higher to avoid excessive body roll. Otherwise, anti-roll bars must be added or increased in size to restrict body roll.

When the designer has prepared a drawing with pertinent dimensions as shown in figure 98, along with cross-sectional views through front and rear suspensions to depict roll centers, tread width, and suspension geometry, he is then ready to start tabulating forces, moment arms, and the moments for various dynamic conditions. The table should include:

1–CG most forward and most aft location due to changes in passenger load, fuel, luggage.

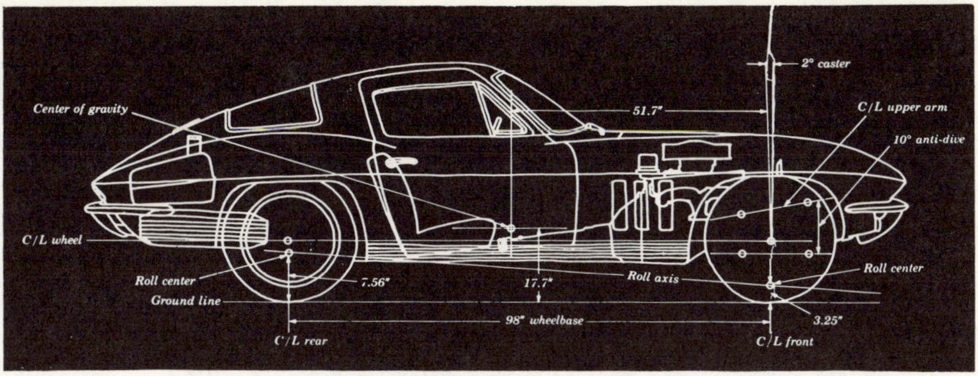

1963 CORVETTE DIMENSIONS
– FIGURE 98 –

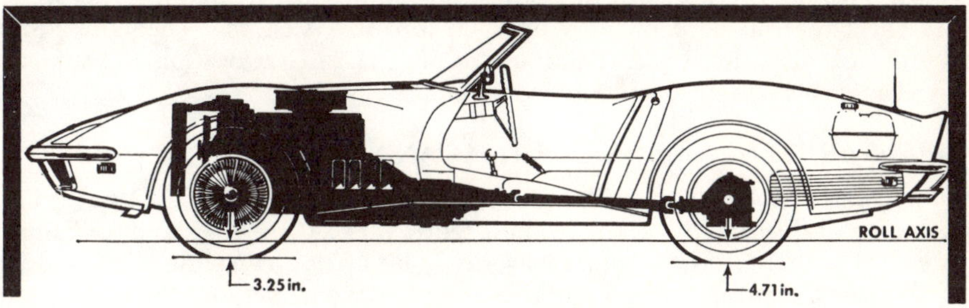

CURRENT CORVETTE ROLL AXIS
– FIGURE 99 –

2–Hard braking – effect on CG cone.
3–Maximum acceleration – effect on CG cone.
4–Centrifugal forces during maximum cornering – effect on CG cone.
5–Aerodynamic forces – moments about CG.
6–Location and weights of major items should also be tabulated if the polar moment of inertia is to be calculated.
7–Any combination of these forces.

At this point, only moment arms in inches can be determined from the drawings, but they should be tabulated.

Vehicle weight, centered at the CG, should next be listed and proportionately divided, front and rear, at the suspensions. Note in figure 94 the dimension "A" is measured from CG height to roll center. This is true at both front and rear. If, for example, the rear roll center is high as with a one-piece "live" axle, the "A" distance or arm will be short, thus reducing the amount of moment taken at the rear.

Weight transfer is another matter. It is always measured from CG height to the ground line for cornering, acceleration, and/or braking.

Aerodynamic forces are first centered at the center of pressure and then measured from the center of pressure (CP) to the center of gravity (CG). These forces then become couples or moments about the CG, both in height and longitudinally. They could be resolved directly into the suspensions and added or subtracted from other forces, depending on force direction. Aerodynamic forces, their center of pressure (CP) and their direction are best determined in the wind tunnel, if at all possible. Otherwise assumptions, sometimes erroneous, must be made.

It soon becomes clear that the table cannot be completed until the forces as well as the moment arms and moments are determined. These forces are dependent on how much adhesion and traction the tires, both front and rear, can offer as reactions. It also depends on the relationship of each tire, front or rear, to its proportion of the load. In many designs, especially racers, the rear tires must be much larger to accept their proportion of the load. This is true of traction as well as adhesion. The extremely large rear tires on a dragster is an example.

If tires are to be identical front and rear, then only tire pressure can be used to alter their characteristics. In this case the CG *cannot* then be very much aft of the mid-point of the wheelbase without developing *serious oversteer.* If a front wheel drive (FWD) is planned, then the CG must be ahead of the mid-point to provide traction. Then, if too far ahead, *excessive understeer* results.

Clearly, high power (applied front or rear) on a short wheelbase vehicle makes the CG location critical for maximum cornering and best handling.

The designer must, at this point, decide what size and type of tires his vehicle will need for its intended purpose, and for acceptable handling and cornering. In the last chapter he was shown how to choose the smallest tire to do the job. If it is a sports car or racer the smallest tire size will not provide maximum cornering power as we shall soon see. The table can be completed once the tire size is chosen. Then alterations in tire pressure, use of anti-roll bars (front and/or rear) and CG relocation and suspension geometry changes may be indicated for further improvements.

"G" FORCES LIMITED BY CHOICE OF TIRES AND CG HEIGHT

The previous chapter began with a discussion of tires and mentioned that (a) Racing cars reach a "g" value of 1.15 g; (b) Modern sports cars .80 to .85 g; and (c) Family sedans about .65 to .72 g. These values may vary a bit up or down with different designs, but can be used to figure "g" forces and thus complete the tabulations for a first approximation.

The reason the family sedan is so low in "g" value is that: (1) the tire size was chosen according to maximum rated capacity; (2) front and rear tires are the same size, although not necessarily loaded equally; (3) the high CG location of a family sedan causes more weight transfer (a larger CG cone), therefore loading outside tires much higher. This will be discussed under the next sub-title.

A modern sports car has: (a) a lower CG, therefore less weight transfer; (b) usually larger tires on the rear to match an aft CG location; (c) tire choice based on optimum cornering ability rather than gross vehicle weight.

Since the racer is primarily concerned with cornering ability, it will use the largest size tires with the greatest width practical at both front and rear for the loads to be carried, and will keep the CG as low as possible for minimum weight transfer. Additionally, racers frequently use aerodynamic devices such as inverted airfoils, both front and rear, to apply more download for increased adhesion and traction, without an increased mass of weight, thus achieving higher "g" forces in a turn and non-skid braking and full throttle acceleration on the straights.

ANALYSIS OF TIRE REACTIONS FOR ADHESION AND TRACTION

During the past quarter century a great deal of research and development has gone into tires and how they react. This has been written up in technical reports for the SAE and repeated many times by writers for trade magazines, and for the more popular magazines for auto enthusiasts, usually without reference to the original works. The serious student who wants or needs the background and complete information might start with references 85, 86 and 87. These references have other references for even more information.

Since the tires must ultimately provide the adhesion for cornering as well as traction for acceleration and braking, their proper choice is of great importance. How the tire reacts while rotating is not the same as during a skid, even though Newton's laws of friction have not been violated.

SLIP ANGLE DURING WHEEL ROTATION VS SKID

The tire is elastic by nature. This permits it to deflect in different directions, not just vertically for ride. If there is a side force due to wind or during cornering, the tire patch deflect sideways slightly in a "creeping" manner while attempting to continue rotating in the plane of the tire. This sidewise motion will cause a divergent (outward) path from the direction in which it is steered. The angle between the plane of rotation and the divergent path is called the "slip angle". The tire is actually "creeping" sideways while it rotates, due to its elastic nature. It isn't slipping. As long as the tire continues to rotate, it is attempting to realign or pull the vehicle back from the divergent path into the path it is being steered.

If the tire is now braked, in addition to being steered, hard enough to stop rotation, then all ability to steer is lost. Pure friction is then being applied by the tire to the roadway and in the direction of the *resultant* of the two applied forces, side force and braking. No matter how the tire is now steered its path will remain along the resultant of the side and braking forces. If hard braking occurs while moving straight ahead on a level surface (no side force), when tire rotation ceases, the path of the vehicle continues straight ahead no matter how steered, because the resultant is straight ahead. Steering reaction can only occur while the tire is rotating.

Accelerating force from the application of power during a turn will also reduce the cornering ability or side force generated while the tire is rotating. This is shown in figure 100. Again the *resultant* of cornering and traction cannot exceed the limit of friction. If acceleration is limited as at "A", cornering is not reduced very much. If high power is applied as at "B", then little friction remains for cornering.

Since the rear tires are not steered directly they can only develop a slip angle, and therefore side resistance, when the vehicle is yawed relative to its direction of travel. In other words, its tail must hang out somewhat. This in itself is not oversteer. It is the only way rear tires can resist lateral forces.

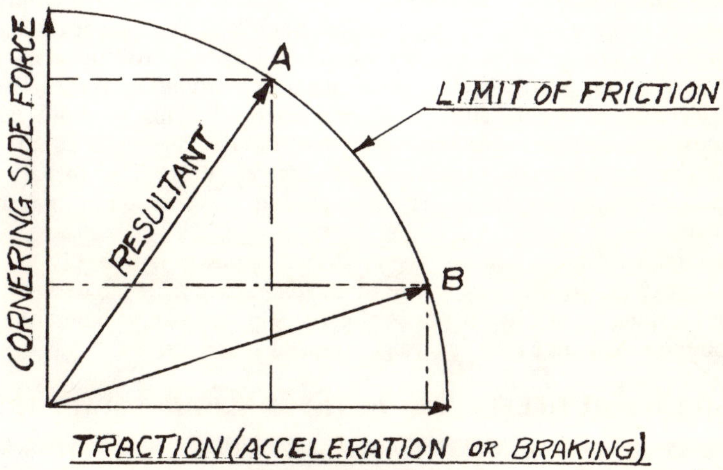

CORNERING VS TRACTION
– FIGURE 100 –

CORNERING FORCE VS LOAD FOR VARIOUS SLIP ANGLES – WHY GREATER TIRE CAPACITY INCREASES CORNERING POWER

Theoretically, as more weight or load is put on a tire its ability to resist cornering or side force should increase in the same proportion. It doesn't work out that way as may be seen in figure 101. Tire characteristics vary from that shown in this figure due to tread design, tire pressure, softness or hardness of rubber, height-to-width ratio of the cross-section and other features of construction, but this depicts in principle, what occurs for all tires.

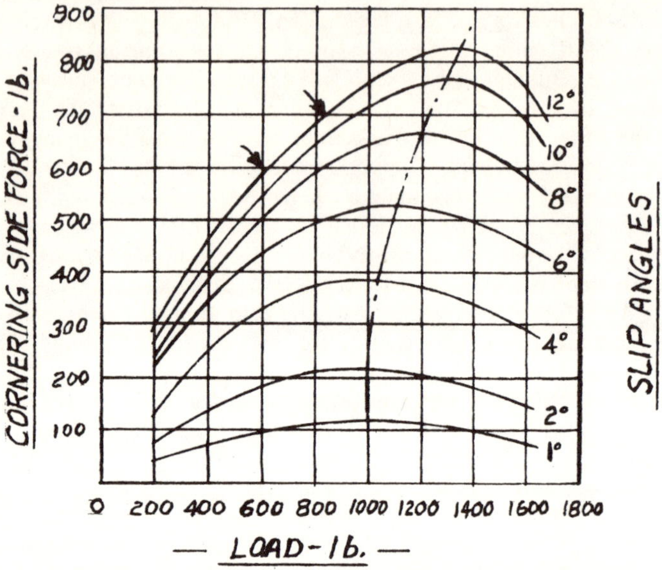

CORNERING VS LOAD FOR VARIOUS SLIP ANGLES
– FIGURE 101 –

Assume the tire is normally loaded to 600 pounds. If the load does not exceed this it will then be capable of about 600 pounds cornering side force at a 12-degree slip angle. If weight transfer increases the download to about 820 pounds, we might then assume we could also achieve 820 pounds cornering side force to retain a 1 g ability. As the figure shows, this is not the case. The value is 700 pounds. The slope of the curve is flattening out. If weight transfer due to acceleration as well as body roll occurs, the download on the outside rear tire might reach 1200 lb. *without considering its loss of cornering power from acceleration.* Then the cornering power reaches 820 lbs. or about .67 g. The vehicle would most certainly oversteer and spin out. Twelve degrees is about the greatest slip angle that can be achieved. As can also be observed, dropoff of cornering power occurs after a peak is reached regardless of slip angle and the dropoff becomes more severe with higher angles.

The obvious way to achieve higher cornering power is to apply less weight (lighter vehicle) to a given tire, or to use a tire of greater capacity for the design weight. Then the weight is working on the lower portion of the slip angle curve where the slope is steeper and much more remains above for weight transfer increases.

RIM AND TIRE WIDTH EFFECTS – RATIO OF HEIGHT-TO-WIDTH

To work on the lower end of the slip angle with old style "narrow" tires would require the use of large diameter cross-sections and large overall wheel diameters. These would affect the differential gear ratio, increase unsprung weight, and be far too large for wheel opening clearances on the body. The center of gravity would also be raised unless the suspension geometry were altered.

The designers of race cars solved the problem by increasing rim widths first to provide a wider cross-sectional base for the tire. This worked so well they asked tire manufacturers to design new wider tires for even wider rims without increasing overall tire diameters. The ratio of height-to-width in cross section became less and less as the years went by and the tire's ability to resist lateral forces continued to improve. Other improvements also occurred but this was something everyone could see.

Sports cars followed with wider rims and tires and soon the practice became a trend for all passenger cars, vans, and pickups.

Today it is generally recognized that tires are the most important single factor in developing high cornering forces on smooth surfaces. Mr. Ron Wakefield, Engineering Editor, Road and Track magazine, reference 88, a decade ago did an excellent series of tests of 9 different cars on a 100-foot circle to determine their speed and the "g" force at the limit of adhesion. One of his conclusions was that tire tread width and "g" forces tend to increase together. Another conclusion was that tread width is more important than contact patch unit loading in obtaining higher "g" forces. Much more can be learned from the article by those who wish to investigate further.

As tires become wider it is important for the designer to review his suspension geometry. Wide tires can be most effective only if they remain vertical with little or no camber change with body lean and/or wheel travel. Just enough swing axle effect to counteract body roll and a low roll center appear to be most desirable.

TIRE PRESSURE AFFECTS CORNERING POWER

Tire pressure differences, front to rear, may be used to assist in improving handling and cornering, but this method has its limits. First do all that can be done by other means such as anti-roll bars, front and/or rear, tire and rim widths, and center of gravity improvements.

Tire pressure influences the slip angle by changing the elastic action of the tire. A soft tire will require a greater slip angle for the same lateral force and cannot generate as much "g" force as when inflated to a higher pressure.

A conventional front engine, rear drive, car usually has the same pressure front and rear with a forward CG to provide some understeer. Sometimes the front tires may have 2 to 4 psi less pressure, or if the rear is fully loaded, the rear tires may be increased up to 4 psi.

A front wheel drive car usually has the same pressure front and rear – however, its understeer may be reduced with more tire pressure in front than in the rear. Again, don't expect miracles and don't change tire pressure more than ±4 psi from the manufacturer's recommendations.

A car with an aft CG location, particularly a mid- or rear-engine design with lots of power, can rarely be helped enough with tire pressure changes only. Rear tires must be wider and/or larger to prevent excessive oversteer. Then tire pressure changes can be used for fine tuning or driver preferences.

ENGINE TORQUE EFFECTS

In tabulating the forces acting on the tires, engine torque effects are often not mentioned, but should be included, especially if the engine is large for the vehicle size. When accelerating, the engine tends to rotate in a direction opposite to its crankshaft rotation. This force is taken through the engine mounts as a couple, loading one side and unloading the other, with resulting weight transfer at the tires. This force can be considerable and is measured in foot pounds of engine torque.

The author learned about engine torque the hard way. During my college days I had an old HENDERSON 4-cylinder motorcycle for transportation. Its crankshaft ran fore and aft. When accelerating from low speed, especially in the gears, it would quickly lean due to engine torque. The rider had to anticipate this to avoid being "laid over", or go easy on the throttle. Needless to say, you remember well after a spill or two. Now-a-days the big motorcycles have their cranks crosswise so this is not a problem. This is the case also for front wheel drive cars with transverse engines.

Engine torque also reacts at the differential on a car. The pinion gear, in effect, wants to "climb up" (or down – depending on direction of rotation) the ring gear. It also loads one side of the differential housing and unloads the other side. These forces can be considerable when the transmission is in first or second gear, so should be calculated by standard gear design methods to determine the reactions.

Reactions at the drive wheels can often be dealt with to some extent in suspension member design. If a high power-to-weight ratio vehicle is being developed the designer should restudy his choice of suspension geometry with engine torque in mind from crankshaft, through the gears, to the drive wheels. Some small changes in angles of a drag link often can provide some compensation.

UNDERSTEER OR OVERSTEER – BOTH REDUCE CORNERING POWER

A review of the tabulated moments and reactions at the tires, front and rear, will clearly show that it is impossible, under all conditions, to always achieve exactly the degree of understeer or oversteer desired during cornering, with or without braking or acceleration. Some compromises must be made and usually testing and modifications determine the best final result.

Normally the quickest way through a turn is to: (1) do braking and downshifting (if required) *before* the turn; (2) apply only enough power through the turn to *maintain* a constant maximum "g" load on the tires. This is easier said than done. Driver skill plays a big part here.

Now we are into under- and over-steer. First, let's say that neither is good or bad and either can be overdone. Understeer occurs when the slip angle of the front tires together is greater than that of the rear tires together. Then more steering angle is required to maintain a constant turn. If all the steering lock is used up without staying in the turn, the vehicle leaves the course head first on the outside of the turn. In oversteer the vehicle will also leave the course on the outside of the turn, this time rear end first, with steering lock in the opposite direction because the slip angle of the rear tires is greater than that of the front tires. In neutral steer, front and rear tires are working at the same slip angle, even though they may be of different size and with different tire pressures. Then when maximum "g" load is taken by the front tires the same is true at the rear. Once the turn is set by the steering wheel it can be returned to its mid-position (straight ahead). To take a turn at maximum speed and highest "g" load capability of the tires, both front and rear tires must work to their limit. Neutral steer is most desirable in cornering. Then it is up to the driver to get the most out of the car. The designer has done his part.

UNDERSTEER IMPROVES STABILITY

If a car oversteers and the driver does nothing the car will continue to swing around until it spins out, rear end first. This it is unstable. If a car understeers and the driver does nothing the car will "wash out" or tend to align itself in the direction it is moving. If left alone it will "aim" itself straight ahead. Thus it is stable. An understeering car requires less attention than an oversteering vehicle. If oversteering is considerable, the driver is kept very busy with constant steering corrections. This can be dangerous if the driver is not alert enough or if the vehicle gets "out of hand" it simply rotates further, beyond the point of safe return (it spins out). This can occur on a straight level roadway as well as during cornering.

Ideally a car should have some small amount of understeer on the straights, and neutral steer for maximum cornering in the turns. Then the driver can "steer" through the turns by applying more (or less) power to the rear wheels.

A front drive vehicle is essentially an understeerer on the straights because of its forward CG and it will understeer even more in a turn as more power is applied. This may even lead to "plowing" of the front end (yaw to the outside of the turn) if excessive power is used in a turn. Then if the driver gets off the throttle the front wheels regain their adhesion and the vehicle rotates (yaws to the inside of the turn) unless the driver quickly corrects the steering angle. This again, is a stable situation, although at times it may be a bit unnerving. Additional tire pressure in the front tires would reduce this condition. If necessary, wider rims and/or tires could be used in front if understeering is excessive.

VEHICLE SIZE VERSUS ROAD WIDTH – SMALLER IS BETTER

Have you ever noticed – really large racers never "make it". Brute power, if it requires a large vehicle, is not enough to get the job done. Conversely, small cars often are more agile and turn in remarkably good lap times. There is a reason for this. Imagine a large, long wheelbase bus or truck in a tight turn. Its front wheels must be steered nearly full lock just to get around at low speed. If the wheelbase is only half as long, or even less, the front wheels need be turned much less for the same curve. More steering lock remains for maneuvering. Width is important too. That leaves more clearance on each side of the vehicle for maneuvering.

A lightweight vehicle can also use wider tires to gain a proportionately higher "g" load in the corners. There is a limit to how wide a heavy vehicle's tires may be. Less weight, of course, requires less power, but the important thing is power-to-weight ratio in either case.

Large heavy vehicles can be made to go fast on the straights, but they are no match for the well set-up small vehicle in cornering and agility. This is one of the main reasons small sports cars are popular and so much fun to drive.

DESIGN RULES

1–Keep the center of gravity (CG) as low as possible to reduce the size of the CG cone and resulting weight transfer.

2–Keep the center of gravity ahead of the mid-point of the wheelbase for some understeer. If power to the rear wheels is so great that a more aft CG position is required for traction, then wider rims and tires are also needed aft. In this case aerodynamic forces at the center of pressure (CP) should be considered. If a front drive is used, then a forward center of gravity is required. In this case tire pressure increase at the front may be justified.

3–Suspension geometry should, preferably, be independent both front and rear. Camber change should be kept minimum for good tire adhesion, although body roll usually requires some swing axle effect to keep the tires upright in cornering.

4–Roll center height is chosen primarily by body roll. If the CG is high then the roll center should also be up a bit. If the CG is low the roll center may be at, or nearly at, ground level.

5–Body roll can also be reduced by the use of anti-roll bars, front and/or rear.

6–If the vehicle understeers too much (usually FWD) then an anti-roll bar should be added to the rear axle or increased in size if one already is used. Front tire pressure should also be increased.

7–If the vehicle oversteers too much add the anti-roll bar to the front or increase its diameter if already there. Rear tire pressure should also be increased. If these do not help sufficiently, rear tire and rim widths may have to be increased.

8–If body roll is excessive but the vehicle is ideally at neutral steer in a turn, then anti-roll bars at both ends, front and rear, should be added or increased in size.

9–If handling is a bit "vague" it may be due to frame torsional weakness, which allows some wandering between suspension geometry and steering linkage. Lateral weight transfer, front and rear, also is affected. A torsionally stiff frame is required to provide a firm foundation for suspension and steering relationships.

10–If handling at freeway speeds or higher is a problem, look into aerodynamic forces at both front and rear. Lift at the front and/or rear may be disturbing handling.

Lift at either end is undesirable. A dam or spoiler at the front may help and an up-turned edge, a spoiler, at the rear may be required to reduce "induced drag" and lift. Directional stability problems are discussed in more detail in the chapter on aerodynamics.

In summing up, it is recommended that only one "fix" be tried at a time, so that testing can evaluate results for each correction before proceeding with next. Document findings thoroughly for constant review as testing continues.

Chapter XIV

THREE WHEELERS

The appeal of three-wheelers goes back as far as the first automobile, but unfortunately they still aren't fully understood, even by those who should know better. It, therefore, seems proper to devote a chapter to them. All the early mistakes of ride, handling, and cornering were made on three-wheelers as well as on four-wheelers, the only difference being that progress was made on the four-wheelers, but precious little was carried over to the trikes. As a result, most people nowadays think of three wheels for slow moving industrial vehicles, particularly in this country. In other countries around the world, three-wheelers are more popular, but even there the full potential has not been reached.

The writer has devoted a number of years in designing and building three-wheelers and feels more strongly than ever about their advantages and that they have a definite place in the motor vehicle marketplace.

ADVANTAGES

The obvious advantages include design simplicity, fewer parts, lighter weight, improved performance, and reduced cost. Since three-wheelers under 1500 pounds are classified as motorcycles, they therefore are subject to Federal and State regulations dealing with motorcycles instead of the more involved automobile regulations.

The disadvantages essentially have to do with limitations of center-of-gravity movement as related to stability. This problem no longer need be troublesome in modern design. Three-wheelers have been built that corner as though on rails. In fact, in Europe some years ago, newly designed racing trikes were entered in side-car races and after trouncing the conventional motorcycle/side-car racers, were promptly outlawed by the revised club rules (FIM). Later they were reintroduced with some restrictions, but in October 1978 the FIM decided to divide them into two classes: (1) Motorcycle/side-cars, and, (2) Three-wheelers, starting in 1979. Reference 89 tells the story.

SINGLE FRONT WHEEL CONCEPT

Where we put the single wheel – front or rear – makes the difference. We will first discuss the single front wheel version. Today it is most popular in industrial applications and has its place, although limited as we shall see.

CENTER OF GRAVITY, EFFECTIVE HALF TREAD, AND WHEELBASE RELATIONSHIPS

Figure 102 depicts the single front wheel design with its typical aft CG location. Note the *effective half tread* is measured from CG location in planview, to the line connecting the front wheel and one rear wheel. As learned in the previous chapter the base circle of the CG cone is determined by CG height and maximum "g" or side force generated by the tires. If the CG is moved further aft, the effective half tread becomes greater, and vice versa. Although rearward CG location improves rear wheel adhesion and traction (as with a dragster), it has an adverse effect on directional stability and promotes oversteer. Clearly a three-wheeler may have as much protection from upset as a four-wheeler if its *effective half tread* is equal to or greater than the half tread of the four-wheeler providing CG height is the same or less, and tire adhesion (lateral "g" force) is equivalent. *CG height* is as important as effective half tread in all cases, as is demonstrated by top-heavy campers, trucks, and semi-tractor-trailer rigs that occasionally overturn, no matter how many wheels they have. Actually, it's the relationship between CG height to effective half tread vs tire "g" force that determines resistance to upset – on any type of wheeled vehicle.

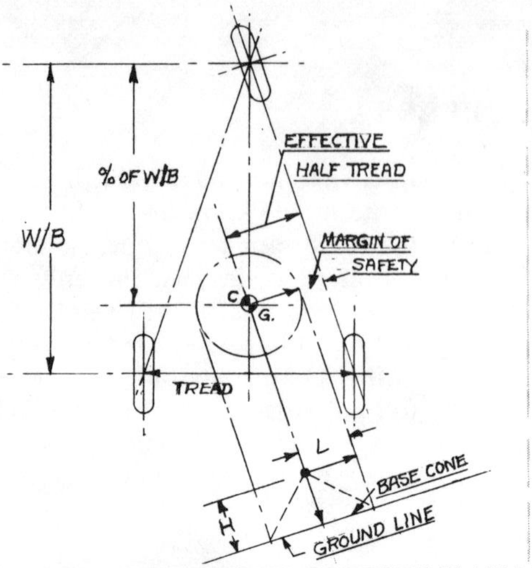

C.G. RELATIONSHIP TO EFFECTIVE HALF TREAD AND WHEELBASE
– FIGURE 102 –

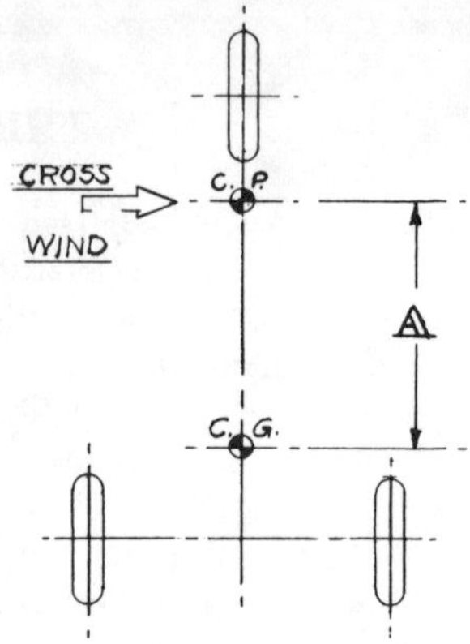

C.P. vs C.G. LOCATION AFFECTS YAW AND PITCH – UNSTABLE
– FIGURE 103 –

THIS DESIGN IS SENSITIVE TO WIND FORCES

Handling is the next problem. Figure 103 shows the effect of side gusts or crosswind on a three-wheeler with single wheel in front and with aft CG. Since the distance "A" is considerable, the resulting yaw and instability can be quite noticeable. Pitch instability is also evident in nose lightness with increasing speed.

OVERSTEER TENDENCIES

Oversteer also occurs due to weight transfer with body roll as during cornering and depicted in figure 104. The single front wheel is incapable of accepting any lateral weight transfer – it must all be taken by the outside rear wheel. This, together with power applied to the rear wheels, can cause excessive oversteer and possible spin-out.

Handling can be improved with wider rims and tires in the rear, but the basic problems are unchanged.

SUCCESSFUL APPLICATIONS

Low-speed vehicles are operating successfully with the single wheel forward configuration in many industrial applications. In England the old "RELIANT" and BOND "BUG" three-wheelers were in production for many years and may still be seen on the highways there. The DAVIS received a lot of publicity in this country after World War II but never really got into production. General Motors, in connection with a feasibility study, built an unusual three wheeled (single wheel front) commuter car, the model 511, which is described in more detail later in this book. This is probably the best looking of such vehicles. It also has other noteworthy features.

SINGLE REAR WHEEL CONCEPT

When the single wheel is located at the rear and the center of gravity (CG) is moved forward, a very significant improvement in three-wheelers occur as we shall see. But first let's look into the background. Most people think the open lightweight sports car idea originated with the British MG four-wheeler. The serious motor enthusiasts know better. The MORGAN three-wheeler is generally credited with being the so-called "original sports car". It began life in 1909 and was built in many varieties up through 1950, although with but few engineering improvements. References 90 and 91 provide a great deal of information on the early MORGANS. Reference 92 goes into three-wheelers

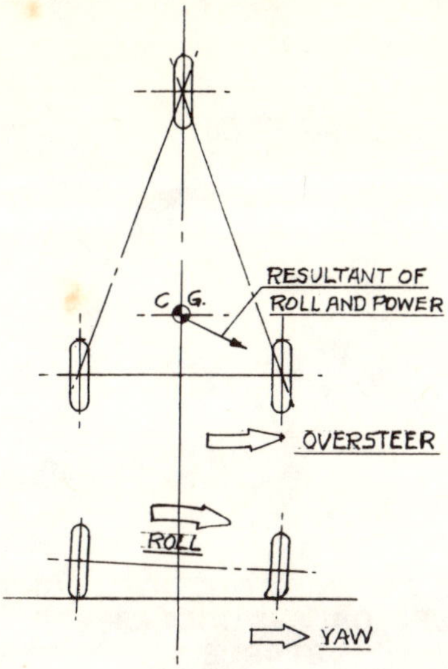

ROLL CAN ONLY BE TAKEN BY OUTSIDE REAR WHEEL – RESULTING YAW CAUSES OVERSTEER – UNSTABLE
– FIGURE 104 –

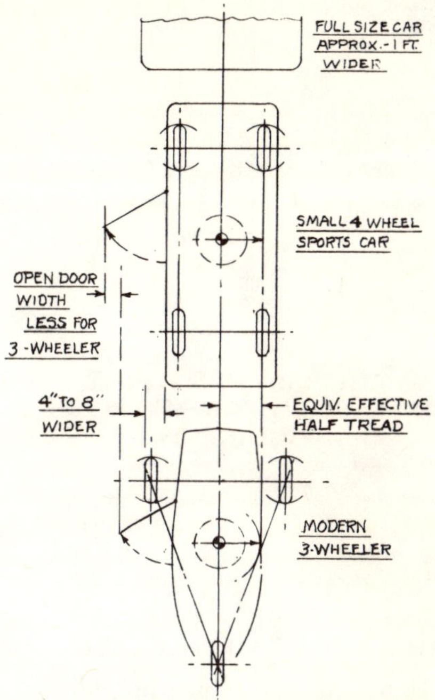

EFFECTIVE TREAD CAN BE EQUIVALENT TO SMALL 4-WHEELER – WITH LESS OPEN DOOR WIDTH
– FIGURE 105 –

created by many others as well as MORGAN. These include the popular MESSERSCHMITT, Buckminister Fuller's DYMAXION, and others.

Today's more expensive fuel and federally mandated smaller cars with much higher miles per gallon requirements, may influence the return of this type of vehicle in the 1980's. Its popularity as a small economical (to buy as well as operate) sporting type vehicle may be reason enough to bring out a modern version. The next few figures illustrates how well this concept comes together.

CENTER OF GRAVITY, EFFECTIVE HALF TREAD, AND WHEELBASE RELATIONSHIPS

Figure 105 compares this type three-wheeler with a small four wheel sports car and a full size car, all in planview. The center of gravity is shown at approximately 1/3rd back on the wheelbase. The CG cone is included along with a margin of safety out to the line connecting a front wheel and the rear wheel. In this case we show that it is quite possible to maintain the equivalent *effective half tread* equal to the half tread for the small four wheel sports car by making the half tread of the three wheeler greater by 4 to 8 inches on each side. This *extra* half tread varies with CG height and its cone base circle. If the height is kept low enough (lower than for a four-wheeler) no additional width is required – as long as there is ample margin of safety between the CG cone's base circle and the front and rear wheels. Of course, the CG could be moved forward more for reduced tread, but this would increase understeer and power to the rear wheel may become limited. A too far forward CG also makes it more difficult for occupants to enter and exit behind the front wheels.

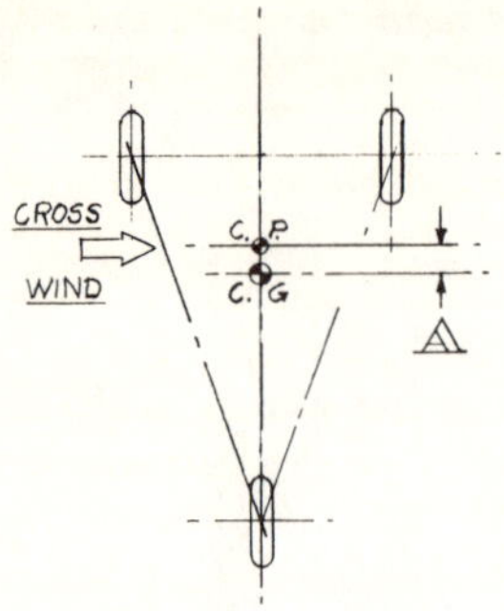

AIR FLOW HAS SLIGHT AFFECT ON YAW AND PITCH
– FIGURE 106 –

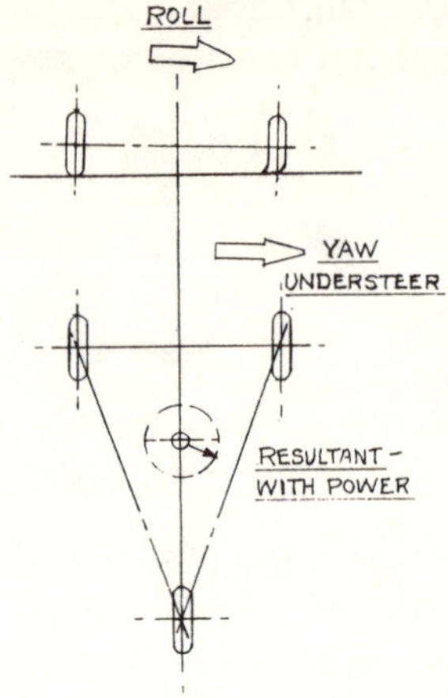

ROLL, TAKEN BY OUTSIDE FRONT WHEEL, CAUSES UNDERSTEER STABLE
– FIGURE 107 –

ALTHOUGH TREAD IS GREATER, DOOR OPENING WIDTH IS LESS

Figure 105 also shows the doors open on both the 3- and 4-wheelers. Since the hinge line on the 3-wheeler is inboard, as compared to the 4-wheeler with its enveloping body, the open door width for the 3-wheeler is clearly less. This is a worthwhile improvement when parking. The extra tread width has never been a problem since the overall width can still conveniently be kept less than for a full size car.

WIND FORCES HAVE BUT SLIGHT EFFECT ON YAW AND PITCH

Contrary to the effect shown in figure 103, the single rear wheel design as shown in figure 106 is but slightly influenced by air forces in either yaw or pitch, because the center of pressure is quite close to the center of gravity. This improvement in stability is very worthwhile at freeway speeds.

A forward CG helps a three-wheeler aerodynamically just as in the case of a four wheel motor vehicle.

ROLL, TAKEN BY THE OUTSIDE FRONT WHEEL, CAUSES STABLE UNDERSTEER

Figure 107 shows this. Roll cannot be taken by the rear wheel; only by the outside front wheel. This in turn causes the front outside wheel to be increasingly loaded due to weight transfer, thus initiating more understeer. Since the rear wheel takes its normal amount of lateral "g" force without any additional due to lack of weight transfer, it (the rear wheel) is always completely predictable. Any application of power to the rear *is assisted by rearward weight transfer but not hampered by lateral weight transfer.* This is a stable condition.

POWER – FRONT OR REAR

The European racers mentioned at the beginning of this chapter have, in many cases, been built with front transverse engines and front drive. The side car passenger is free to move about on his platform to counteract weight transfer during cornering, braking, and acceleration to achieve higher lateral "g" forces and better lap times. These small FWD's are very successful and often utilize stock FWD components in their design. A more conventional FWD three-wheeler with single wheel aft could do a fine job as a small sports car also. Proper understeer, without becoming excessive, can be designed in by CG location, tire and rim choice, and attention to all the usual details of suspension and other features.

Rear wheel drive is relatively easy because no differential is required. This too, can be quite successful with normal power-to-weight ratios. *Aft weight transfer* during straight line acceleration, can in many cases, be sufficient to achieve 0.5 "g" force with a static CG position at one-third of the wheelbase. This is more than ample for urban and highway use. Contrary to rumors, tire wear is not excessive. As mentioned earlier, acceleration in a corner is not jeopardized by *lateral* weight transfer on the rear wheel. Transition to oversteer is gradual, smooth, and completely predictable. A spin-out is next to impossible and if imminent can be avoided by less throttle during acceleration out of a turn.

MAJOR COMPONENTS DISTRIBUTION FORE AND AFT OF CENTER OF GRAVITY

A CG location at about one-third of the wheelbase is desirable. Changes in disposable load such as cargo aft will move the CG aft also. We have tested with a CG as far back as 40% with no adverse effect on our DUO-DELTA. We have also tested with the CG as far forward as 27% with only slight increase in understeer. It is important that CG height be kept low to minimize weight transfer in all cases. A high CG is not conducive to as much fore and aft CG movement as mentioned above.

Once the CG location is chosen the distribution of major items fore and aft can be made. Variable weights such as passengers, fuel, and cargo should be kept as near the center of gravity as possible to minimize CG changes.

A difficulty arises if the passengers must be ahead of the CG by very much for balance. In this case, entry and exit sidewise behind the front wheels becomes limited and more difficult. Figure 108 shows the general arrangement on the MINI-K. Figure 112 shows the arrangement we used on the DUO-DELTA. In this case the CG is at approximately 35% with two people aboard. With driver only, it is at about 38%. With cargo added in front, and two occupants, the CG moves forward to about 32%. Incidentally, the battery is forward of the passenger's feet. It is also necessary to keep the powerplant as far forward as possible – immediately behind the seats.

Balance could have been improved a bit more by putting the fuel tank forward instead of aft.

In the case of the MINI-K we used smaller diameter wheels and moved the rear wheel farther aft so the front wheels could also be moved forward a bit also. This is evident in figure 108.

– FIGURE 108 –

If the powerplant is moved forward, ahead of the passengers, into a front engine position, entry and exit is not critical and balance is satisfactory. Cooling is easier but the exhaust is longer and the powertrain has to go from front to rear.

The MINI-K used a Yamaha 250 cc two-cycle engine behind the seat. It was trouble free and never overheated. Cooling air came from underside of the smooth floor pan.

The DUO-DELTA used a 750 cc snow-mobile engine that included a built-in fan on the right side and ducts enclosed the engine. It too, never overheated and was trouble free.

Both engines were two-cycle with short tuned exhaust systems. If either engine were front mounted the exhaust system for a two-cycle becomes more difficult, especially if a tuned length is used and the outlets must be behind the passengers.

MINI-K DESCRIPTION AND SPECIFICATIONS

This three-wheeler was developed as a research project by the author's company in 1968. Manufacturing facilities did not exist, so the vehicle never reached the production stage although early in the program we considered selling KITS for home builders. As the project developed, a patent was applied for. U.S. patent number 3,610,358 was issued with 14 claims. The MINI-K was tested and favorably reported on by several magazines and later sold along with parts and limited tooling.

We described the MINI-K as "The new idea motorcycle – seats two like a sports car". It was designed to be powered by a motorcycle engine, carry two people as in an open sports car, be superior to a motorcycle/side-car combination, and be quite reasonable in cost. As the test reports show, ride, handling, and cornering were excellent and everyone enjoyed driving it.

Creature comforts were less than in a car, but better than on a motorcycle. Motorcyclists were accustomed to and didn't mind: (1) the noise level of the engine, (2) the lack of a reverse, (3) lack of foul weather protection. They were very enthusiastic about the vehicle and this encouraged us to consider marketing it ourselves as a KIT, although we soon had to face up to the problems of larger facilities and ample financing, which in turn led to its sale.

Figure 108 shows it in side view and figure 109 the Yamaha 250 cc engine in the rear. The powerplant was complete in an aft frame of our design. It was bolted on at the rear of our forebody. As figure 110 depicts, we later modified the standard motorcycle frames by cutting off the steering post and adjoining tubes and adding side plates for attachment. This placed the engine weight up closer to the center of gravity. The picture does not show the diagonal tie-rods on each side between forebody and powerplant frame.

Figure 111 is of sketches made for a body of improved appearance with a conventional door on each side. This never got beyond the idea stage.

SPECIFICATIONS

General

- Licensed as three-wheeled motorcycle; engine ahead of rear wheel.
- Side by side seating slightly *ahead* of center of gravity for balance; two wheels forward and center of gravity at 40% of wheelbase for superior handling.
- Racing type roll-bar integral with frame; lap belt and shoulder harness attachments provided.

SPECIFICATIONS (Continued)

Chassis

- W/b min. 75"; tread 56".
- Clearance 5½"; height at R/B 43½".
- Width at seat 40".
- Tire size 5.20 x 10; rim 4½"
- Brakes – Airheart Disc 7" dia.
- Steering – rack & pinion.
- Suspension – fully independent, may be adjusted as preferred.

Powerplant

- Wide choice of motorcycle engines 200 cc to 750 cc – 2 or 4 cycle – electric or kick starter.
- Motorcycle instruments, controls, fuel & oil tanks, and most accessories may be used.
- Finance 2 wheel bike, new or used, through dealer; resell unused parts; or, buy engine and parts as needed.

WEIGHT empty – 485 lbs.
PERFORMANCE
Top speed 68 mph with 26 hp (250 cc Yamaha 2-cycle)
Excellent acceleration – due to sprocket change and smaller wheel size.

– FIGURE 109 –

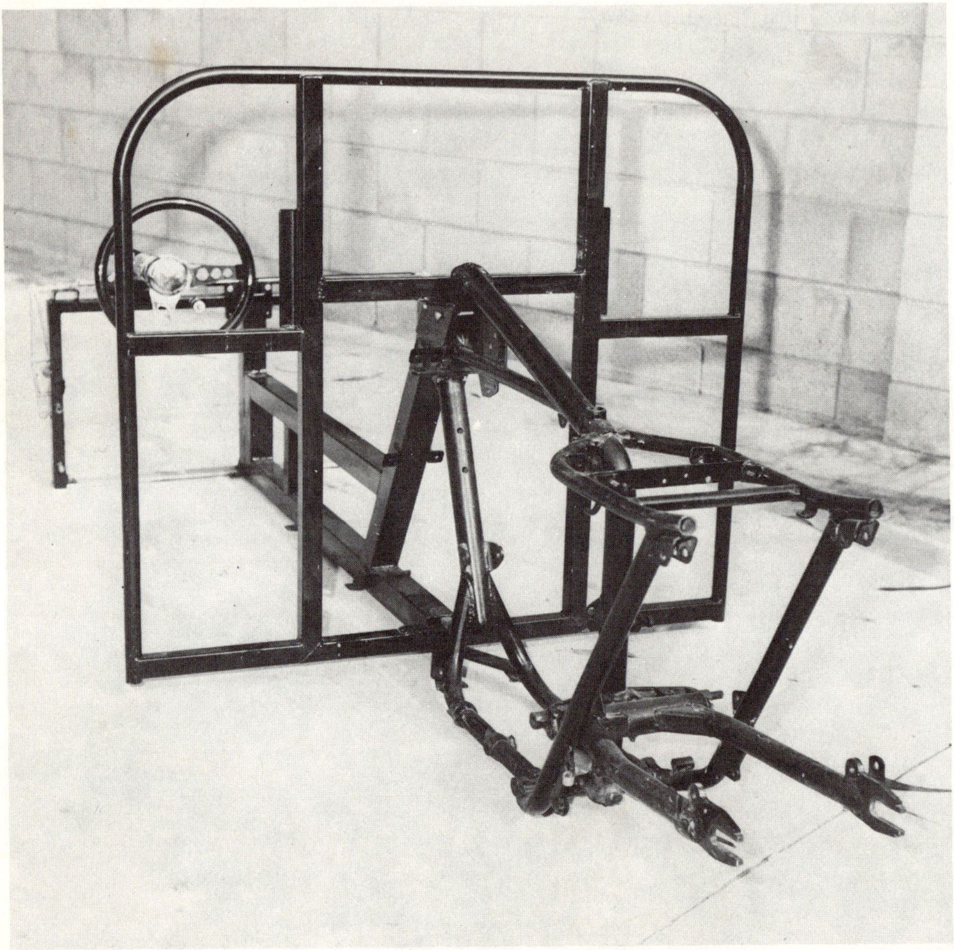

– FIGURE 110 –

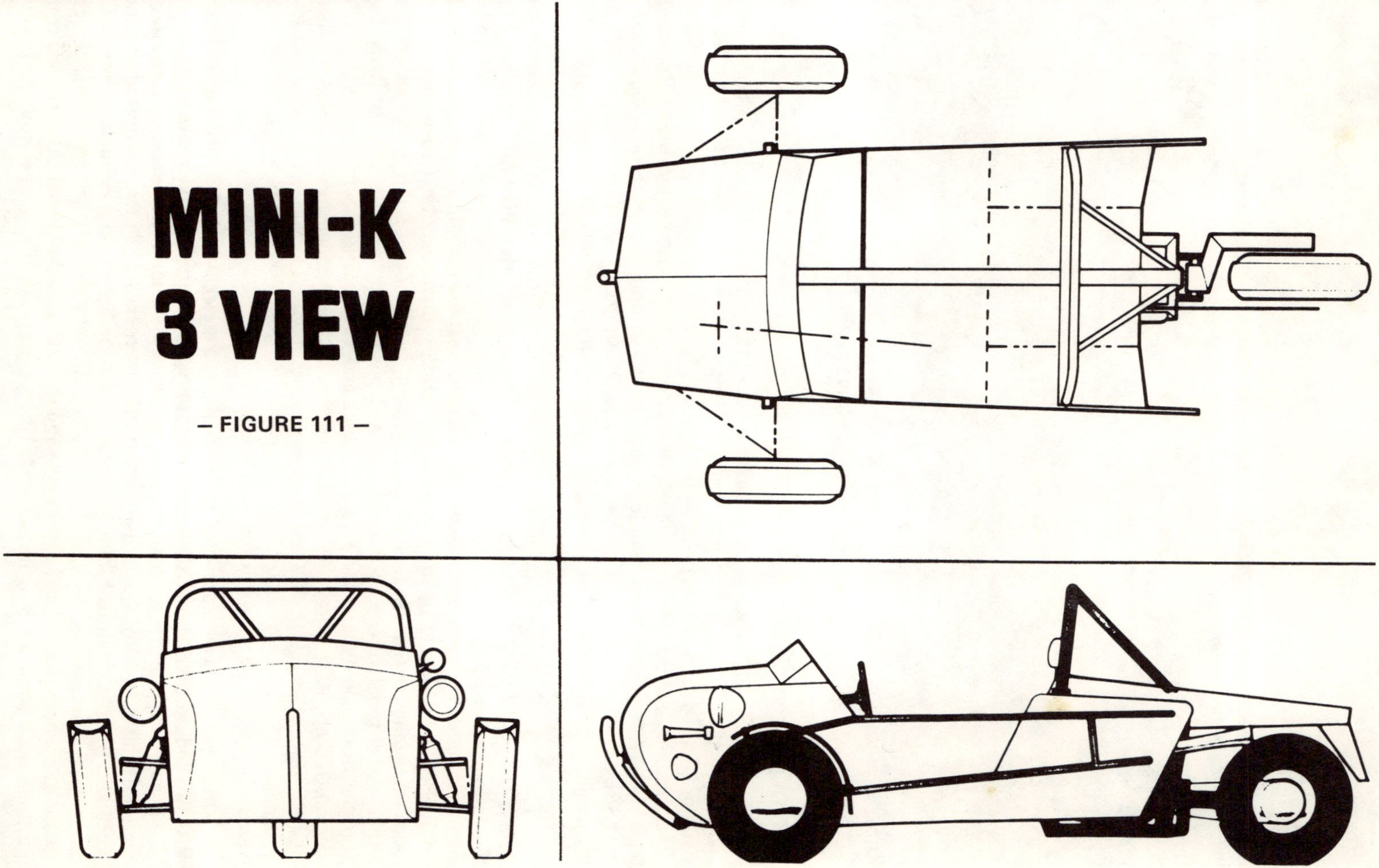

MINI-K 3 VIEW

– FIGURE 111 –

— FIGURE 112 —

DUO-DELTA DESCRIPTION

In the early 1970's we again began to think of three-wheelers because forecasts were made of depleting supplies of petroleum and resulting higher prices. By early 1974 we had our DUO-DELTA running, just in time for the "energy crunch".

This machine had the same wheelbase and tread as the MINI-K but used a Rockwell-JLO 744 cc two-cycle snow-mobile engine. It had a more complete body and used a SALSBURY V-belt automatic torque converter transmission. A reverse was provided by tightening a narrow V-belt to an electric starter motor somewhat as described in the ICE powertrain chapter.

The DUO portion of the name referred to its dual purpose. It has a convertible power train (CPT) which can be quickly disconnected from the DELTA three-wheel configuration and used with a front fork to provide a two-wheel motorcycle at minimum additional cost, for cycle enthusiasts who like the dual feature. Mr. Charles Slover could, by himself, with the aid of the special jacks we built, disconnect the CPT from the forebody in 4 minutes flat. He then could attach the front fork assembly in 3 minutes. No other special tools were needed and lifting and positioning was easy with the jacks. The transfer from one type vehicle to the other and back again occurred many times in demonstrations. A motorcyclist could use the two wheel version in warm, dry weather, and the three-wheeler DELTA in cold, wet weather. Figure 112 was taken during our test program at El Mirage dry lake. The CPT, when the front fork assembly was added, provided the motorcycle shown at the top of page 176.

Other versions were conceived as depicted in the sketches of the BEARCAT, SPORTS RACER, and ski adaptations for winter sports shown in figures 114, 115, and 116.

Figure 117 provides an indication of the chassis general arrangement and its safety features. This view also shows the convertible power train (CPT) in relationship to the forebody. It has a small ball joint at the bottom – similar to a trailer hitch. At the top a single 3/8" diameter bolt was used. A diagonal tie rod on each side took lateral forces. Quick disconnect joints for the rear brake and throttle were developed. All electrical connections were standard plug-ins. Joints, fittings, and plug-ins on the front fork were mates.

The convertible power train is shown in figure 113 mounted on its special two-wheeled jack for easy positioning.

– FIGURE 113 –

IN RETROSPECT

The DUO-DELTA was designed to appeal to car owners as well as motorcyclists by providing a roadster type body with doors, windshield, and roof. The automatic torque converter eliminated the clutch and gear shifting, which would appeal to the ladies. The small size would appeal to nearly everyone interested in an urban car.

Unfortunately, our restricted budget did not allow for expensive body molds and other tooling to provide beautifully contoured body panels. We settled for simple flat-wrap single contour lines. The material is ABS sheet on wood frames. Since the sheet material was not formed in a mold to include flanges, all stiffening and shaping had to be obtained from the wood frames. Wood construction is labor intensive and adds weight. It is nice for experimental projects and one-off home builts, but has definite limitations, especially for production.

The project was shelved for the following reasons:

1– OVERWEIGHT – The vehicle weighed slightly over 900 pounds. It is 150 pounds overweight due primarily to the wood framing and larger steel wheels.
2– UNAVAILABILITY OF CHOSEN ENGINE – The engine was no longer available in this country. We did not want to initiate our own importation program.
3– REDESIGN FOR PRODUCTION – Redesign would be a major undertaking in view of the body work required as well as a powerplant change.
4– ECONOMY DOWNTURN AND UNCERTAINTIES – It was not feasible for our small company to risk its limited assets further in view of the uncertain business climate at the time.

Although this project was shelved, our interest in three-wheelers has never waned. The concepts depicted in chapter V provide some indication of the author's thinking on the subject at this time.

THE BEARCAT

Add hood - doors - windshield later.

— FIGURE 114 —

THE SPORTS RACER

For Club racing Remove windshield and roof

— FIGURE 115 —

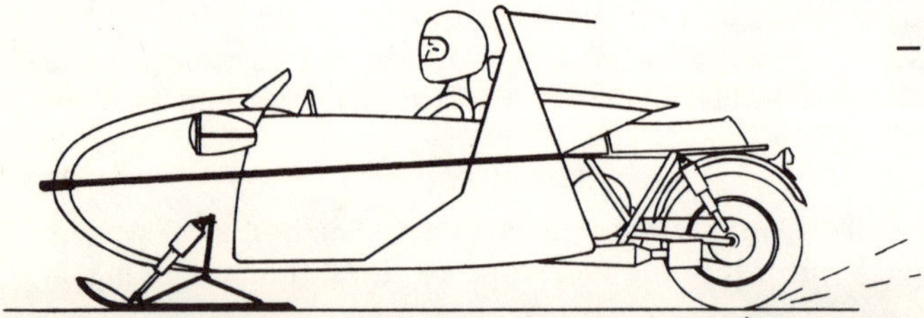

WINTER SPORTS

— FIGURE 116 —

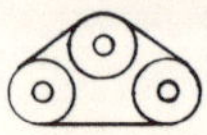

or TRACKS

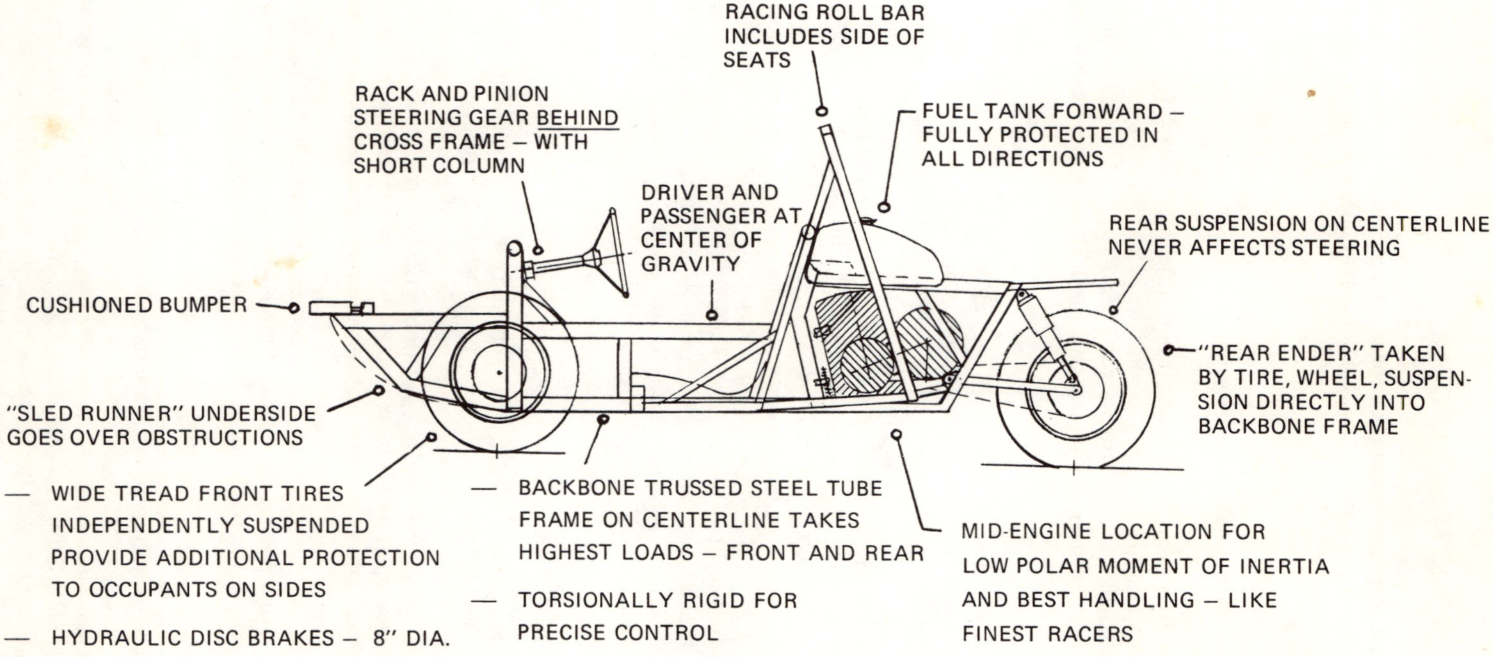

SAFETY

– FIGURE 117 –

Chapter XV

AUTOMOTIVE AERODYNAMICS

Before discussing what happens when air flows around an automobile body, it might be well to get acquainted with the air itself, its inherent characteristics in motion, and the effect of temperature and pressure on its density.

PHYSICAL PROPERTIES OF AIR

Air is a fluid, and since its internal friction is small, it is usually treated as if it were a perfect fluid, which it isn't.

Air is often considered as a mixture of four-fifths nitrogen (actually 78%) and one-fifth oxygen. Minor additions such as Argon, Carbon Dioxide, and Hydrogen are also in the mixture. Close to the earth there are usually particles of dust suspended in the air. Water vapor is also generally present in varying amounts up to about 5 or 6 miles altitude.

AIR DENSITY

The air or atmosphere surrounding the earth extends upward (outward) for about 500 miles. The air in our atmosphere is subjected to pressure due to the weight above it.

Since air is invisible we will describe it with an analogy. To visualize air think of particles – such as grains of sand – floating in an invisible foam. These particles have mass. Air at higher altitudes is "thinner". There are fewer particles in a cubic inch of air at the top of a mountain than at sea level. Now envision a "chimney of air". Stack cubic blocks of air vertically – one on top of another – as you would stack bricks. But remember that air is compressible. Since air particles have weight, the bottom cubic block of air is supporting the weight of all the blocks of air above it *for some 500 miles.* Therefore, it is compressed by the weight of all the blocks above it. Since it is under pressure (weight) the particles are pushed together so there are more air particles in a cubic inch at sea level. This is density. At sea level the density or weight is 14.7 pounds per square inch (psi). At one mile altitude (5,280 feet) the pressure is 12.1 psi. At two miles 9.9 psi and at three miles 8.0 psi.

EFFECT OF TEMPERATURE

As air is compressed it becomes warmer and conversely as it expands (as with altitude) it becomes cooler. The temperature drops approximately one degree F for every 280 feet increase in altitude until the temperature has decreased to -67 degrees F, at which point it is presumed to get no colder. Altitude affects the temperature of air.

The effect of increasing temperature is to *expand* air – move the particles away from each other – make it thinner. This is because heat speeds up the random motion activity of the molecules.

Fewer particles due to altitude or increasing temperature means the air has less resistance to, or drag on, a shape passing through it. Fewer particles also means there are less oxygen molecules in the air for combustion in an engine or respiration in a person.

STANDARD ATMOSPHERE

If test results in wind tunnels, or road tests, are to be compared in a meaningful manner, the temperature and density of air must be converted to values that would be correct for so-called "Standard Atmosphere". The National Advisory Committee for Aeronautics (NACA) has established Table VII for the convenience of aeronautical engineers in their tests. Since the same wind tunnels are often used for automotive testing, these values remain for all testing. A slug (listed in the table) is a British engineering unit of measurement expressing air density in terms of mass rather than force units.

TABLE VII

Standard Atmosphere Based on N.A.C.A. Report No. 218

Altitude (feet)	Temperature (°F.)	Pressure inches of mercury	ρ Slugs per cubic foot
0	59. 0	29. 92	0. 002378
500	57. 2	29. 38	. 002343
1, 000	55. 4	28. 86	. 002309
1, 500	53. 6	28. 33	. 002275
2, 000	51. 9	27. 82	. 002242
2, 500	50. 1	27. 31	. 002209
3, 000	48. 3	26. 81	. 002176
3, 500	46. 5	26. 32	. 002144
4, 000	44. 7	25. 84	. 002112
4, 500	43. 0	25. 36	. 002080
5, 000	41. 2	24. 89	. 002049
6, 000	37. 6	23. 98	. 001988
7, 000	34. 0	23. 09	. 001928
8, 000	30. 5	22. 22	. 001869
9, 000	26. 9	21. 38	. 001812
10, 000	23. 3	20. 58	. 001756
11, 000	19. 8	19. 79	. 001702
12, 000	16. 2	19. 03	. 001648
13, 000	12. 6	18. 29	. 001596
14, 000	9. 1	17. 57	. 001545
15, 000	5. 5	16. 88	. 001496
20, 000	−12. 3	13. 75	. 001267
30, 000	−48. 0	8. 88	. 000889
40, 000	−67. 0	5. 54	. 000582
50, 000	−67. 0	3. 44	. 000361

For automotive aerodynamics "Standard Air" occurs when the barometric pressure is 29.92 inches of mercury (sea level) and the temperature is 59 degrees F. It is equivalent to an atmospheric air pressure of 14.7 pounds per square inch. Standard air density is .07608 pounds per cubic foot. Thirteen cubic feet of standard air weighs about one pound.

Now let's look at the effect of velocity (V).

EFFECT OF RELATIVE MOTION (VELOCITY)

We say "relative motion" because it doesn't matter if the air is moving as in a duct or wind tunnel or if a shape is moving through air.

When speed (velocity) is doubled the air drag on a shape increases four times – it is squared. Visualize it this way: when a shape is moved through air twice as fast, the shape (a) hits twice as many particles of air due to doubling the distance traveled, and (b) hits each particle twice as hard due to doubling the speed. This is illustrated in figure 118. If an automobile requires 75 pounds force to overcome wind resistance at 40 mph, then it will require 300 pounds force at 80 mph. *The horsepower needs to go up eight-to-one!* If 5 hp overcomes wind resistance at 40 mph, then 40 hp is required at 80 mph. The power requirement is cubed.

REYNOLDS NUMBERS AND SCALE EFFECT

Before leaving the discussion of particles we should include a brief description of Reynolds Numbers. In any comparison of test results in which the size of models varies greatly, and/or air velocity of tests differs greatly as well as possibly other factors, we may not have a fair comparison because the Reynolds Numbers are not comparable. One set of test results might even be in a critical range. This calls for a correction factor to account for the effect of scale.

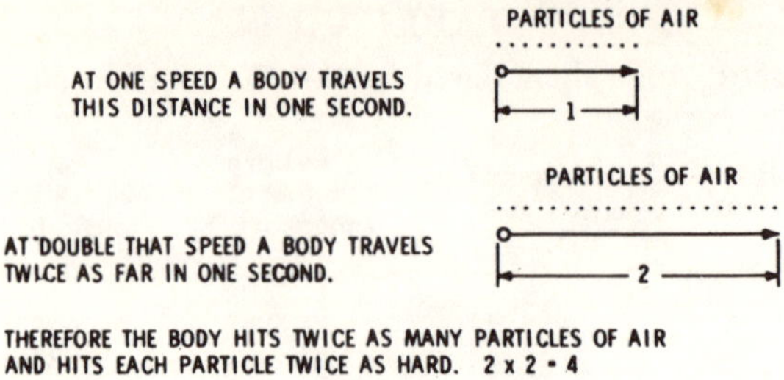

Effect of speed

— FIGURE 118 —

In the 1890's Osborne Reynolds, an English physicist, while studying the flow of liquids in pipes made his discoveries. When he injected a dye into water with both the water and dye flowing at a certain speed in a tube, the dye would remain intact like a colored thread, without mixing with the water even though the speed of the flow was decreased or increased somewhat. Then, as the flow rate was increased beyond a certain critical speed the flow changed from smooth to turbulent and the dye mixed completely with the water to make it cloudy. The point at which it changed its flow pattern he called the "critical Reynolds Number". This condition is true also when the fluid is air instead of water and smoke is used instead of dye.

The Reynolds Number is a nondimensional coefficient used as a measure of the dynamic scale of a fluid flow. It is expressed as $P \frac{VL}{\mu}$ in which P is density of the fluid (slugs per cubic foot), V is the velocity of the fluid, L is the linear dimension of a body in the fluid (usually a length dimension), and μ is the coefficient of viscosity of the fluid (standard air) in pounds seconds per square foot.

In more general terms it can be considered as an expression which may be expressed as VL/μ which is used in wind tunnel testing of scale models. It is a correction factor which takes into account the linear dimensions of the model, the speed at which tests are run, and allows for increasing the density of the air in the wind tunnel in order that the results will compare favorably with those that would be obtained if the tests were run with full-sized vehicles and at their proper speeds.

Reynolds Numbers are very important when comparing a small scale model airplane at low speed with a large airliner at several hundred miles per hour. The scale effect is tremendous.

It is still important for comparing high-speed road racing and record breaking vehicles with their small models in a low-speed wind tunnel.

It is of less importance for more normal vehicles at reasonable speeds when compared with their models.

There are differences that will now be described, again with particles. This time envision a small model of a wing and a full size wing surrounded by uniformly spaced dots or air particles disposed in all directions. Let's make the chord of the full size wing 15 feet and its identically shaped model ½ foot (6 inches) long. The model scale is then 1/30th. Its width is also 1/30th as wide as the full size wing. Therefore the skin surface area becomes 30 x 30 = 900 times as great for the full size wing as for its model. For this reason the "relative density" or number of air particles surrounding the full size wing is 900 times as great as for the model, rather than just 30 times as great, like the scale.

The above is true only when the model and full size wing are at rest, or when they are moving at the same speed. If the model were in a very high speed air flow compared to a low speed air flow for the wing, it would have far more air particles passing around it so that its "relative density" would be increased, makiing its results more comparable. The density times its velocity is a measure of its momentum (strength). *Flows with high momentum tend to remain established around the surfaces for a greater distance downstream.*

Adverse pressures and the viscosity or cohesive nature of air tend to impede the flow downstream. Viscosity is the "stickiness" of say, honey or motor oil. Water has viscosity also but much less than oil, and compared with water, air has very little viscosity, but still it must be recognized and dealt with.

In review, we now see that the air flow or momentum (strength) has to do with density, velocity, and size or length downstream. The resistance has to do with viscosity or "stickiness". We may therefore say that Reynolds Number is a ratio of $\frac{\text{Momentum}}{\text{Viscosity}}$. High values of Reynolds Numbers (RN) are good and low values not so good.

Looking back at our RN formula we see P = slugs per cubic foot (mass density) and μ is viscosity of standard air in pounds seconds per square foot. Therefore P/μ is equal to 9378 under standard air conditions. From this we can now say RN=10,000 x mph x length in feet. Some automobile companies use cross sectional frontal area instead of length as stated above.

We now know a few things about Reynolds Numbers and how they are figured. The important thing to recognize is that test results can be a bit misleading if the RN is too far below actual operating conditions for the full size vehicle. Attempts to study the effect of certain changes downstream on a body shape may be discouraging if, due to a low RN, the air flow is too weak to obtain meaningful results. This is especially true of very efficient low drag shapes.

EFFECT OF SIZE

Assuming the Reynolds Number is not in the critical range, the size of a vehicle body has a direct effect on drag. In fact, the scale model, when properly tested, will accurately depict flow patterns, pitch and yaw moments, lift (if any) and the coefficient of drag can be determined with considerable accuracy. The size dimension, in this case is not the length, but rather the cross sectional dimensions of width and height. This is referred to as frontal area or FA. A frontal area of 10 sq. ft. will have half as much drag as a frontal area of 20 sq. ft. at the same speed and providing the shapes are otherwise identical. In this case the wind drag of the cross section or frontal area varies directly with its size (at the same speed).

EFFECT OF SHAPE

The shape of a body is most critical. To demonstrate, try this in your bath tub or swimming pool. Tie a string to a square block of wood and drag it down under water. Watch the turbulent flow of water around it. Now do the same with a fish-like shape and notice how much easier it is pulled and how little disturbance is created. The paths of water (air also) bend gently at the nose of the fish and flow smoothly and equally to the tail and rejoin with the least resistance or disturbance. But not so with the square block – the water is disturbed violently – it simply cannot flow around the sharp corners. Air becomes turbulent like water, thus causing the tremendous increase in wind resistance.

STRUT SHAPES

In considering shapes let's start with five long rods, all 100 feet in length and one inch high (thickness). The frontal area (FA) is then 1" (1/12th foot) x 100 ft. = 8.333 sq. ft. in each case. Each shape is depicted in cross section in figure 119 and include a square at the top, next a round shape, third is elliptical, fourth a streamlined shape 1 x 2 inches, same as the elliptical, and the fifth shape is 1 x 3½ inches and also streamlined.

At 100 miles per hour the square shape has 433 pounds of drag and the last streamlined shape but 18 pounds of drag – only 1/24th as much as the square shape. Even the

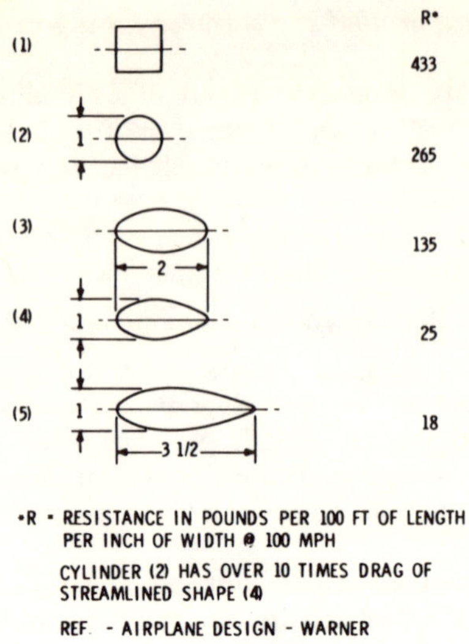

— Drag of strut shapes

— FIGURE 119 —

cylinder (2) has over ten times as much drag as the short streamlined shape (4). The source of this information is reference 93.

These strut shapes, if on an automobile, might be a protruding front bumper, an external radio antenna, a strut extension for an outside mirror, or suspension members under the car including tie rods, linkages, and brake hoses. While the drag may not be as high as indicated above, due to interferences, it could be higher in the case of the mirror and antenna due to air flow speed-up from around the edge of the windshield.

FLAT PLATE VS STREAMLINED SHAPE

These are the air drag extremes in shapes. See figure 120. The flat plate has two values based on the differences in Reynolds Numbers and should be explained. The drag of flat plates has been studied for over two hundred years. Early on it was decided that the coefficient of drag of a flat plate should be unity (1.00) since every other shape would have a lower coefficient. But the flat plate tests were run in very low speed wind tunnels and before they understood about Reynolds Numbers. Much later, when the same tests were run again, but at higher RNs, the coefficient was found to come out to approximately 1.28. Rather than try to change all other test data proportionately, it was easier to accept the drag of a flat plate as being greater than unity when tested at proper RNs.

Incidentally, the drag of a circular shaped flat plate is about 5% less than for a square shape.

The fish-like streamlined shape disturbs the air much less and has a drag coefficient of but .04, just 1/32nd as much as the flat plate. Actually, it would be even less if, like certain fish, its skin could ripple slightly to reduce skin friction.

Much of the following information has appeared before in articles and papers by the author such as reference 94.

SKIN FRICTION AND SURFACE CONDITION

We learned earlier that air has viscosity or a small amount of "stickiness". This in turn results in some friction or drag along the skin surface of a shape as the air flows rearward. This slows it down next to the surface, like water along the banks of a stream. Out in mid-stream the flow is unrestricted if the stream is relatively deep. So it is with air flow around a shape. A smooth, continuous surface keeps skin friction low. A polished surface is better. While this helps, this element of wind resistance is of itself considered a minor factor on an automobile, except for highest speeds with streamliners. Low skin friction can, however, be quite important in critical areas to delay flow separation.

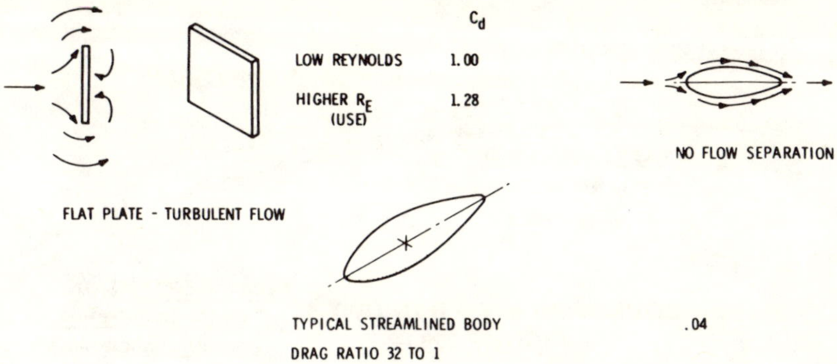

— Air drag extremes in shapes

– FIGURE 120 –

THE BOUNDARY LAYER – LAMINAR FLOW AND TURBULENT FLOW

The boundary layer theory, conceived by Prandtl in 1904, opened up the study of lift and drag and is described by nearly every book on aerodynamics since, and will be repeated here to show what happens as air flows downstream along a surface. Reference 95 provides an excellent in-depth analysis for the serious student. Reference 96 provides a similar analysis, particularly as it applies to automobiles.

According to this theory, the viscosity of air must be considered only within the limited sheet of air adjacent to the body surface. Outside this boundary layer, considering the motion of the air particles against each other, the viscosity does not have any noticeable effect.

While a thin boundary layer seems at first unimportant on an automobile *it should be realized the boundary layer grows in thickness as it moves downstream,* quickly changes from laminar to turbulent flow, and then easily separates or becomes detached from the surface at which time the air drag increases remarkably. *Separation is what we want to avoid.*

Figure 121 depicts, but not to scale, what happens to the boundary layer as it moves downstream on the upper surface of a wing or possibly the roof line of a streamlined sports car. Remember, earlier we discussed the viscosity of air. Now visualize a thin sheet of air made up of layers flowing over the body surface. If, due to the viscosity of the air, the layers can remain intact, the flow is said to be laminar – consisting of laminates. In laminar flow *the drag is directly proportional to the speed.* As long as the air is accelerating on a rising or convex, very smooth, (even polished) surface it has a very good chance to remain laminar. This condition can remain for only a relatively short distance even under the best of conditions, except on laminar flow airfoil sections.

When the air is no longer accelerating beyond the point of maximum height of the surface the speed drops and the pressure increases until the weak viscous forces are no longer able to hold the laminates or layers together and keep the flow smooth over the surface. This condition may be visualized as tiny "rollers" along the surface. These tiny rollers, or disturbance, is the first sign of turbulent flow and is the transition point from laminar to turbulent flow. At this point higher velocity flow from free air beyond the boundary layer begins to mix with the layer next to the surface. Then air drag *rises to a value proportional to the square of the speed.*

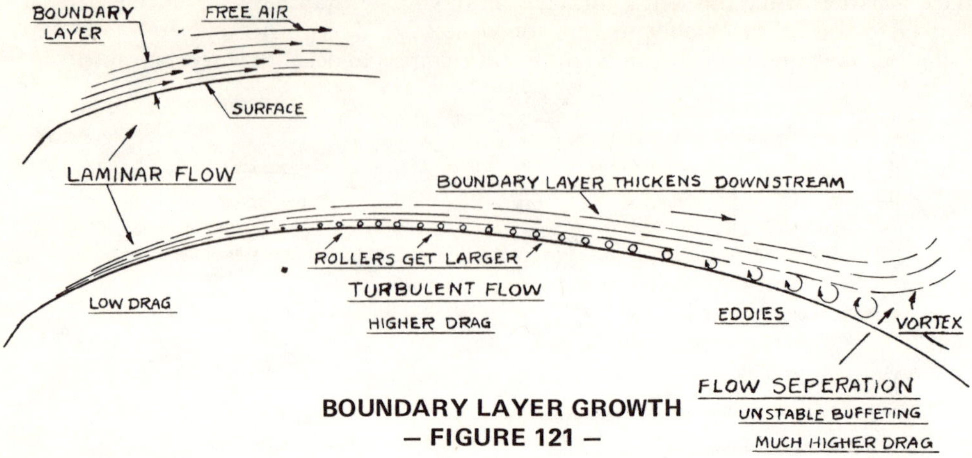

BOUNDARY LAYER GROWTH
– FIGURE 121 –

BOUNDARY LAYER BUILD-UP DOWNSTREAM AND FLOW SEPARATION

As the boundary layer progresses downstream the rollers get larger and the layers thicken. It becomes increasingly more difficult to keep the boundary layer next to the body surface. Eventually the rollers become small eddies and begin to lift the layers. This is the beginning of separation. It is an unstable condition and may move fore and aft a bit. Next the eddies enlarge and become a vortex in which separation is complete.

AT WHAT ANGLE DOES FLOW SEPARATION OCCUR ON ROOF AND SIDES?

Dr. W.I.E. Kamm, while developing his streamlined cars in the Automotive Institute of Stuttgart wind tunnel, determined that a 10-degree downward slope of the roof line after a well curved forward portion is, under normal conditions, about the maximum that will still prevent the boundary layer from separating. Obviously, many "fast back" sports cars slope downward too quickly to avoid separation. The author tried *curving the sides* of his racer design, which in model form was tested at Mississippi State University. It was found the 10-degree angle is also approximately true for the sides, providing the nose is nicely curved and wheel wells are covered or properly faired after an opening. The sides must also be kept smooth with easy curves. The 10-degree angle assumes zero degree yaw and can only be exceeded (as with a cross wind) when the underside and topside have a nice radius with the sides so that the air may move laterally as well as rearward.

When we tested the GOLDENROD model at Cal. Tech., we were concerned about side winds and stability in yaw in case of a skid. The model was therefore progressively yawed from 0 to 15 degrees. The drag coefficient slowly increased from 0.1165 at 0 yaw to 0.1210 at 15 degrees yaw as may be noted in the tabulated data in the last chapter. This is an increase of but 0.0045 or 3.86 percent. Clearly, no separation occurred even with this long shape. The sides had about 2 degrees taper (after a smooth curve from the nose plus wheel covers) between front and rear wheels. Aft of the rear wheels the taper was greater. The well rounded corners had much to do with permitting air to flow from the underside and over the long hood or topside to the negative side when the model was yawed.

Drag typically increases much faster with side winds on most motor vehicles because of sharp corners and a rough underside, both of which restrict cross flow. *This is dramatically shown later when large trucks, vans, and trailers (with square corners) have their already high drag coefficient nearly doubled in a cross wind.*

THE SNOW PLOW EFFECT ON SOME HOODS DUE TO THE WINDSHIELD

Separation also typically occurs on a relatively flat hood about halfway back to a rather steeply slanted windshield. The free air stream above the boundary layer hits the steep windshield and splits its flow with some flowing downward to lift the boundary layer. This is similar in appearance to the snow being lifted ahead of the blade on a snow plow. Three different changes would improve the air flow: (1) slope the hood more into a wedge shape with a radius at the nose; (2) tilt the windshield back more to reduce the angle between hood and windshield; (3) a gently curved transition instead of a sharp notch between hood and windshield.

Exposed windshield wiper blades also trip the flow and increase the drag. On a very "clean" hood/windshield contour exposed blades have been known to increase the drag coefficient of the complete body as much as 10%.

WHAT HAPPENS ON THE UNDERSIDE?

The rugged exposed underside of an automobile somewhat follows the frame but can hardly be called a surface as such. The frame and its cross members, outboard brackets, springs, bouncing axles, suspension members, steering rods, levers, hose lines, cables, and other exposed parts are real wind catchers and tear the flow so badly that the underside of a car is a major source of wind resistance.

Studies have shown that the boundary layer thickens so much, its velocity is slowed down so much, and it is hemmed in by the ground below and wheels front and rear, that the resulting blockage greatly limits flow along the underside. This blockage also varies with height above the ground.

DAM THE AIR FLOW OR SMOOTH IT WITH AN UNDERPAN

Two approaches have been used to overcome the problem: (a) a smooth underpan to cover the exposed rough surface, or (b) a "dam" at the front below the nose to block, to some degree, the flow of air to the underside.

A great deal of misunderstanding has resulted; followers seem to have divided into two camps, favoring either (a) or (b) above and insisting their approach is best. Actually, both methods work (or don't work) depending on the skill of the designer and his understanding of the problems with a specific application. A good dam will reduce drag, usually about 6%; will increase flow through a radiator by creating a low pressure area under the engine; and this same reduced pressure tends to reduce lift over the nose. Some racers have the dam curve around the front wheels to direct air to the sides instead of over the hood. This type is probably the most efficient. A street automobile requires more clearance and often has a dam a bit farther back and not nearly as wide, with considerable ground clearance for practical reasons. Clearly, it cannot possibly be as effective. It is usually more cosmetic than functional. Manufacturers the world over have embraced the dam rather than accept the more difficult (usually) problem of installing an underpan. Servicing the underside of a car can be a problem, admittedly, unless the problem is dealt with from the inception of design rather than treat it as an "add-on".

Dr. Joseph Cornish, reference 97, provides an excellent study of the problem and introduced the spoiler, or dam on the underside, along with its effects on both lift and drag with different locations. Dr. Cornish, in other studies, also recognizes the work done with underpans and the need to analyze the problem as related to a specific application.

The case for underpans is about as follows: Dr. Kamm's excellent streamlined sedans, K-1 and K-5, had underpans as early as 1935 with a drag coefficient of but 0.25. He certainly achieved excellent underside flow to get the drag coefficient that low. Reference 98 more recently achieved a design, with underpan, in which the drag coefficient was 0.23 along with other features. The GOLDENROD, with its very long narrow body and smooth underside achieved a drag coefficient of 0.1165, download (negative lift), and complete stability.

The disappointments with underpans can usually be traced to: (1) lack of a curved entry at the nose to initiate smooth flow, (2) lack of a gently increasing exit downstream with a venturi-like flow pattern. Additionally, small "fences" around suspension members and rear wheels may be helpful. Another important feature is to provide well rounded rocker panels to assure crosswise flow with cross winds.

Rather than remove the complete panel for underside servicing, lubrication, and oil changes, only small hinged covers with quick acting fasteners, as used on aircraft, would be incorporated in the design.

A good underpan can usually reduce the drag of the complete body by about 15 to 17 percent as well as assist in reducing drag due to cross winds. This much improvement justifies the added expense (over a dam) in many cases. The cross wind improvement alone is worth a great deal. Good underside flow can also reduce lift – especially if the nose and hood lines are sloped properly.

PROTUBERANCES ON THE SURFACE

Unfortunately, an automobile has quite a few protuberances from its body surface, many of which can trip the boundary layer, cause flow separation, and thus indirectly cause more wind resistance out of all proportion to the size of the protuberances.

The wheels extend below the body and the front ones must be steered – thus requiring additional clearance on the underside and to the sides of the body. Other items are bumpers, outside mirrors, door handles, fender mouldings, headlight bulges and bezels, rain gutters, windshield joints and wipers, as well as outside trim strips.

Additionally, there are many openings and gaps such as joints between doors and their frames. Wheel openings often are larger than need be, and designed to do nothing to restore airflow downstream. Recessed windows are also wind catchers.

Reference 99 lists the drag of typical protuberances, gaps, rivet and bolt heads, strips and grooves, as well as a great deal of other information. References 93 through 96, as well as most books on the subject of aerodynamics have similar data, usually as measured on aircraft surfaces.

The surface of an automobile, especially the front half or two-thirds, should be nearly as smooth as a "clean" airplane, and the aft portion should have very limited protuberances, if any.

INTERFERENCE DRAG

As the air flows around, or over, an object it must speed up to cover the greater distance of the curved surface in the same time that adjacent air flows along a straight path. Any object in the path of the curved air flow will then have more wind resistance because of the relatively faster air flow. A well-streamlined rear vision mirror located near the curved side of the windshield will be in the faster moving air around the side of the windshield corner. This is called interference drag and might easily cause the mirror to have double the amount of drag it has when tested alone.

Air that is channeled between humps such as the hood and fender bulges on some racers usually add interference drag.

Exposed wheel racers of course have many examples.

There are many items subject to interference drag. In fact, anything exposed to the wind stream and located near another part or surface. Interference drag is one of the worst offenders and deserves detailed analysis.

WIND RESISTANCE DUE TO LIFT (INDUCED DRAG)

The wing on an airplane deflects air downward while supporting (lifting) its weight. This can be observed by attaching a thread to the trailing edge of the wing as shown in figure 122. This downward forcing of the air causes additional turbulance or resistance in the surrounding air, hence the name "induced drag". Power is consumed by this action.

Air forced upward by a windshield and then curved back downward at the rear by the roof and rear fastback deck, lifts in the same manner as the wing and creates the same kind of "induced drag". Fenders and other airfoil-like shapes create lift and its accompanying induced drag as shown in the lower portion of figure 122.

Induced drag on this type of automobile may easily lift as much as the pressure drag only of the body shape. Conventional domestic sedans of a few years back generally "lift" about 300 pounds at 100 mph. It is divided somewhat equally between front and rear axles. The same car may have about 300 pounds air drag also at 100 mph. A lift of 300 pounds accounts for a serious loss of weight on a "light" car and will adversely affect stability, controllability, and braking, at high speeds. It is obvious that proper aerodynamic design is a must for these reasons alone, as our cars are getting lighter and for all cars at high speeds.

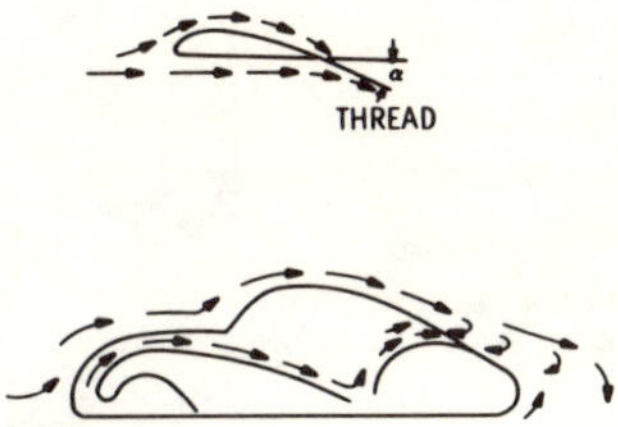

Lift and induced drag

– FIGURE 122 –

DRAG TRADE-OFFS DUE TO "BOB-TAILING"

Aeronautical engineers have known for decades that the pointed tail of certain streamlined shapes could be cut off with but minor effects on drag. A good example is WW II bombers on which the tip of the fuselage tail was "bobbed", or cut off to provide room for the tail gunner. Tests have shown that a surprising amount of the rear portion can be removed, in certain cases, with relatively small increases in drag. This is important to the problem of the streamlined automobile in keeping the length reasonable for parking.

Bob-tailing also affects a lifting shape (induced drag) if the "bobbing" occurs some distance after the start of the downward curve of the roof line on the automobile. Since the cut-off *increases* the base drag slightly and *reduces* the induced drag, an optimum cut-off point can be determined for a motor vehicle at which the two effects essentially cancel each other – a reasonable "trade-off". Then the motor vehicle with this blunt, or cut off portion, will have no more drag than a full length streamlined body that curves down from the roof line to the belly pan. Dr. W.I.E. Kamm of Germany was the first to apply this knowledge to the design of a streamlined sedan body. The Kamm effect is well known now.

Reference 95 indicates the optinum cut-off point occurs where the cross section of the blunt aft body equals 50% of the maximum cross section. This is depicted in front and plan views in figure 123. Typical car designs would have some difficulty in avoiding separation on the aft roof line if all the 50% were to be taken off the upper air flow. They must settle for an aft body cross section greater than 50%. The point of separation is usually chosen by placing a "spoiler" on the roof line. Some are more effective than others since some hardly reduce the aft body cross section at all.

In figure 123, similar to figure 11 of chapter V, the writer has chosen to direct about half of the air around the sides instead of nearly all over the top, to reduce drag and achieve a smaller aft body with less base drag. In this design it was easy to reduce the aft body to 50% of the maximum cross section without causing separation of the air flow along either the sides or the top. This is the subject matter of a patent as mentioned later.

In the GOLDENROD design we avoided induced drag and nearly all base drag by having the aft body taper in plan view to a thin vertical wedge.

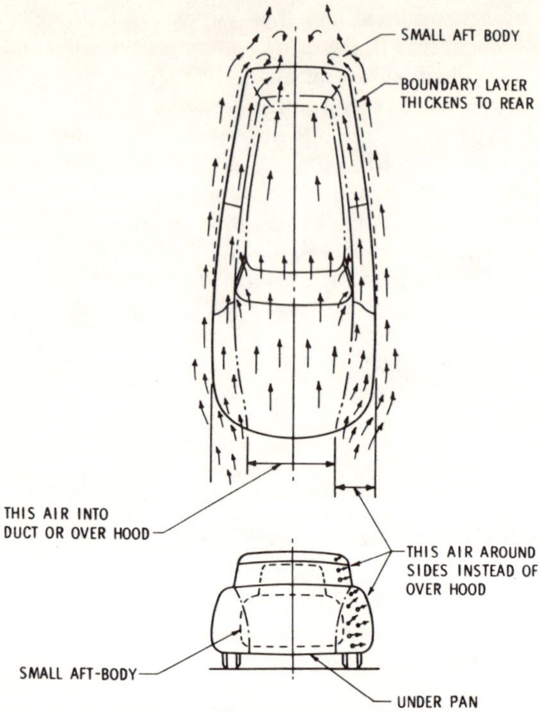

— Flow paths around a streamlined body

— FIGURE 123 —

INTERNAL AIRFLOW AND DUCTING

Internal air flow to the radiator and engine compartment represents a considerable source of drag. Reference 99 indicates 12%. Again, in the case of aircraft, excellent duct designs of very low drag and superior cooling ability have been developed. The principles are shown in figure 124. Everything must be sealed to regain energy.

Obviously, a radiator immediately in front of an engine would lack space for the aft duct (the one behind the radiator); maybe lack space for much of a forward duct as well. Dual radiators, one on each side of a smaller engine might work. Another idea would be to use a single radiator parallel but to one side of an offset engine. Mid- or rear-engine designs, of course, have ample room for a properly designed duct system in the nose of the vehicle.

The one-sixth opening at the nose (figure 124) is typical for fast moving vehicles such as racers and aircraft. A vehicle in slow moving traffic, or idling while parked for long periods, should have a larger opening, possibly a third or even half or more. This can only be determined by testing.

When a separate electric motor driven fan, temperature controlled instead of belted from the engine, is used the opening need not be so large.

DRAG COEFFICIENTS OF SOME VEHICLES

It is always nice to compare values to determine progress and to discover opportunities for improvement. Table VIII does this and goes back several decades, as well as include the latest available data. Unfortunately, this information cannot always be compared exactly because of the following:

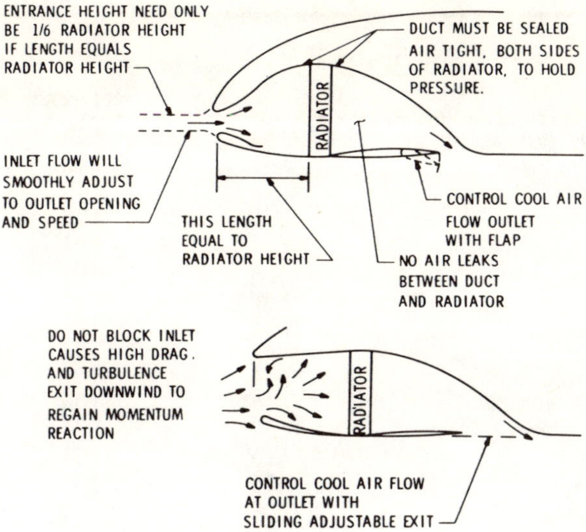

— Duct design

– FIGURE 124 –

a) Model accuracy and details vary;
b) All tests were not run in the same wind tunnel and under exactly the same conditions;
c) In some cases models were tested and in some other cases full size vehicles were tested. Models rarely have all the detail of a full size car.

Clearly there is room for some discrepancy. We believe this information is still useful, especially to students. Opportunities for improvement still remain.

Much data also exists on other types of motor vehicles such as motorcycles, vans, trucks, tractor rigs as well as racers and experimental vehicles. To include it all would not necessarily be of general interest to most readers. SAE papers and references 107 and 108 provide such data for those who wish to investigate further.

As this was being written reference 109 arrived in the mail. Sixteen electric, hybrid, and sub-compact vehicles were tested, full size, in the same wind tunnel, under identical conditions. The Lockheed-Georgia low-speed tunnel was used. Table IX lists the vehicles and provides some description of each. Most are rather well known.

The IC engine vehicles, except the G.M. 1978 Oldsmobile Delta 88, were tested with their grills covered to be more representative of electric vehicles, which would not have radiator airflow losses. The Olds 88 was considered equivalent to a Hybrid with both electric and IC power.

A photographic setup was used to take front view pictures at a fixed distance to accurately determine frontal area from enlargements with the aid of a planimeter.

In Table IX the CD_O column lists the drag coefficient for each vehicle at zero yaw. The $CD_{max.}/CD_O$ column lists a ratio or multiplier to be used in finding the drag coefficient in the worst condition, with added factors such as a cross wind, windows open, and in some cases, with pop-up headlights.

All vehicles had a rapid increase in drag with yaw (cross winds). Yaw angles up to 40 degrees, except for two long vehicles, were recorded.

Surprisingly, four vehicles, Honda Civic sedan and wagon, Hevan, and the Corvette,

TABLE VIII

Vehicle	Coef. of Drag	Reference	Notes
1959 Chevrolet Bel Air	0.44	100	
1959 Chevrolet Impala	0.48	100	
1960 Corvair Sedan	0.43	100	1
1960 Corvair Coupe	0.40	100	1
1960 Volkswagen	0.52	101	
1960 Falcon	0.54	101	
1960 Plymouth	0.48	101	
1960 Cadillac	0.49	101	
1960 Ford	0.55	101	
1978 Ford Fairmont/Zephyr	0.54	102	
1966 VW Rabbit (before)	0.46	103	2
" " (after optimized)	0.38	104	2
1980 G.M. X-cars	0.417 to 0.466	104	
1977 Porsche 924	0.36	105	
with headlights extended	0.38		
1976 Copper Electric Town Car	0.3002	106	
Target with development changes	0.27		
1976 Pininfarina Body of min. drag	0.177	98	
with cooling, gaps, unevenness, etc.	0.230		
1935 Kamm Sedans K-1 & K-5	0.23	95	3
1979 Dodge Omni/Plymouth Horizon	0.395		
1979 ETV-1 (Elec. Test Vehicle-One) GE/Chrysler–D.O.E. Contract	0.295	116	
1979 Ford Probe-I (Experimental)	0.25		

NOTES:

1–Corvair's lower drag is possibly influenced by air cooled rear engine instead of water cooled radiator.

2–The reference report describes how body details such as nose, dam, rain gutters, aft roof line were changed slightly without changing basic body lines, to reduce drag nearly one-third. These changes delayed bounday layer separation.

3–Remarkably, the 1935 KAMM experimental cars had as low a drag coefficient nearly half a century ago as experimental designs of today with our sophisticated computer programs.

Our domestic production sedans still have not equalled the Corvair Coupe of 1960.

actually had a lower zero yaw drag with their front windows open than when closed (almost 4% lower on the Civic wagon).

The Chevette and the Corvette were tested with their radiators both open to airflow and blocked. The Chevette exhibited about 7 to 8% higher drag with the radiator open. The Corvette 6½%. These values would differ for each vehicle but probably remain in the 5 to 10% range.

The Corvette with headlights up had about 3% more drag. A very low drag car would have a higher value.

Since the wind tunnel floor has a natural boundary layer (velocity gradient) growth of about 6 inches, a test on the Chevette was made by raising it about 3 cm higher (+ one inch). This increased the drag by from 1% to 2% over the entire range.

All tests were run at a wind speed of 55 mph.

TABLE IX

		CD_O	CD max/CD_O
Otis Elevator Co.: Otis P-500A Van[1]	Electric delivery van	0.581	1.30
General Motors Corp.: 1978 Chevrolet Chevette 4-door	Internal combustion engine	0.502	1.14
Chrysler Corp.: 1978 Plymouth Horizon 4-door	Internal combustion engine	0.411	1.32
American Motors Corp.: 1978 Pacer Station Wagon	Internal combustion engine[2]	0.406	1.27
American Motors Corp.: 1978 Pacer Sedan	Internal combustion engine	0.450	1.24
Kaylor Energy Products: Kaylor GT	2-passenger hybrid–electric open roadster	0.583	*
General Electric Co.: Centennial Electric[3]	4-passenger electric commuter	0.337	1.12
General Motors Corp.: 1978 Oldsmobile Delta 88 2-door	Internal combustion engine[4]	0.558	1.46
Honda Motors: 1978 Civic Sedan	Internal combustion engine	0.503	1.28
Honda Motors: 1978 Civic Wagon	Internal combustion engine	0.514	1.22
Energy Research and Development Corp.: HEVAN (Hybrid Electric Van)	Hybrid-electric delivery van	0.497	*
Ford Motor Co.: 1978 Fiesta	Internal combustion engine	0.468	1.22
General Motors Corp.: 1967 Chevrolet Corvette	Internal combustion engine[5]	0.490	1.10
Copper Development Association: Town Car	2-passenger electric commuter	0.367	1.16
Sebring-Vanguard: Citicar[1]	2-passenger electric commuter	0.541	1.20
Zagato-Elcar Corp.: Elcar[1]	2-passenger electric commuter	0.490	1.37

TABLE NOTES:

[1] Loaned by NASA-Lewis Research Center, Cleveland, OH.

[2] This production IC engine Pacer Wagon represented a reasonable facsimile of the Electric Vehicle Associates "Change of Pace" converted electric Pacer Wagon.

[3] When these tests were performed, the Centennial was identified only as the GE "Reference Electric Vehicle"; that designation therefore appears on the data sets in Appendix B.

[4] This production IC engine Delta 88 was a reasonable facsimile of the National Motors Hybrid-Electric Gemini II. Here the radiator was not blocked since the hybrid vehicle retains its V-6 engine and cooling system.

[5] This production IC engine Corvette represented a reasonable facsimile of the Cutler-Hammer Electric '67 Corvette of Santini. The front grille was blocked in order to eliminate the radiator losses, which are not present in the electric version.

*Maximum CD was not determined since test yaw angle was limited to 20 degrees.

It is interesting to also compare the 1976 Copper electric town car model test (Table VIII) with the actual car (Table IX). The model C_d was 0.3002 and the actual car C_d 0.367. This may be due to wind tunnel differences, insufficient detail on the model, and possibly other factors. If data from coast down tests were available, they too might be different, because the wheels would then be rotating and there would not be a boundary layer effect from the road surface.

All this does not mean model and full-size wind tunnel tests are unreliable and/or that coast down is better. Experienced testers recognize the limits in each case. Each method offers definite advantages and is an excellent research tool.

WIND TUNNEL TESTS OF THE AUTHOR'S DESIGNS

In chapter V the full-size fiberglass body of the author's little streamliner is shown in figures 13 and 14. The scale drawing, figure 12, is used as an example, along with the pictures, of a project in this book. To continue the project example it seems appropriate to include the wind tunnel work on the model by the Aerophysics Department of Mississippi State University. I am indebted to Dr. Joseph J. Cornish III, the University, and the students who participated in this study. Their enthusiastic and thorough efforts are deeply appreciated.

In the early 1960's the Aerophysics Department, under the leadership of Dr. Cornish, was engaged in advanced research on numerous aircraft projects. As these projects were completed, Dr. Cornish directed the department's efforts into automotive aerodynamics and a number of very interesting and valuable papers such as reference 97 followed. This was also about the time the author's writing efforts mentioned earlier began to appear. It was only natural that we became acquainted and corresponded. My streamliner design was of interest and Dr. Cornish's department built a model and tested it as a class project about the same time the fiberglass moulds were under construction by Mr. Rudy Makela.

Shortly thereafter the Summers brothers approached me regarding the GOLDEN-ROD design. So, of course, the little streamliner was put on the shelf for a few years. Then chassis design, construction, and testing finally got under way, part time, and with many typical changes. After that the three-wheelers and the electric came along, all part time. Now that I am semi-retired I am beginning to again look at this half-finished project and hopefully, when this book is finished, turn my attention back to it. Possibly this background is of some interest, inasmuch as it is rather typical of what happens with many spare-time hobby projects.

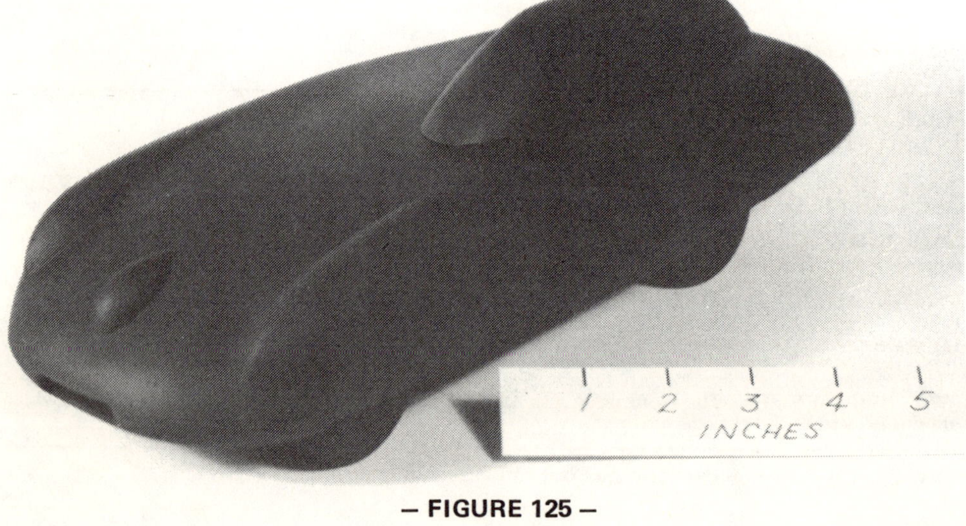

— FIGURE 125 —

By way of design review, it should be mentioned that I intended that about half of the air would be directed around the sides to further reduce *induced drag* and that the side would taper back to a narrow aft body to reduce *base drag.* Both features are depicted in figure 123 and accounts, in part, for the low drag coefficient. Now for the tests by Dr. Cornish and his group, along with his recommendations.

The size of the model, along with a scale, appears in figure 125. As the other pictures indicate, two different testing methods were used. Dr. Cornish's remarks will accompany each picture.

Figures 126 through 129 were taken with smoke (remember Renolds tests mentioned earlier). Figure 130 used the second method, a supersaturated naphthalene solution sprayed on the model and described more completely later.

– FIGURE 126 –

Figure 126 is interesting in that it shows:
1) Turbulance along the ground plane of the wind tunnel.
2) The stagnation line (lower straight line). Dr. Cornish points out that the cool air intake could be raised a little.
3) Note the disturbed air flow above the aft portion of the "blister".

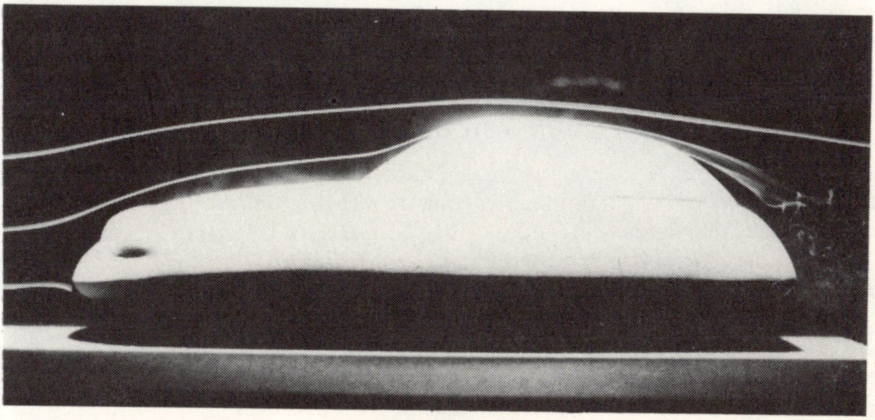

– FIGURE 127 –

Figure 127 is interesting for:
1) Boundary layer weakness over the hood.
2) Dr. Cornish mentions "fairly good flow behind the racing canopy". This clearly shows the classic build-up of the boundary layer over the canopy as depicted earlier in the sketch, figure 121.

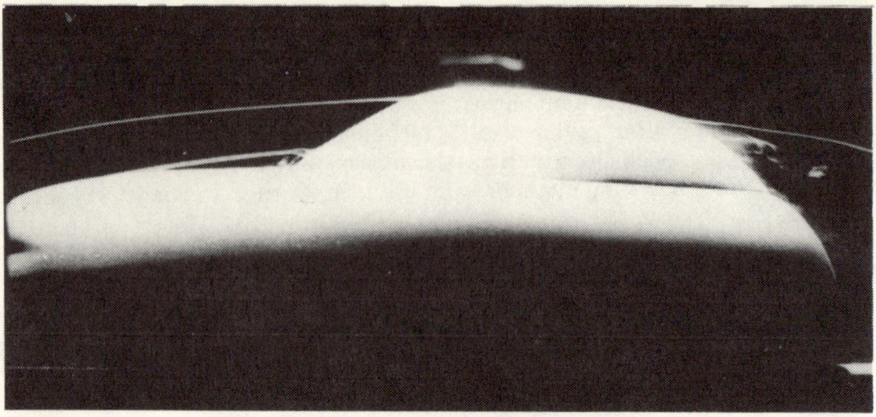

– FIGURE 128 –

Figure 128 is important for showing the "snow plow effect" in raising the boundary layer at the base of the curved windscreen. Even though the slope is good, a gentle curved transition from hood to windscreen would help to avoid this.

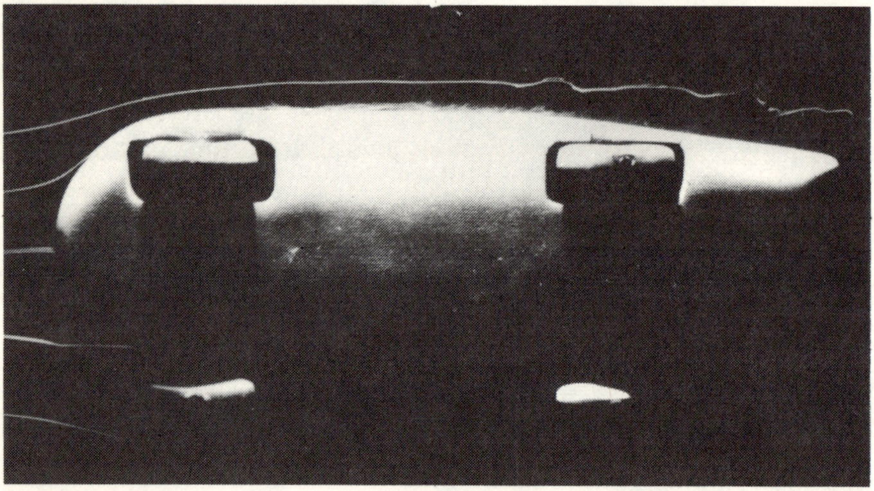

–FIGURE 129 –

Figure 129, taken from the underside, clearly shows how the air flow can be bent around the nose without separation. Remember, this is air that otherwise would have to go over the hood, the windshield, and roof line to create additional induced drag.

The boundary layer begins to weaken behind the front wheel and gets thicker as it moves downstream toward a narrow base width at the rear. Hopefilly, we will be able to improve this a bit more with some minor modifications. It may also be weak and separated partially because of a low and critical Reynolds Number.

Now for the second method of testing as shown in planview in figure 130. Dr. Cornish provided a series of four pictures, all similar but differing only in darkness with time in the tunnel. Figure 130 only is used since it most clearly shows what happens. His description of the method follows.

"The model was painted flat black. Then was sprayed with a supersaturated solution of naphthalene dissolved in petroleum ether, i.e., mothballs in white gasoline. This deposits a frosty coating of naphthalene on the model. It was then placed in the wind tunnel. As the air flowed over the model, the naphthalene is sublimated off in regions of

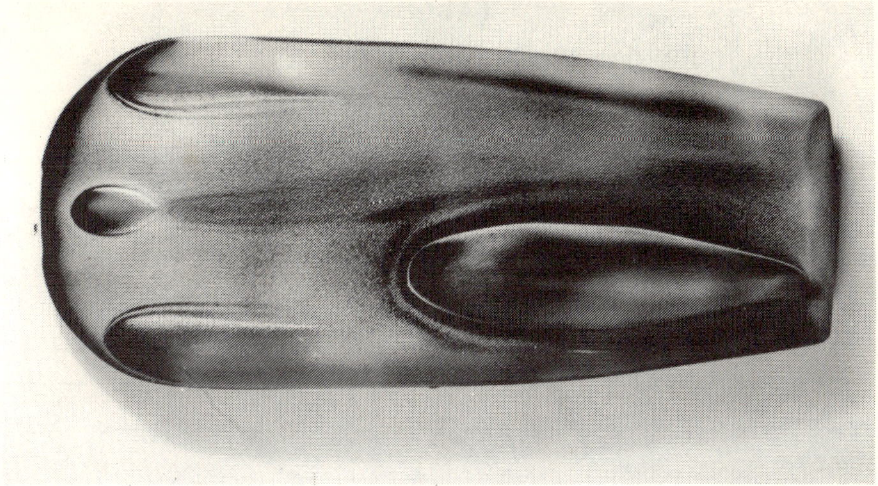

– FIGURE 130 –

high shear more rapidly than in areas of low shear. If progressive pictures are taken of this sublimation process, the regions of high shear can be isolated".

"These pictures were taken after the model had been in the tunnel for 15, 30, 45, and 60 minutes. Notice in picture (figure 130) (a) the high shear near the forward parts of the car; (b) the twin vortices inside each fender hump; (c) the stagnation vortices ahead of the "blister" and the canopy (very strong ahead of the canopy); (d) the wake behind the "blister". The general picture remans the same in time. In the full-size car you could expect to get laminar flow back just aft of the front wheel-well if the forward parts were very smooth".

Table X was provided by Dr. Cornish and indicates the estimated drag coefficient of the model as tested to be 0.02060. Faired and sealed this drops to 0.1963. With a new rear end this drops further to 0.1794. In each case these figures were based on the frontal area of the body alone (8.75 sq. ft.). Our calculations in chapter VI include the wheels as they extend below the body for a frontal area total of 9.62 sq. ft.

As to changes we can incorporate: it is expected the following can be accomplished without drastic changes to the mould already completed.

1) The blister will either be eliminated by another choice of engine or re-arrangement or the blister will be redesigned as recommended.
2) The canopy will be redesigned.
3) Possibly some minor changes can be made in the moulds in the fender notches and doubtfully in the rear aft body area. These will require further investigation.

In summary I am pleased with the results and grateful to Dr. Cornish, those who assisted him, and to Mississippi State University.

HOW LOW SHOULD THE DRAG COEFFICIENTS OF VEHICLES BECOME – AND HOW DO WE GET THERE?

Over 17 years ago, January, 1963, the writer presented reference 94 at the SAE Automotive Engineering Congress in Detroit. It discussed the reduction of frontal area and drag coefficients. Much has been done with frontal area reduction, thanks mostly to the downsizing brought on by the federal CAFE requirements. Precious little has been accomplished by domestic manufacturers in the real work of automotive aerodynamics –reduced drag coefficients–as evidenced by the data in Table VIII.

The writer will stick to his reference statement of 18 sq. ft. of frontal area for a

TABLE X

DRAG ESTIMATE AND POSSIBLE CLEAN-UP
of
KORFF "STREAMLINER"

C_d based on frontal area of body alone

Item	Present	Faired & Sealed	New Rear End
Basic Body	0.0400	0.0375	0.0375
Blister	0.0010	0.0008	0.0008
Fenders	0.0030	0.0021	0.0021
Wheels & Wells	0.1180	0.1180	0.1180
Canopy	0.0150	0.0123	0.0120
Base Drag	0.0290	0.0256	0.0090
Total	0.2060	0.1963	0.1794

$\Delta\ C_d = 0.0097$ | $\Delta\ C_d = 0.0170$

$\Delta\ C_d = 0.0267$

Frontal area = 8.75 sq. ft.

Final version has approximately 50% of the drag of Volkswagen (C_d x 0.37) and less than 40% more than extreme streamlined shape tested by A.V.A. (C_d x 0.13)

(J. J. Cornish)

---

streamlined compact sedan. This is not difficult to accomplish. The reference figure for a drag coefficient of 0.21 was stated - - "can be achieved". Reference 98, the Pinnfarina body of minimum drag, shows an experimental value of 0.177 and a total value estimated to be 0.23 as shown in our Table VIII. I still feel 0.21 "can be achieved", although I certainly agree it will not be easy.

As to "how to we get there", let's review a bit. Reference 103, "The Optimization of Body Details" – A method for reducing the aerodynamic drag of road vehicles", shows how the VW RABBIT went from 0.46 (our domestic values) down to 0.38 after optimization. All this was accomplished with practically no basic change in appearance of a production vehicle that hardly looks like a streamlined vehicle with its straight lines and creased sharp edges!

Next, back in 1935, Dr. Kamm's experimental cars got down to a drag coefficient of 0.23 as our table shows. This was due to:

1) The Kamm "cut-off" at the rear.
2) Well rounded edges.
3) Smooth surfaces with blended transitions for air flow without separation over most of the body.
4) A good underpan.

Now, how do we get a little lower? The writer's work with his little streamliner as tested at Mississippi State University and the GOLDENROD'S model as tested at California Institute of Technology shows clearly that, in planview, a well rounded nose gets much of the air (about half) around the sides instead of over the top, to thus reduce drag further. Then tapering the sides slightly, further reduces the aft body cross sectional area. Both of these factors will lower the drag coefficient as the tests indicate.

The underside air flow can also be re-energized by flow from the cooling duct(s) as indicated in Patent No. 3,425,741, "Land Vehicle having minimum air drag" issued to the author. The sixteen claims and drawings cover a number of other features as well – including features for open wheel racers.

Now for the big rigs. The drag coefficients for large tractor-trailer rigs, trucks, and vans are atrocious, as indicated in reference 110. Values as high as 0.80 with no cross wind and up to 1.50 with cross winds are shown.

Add-on devices reduced drag at zero yaw by 30 to 35% (reference 114) and half that much at a yaw angle of 10 degrees. References 111 through 115 are also of value.

A box-like truck or van shape had its edges rounded, front and rear somewhat rounded, and an underbody seal added, all of which dropped its C_d from 1.13 with sharp corners to 0.440 – a drop of 61% (reference 112).

All of the above was accomplished without changing the essential box-like cargo area or the cab-over-engine conventional cab area. When efforts include what can be done here also, another drop of 25% of the remainder may occur, especially when internal cooling is improved as with the adiabatic diesel.

Since fuel economy is so directly affected by the high drag of vans, trucks, and tractor-trailer rigs, it is reasonable to expect major changes in the next generation of these vehicles.

Drag reduction is also important from a safety standpoint. Passing ability for any vehicle, automobile, van, truck, or racer depends on excess power available for acceleration as detailed in chapter VI. Lower drag provides additional excess power at passing speeds, thereby reducing the time of extra risk.

If acceleration is considered adequate, engine size can be reduced further with lowered drag to improve fuel economy, beyond that for reduced size and weight, along with lowered initial cost.

Automotive aerodynamics still remains one of the best opportunities for improving the efficiency of motor vehicles of all types.

Chapter XVI

FRAME AND FRAME/BODY STRUCTURES

Automotive design and development has always been logical. This is quite evident in the development of frames and bodies. The first horseless carriages were small and light with little more than a seat on a very short frame, not unlike a buggy. Although the earliest cars used wire wheels from the bicycle industry the frame and rudimentary body was of wood as used by the carriage builders. As cars became larger, longer, and heavier, wood became more prominent in construction, even to artillery wood spoke and rim wheels. The FRANKLIN air-cooled was the last hold-out on ash wood frame rails.

ORIGINALLY THE CHASSIS AND WOOD BODY WERE UNRELATED

The chassis and body were definitely unrelated and treated separately in those early days. It was quite normal for certain prestigious motor car manufacturers to concern themselves with, and to build, the chassis only. You then ordered the chassis delivered to the custom body builder who submitted sketches and ideas for your approval much as an architect would design a home for you. Those early custom body builders were originally carriage makers, artisans with wood. To this day our body nomenclature is a carryover from carriage design.

The body's wood framework provided the structure and sheet metal panels (steel or aluminum) were hand formed by planishing and fitted to the wood frames. The metal was not used to carry any structural loads, only to serve as skin. It was, and still is, possible to build a beautiful body by hand, completely without body panel dies. Naturally, it will be quite expensive because wood construction is very labor intensive. If you are into restoring an early motor car, a visit to a large public library to study the early books on wood body design and construction is a must. There are excellent books that go into the details and provide much needed design information.

PROGRESSIVE DEVELOPMENT IN FRAMES

The early cars with beam axles and stiff springs depended on some flexibility in the frame. In fact, designers of that era believed the frame should be strong in bending between axles but allow some flexibility laterally to assist in improving the ride. After the earliest wood frame rails the frames were made of steel channel sections with cross members riveted in place so that in planview the frame looked somewhat like a ladder. Electric welding was not yet used. Gas welding was too slow for production. These open channel frames continued for many years on trucks after they were no longer used on cars.

THE NEED FOR TORSIONAL RIGIDITY AND A LOWER CENTER OF GRAVITY

In the early 1930's, with the introduction of independent front suspension (IFS), frame rigidity in torsion was recognized as a requirement.

About this time ride, road holding, and handling, were becoming more important and the desirability of a lower center of gravity was recognized. Seats were lowered closer to the top of the frame and the floor pan had deep-dish offsets so feet were *between* frame rails instead of above. Side rails were lowered too by kick-ups over the axles.

THE "X" BRACED FRAME AND THE "X" TYPE FRAME

The next design feature was the use of "X" bracing between side rails to improve torsional stiffness. Another design bent the side rails at the mid-point to bring them together to make an "X" in planform for torsional stiffness. Electric arc and spot welding were becoming production methods by this time.

The best way to achieve torsional rigidity is by having an enclosed section like a round or square tube. Boxing in the open side of the channel by electric welding another channel to it to make a box section was the next step in improving torsional rigidity. This was a big step because torsional stiffness of an enclosed section is of the order of about 500 times as great, with the same weight of material, over a channel section.

THE PERIMETER FRAME

With torsional stiffness greatly improved by the use of enclosed sections the side rails could once again move outward and the "X" feature lost its importance.

This time the frame side rails were bent *outward* in planview, around the wheel openings, so that the rails somewhat followed the perimeter of the body to allow more space for feet. In effect you stepped over the side rail to enter the car and the floor pan was at the lower side of the frame rails.

To maintain torsional rigidity it is necessary to have the cross members enclosed also and firmly attached to the side rails, usually by welding.

THE SPACE FRAME

Race cars need torsionally rigid frames also and lightness is of utmost importance. So the space frame was created. This essentially is a bridge-like structure of small diameter steel tubes welded together. It is lightweight and by careful triangulation can be made very rigid torsionally and in bending. The MERCEDES 300 SL was an example of a mass-produced car with such a frame. The VIKING described in the last chapter depicts this construction as does FORD's mid-engine MUSTANG experimental sports car – also described in the last chapter.

This is still a favorite frame design with most race car builders.

THE ENGINE BLOCK AS A SOURCE OF TORSIONAL RIGIDITY

Race cars also save weight by using the engine block as a portion of the frame. The engine block is a very rigid casting both in bending and torsion. When solidly bolted to the remaining small space frame it adds great strength without adding frame weight.

The GOLDENROD, also described in the last chapter, with its four engines in line provide great torsional rigidity when bolted to the simple closed section side rails, which take the bending.

THE BACKBONE FRAME

A number of small cars used a so-called backbone frame down the centerline of the car. In a few cases it is no more than a large diameter thin wall tube with attachment points as needed.

Some electric cars, with the batteries located in line at the center, have a backbone frame enclosing them and a bulkhead at each end for suspension attachment points. This is an efficient approach to frame design.

The writer's three-wheelers also used a backbone frame of square steel tubes for torsional rigidity and with an upper and lower rail tied into a bumper at front, a cross frame at the front suspension, a roll-bar behind the seats, and the rear suspension aft. This is shown in figures 110 and 117.

THE ALL-STEEL BODY IS BORN – FOR MASS PRODUCTION

With the advent of spot-welding as a production tool, the all-steel body became a reality. Body panels could be stamped out quickly on large presses in a manner that included offsets and flanges ideally positioned for access of the welding electrodes. The inner body and the outer skins, spot-welded together, provided great strength and added

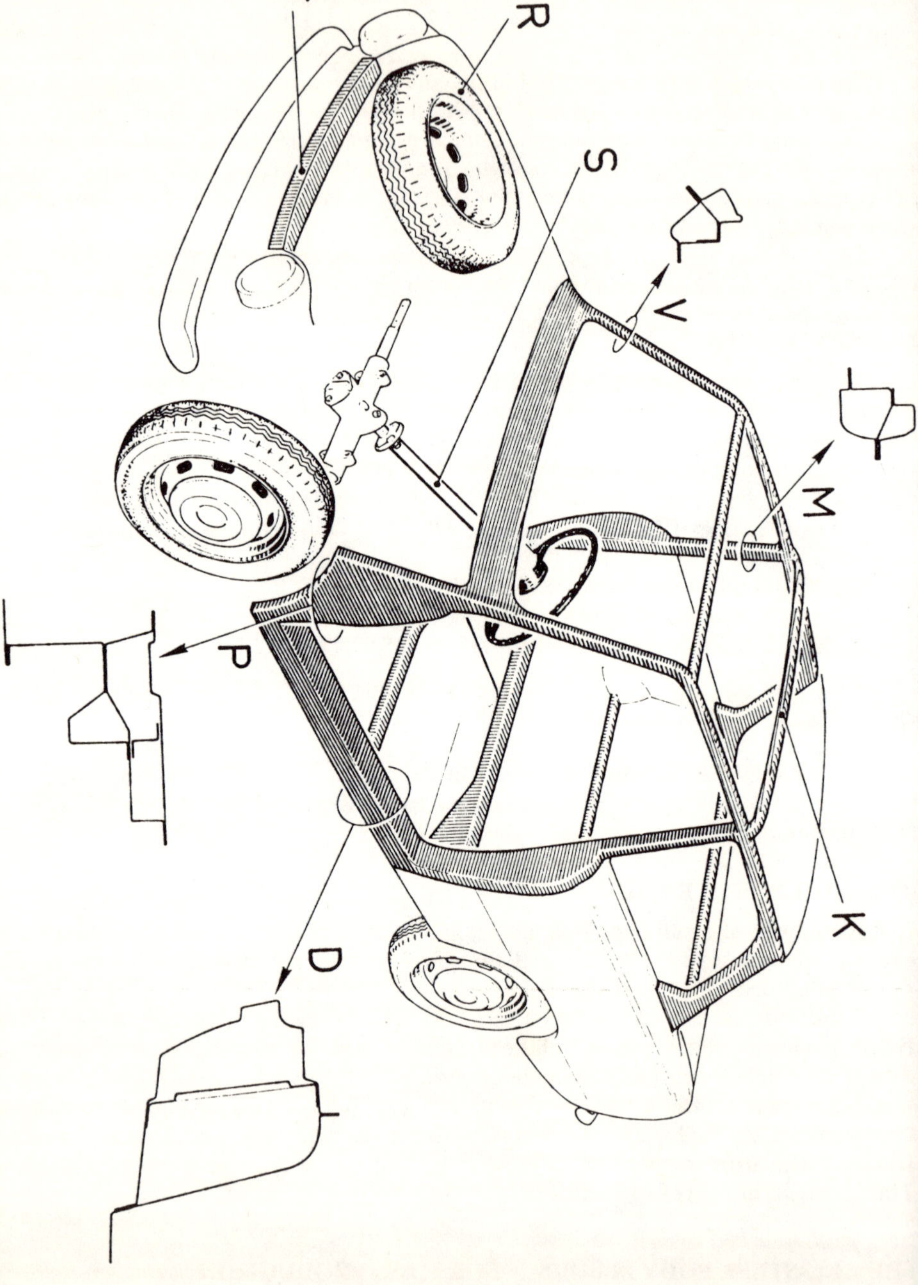

– FIGURE 131 –

safety for the occupants. Sub-assemblies were created and built separately to be combined later in a master body jig for final assembly by massive electric spot-welders – now of robot design. This became the "body in white".

Figure 131 depicts the DAF all-steel body shell with its various members shown in cross section. These are indicative of the shapes and sub-assemblies, and clearly show flanges that are spot-welded together. The DAF is a small European car now manufactured by VOLVO. The lettered parts are described as follows:

The cage (K) consists of box-section members (thick lines in drawing).

The front pillars (V) and centre pillars (M), although kept slender for good vision, are of extra strong double box-section. Door sill (D) is of similar construction, so as to stand up to side impacts, while the door hinge pillar (P) is of treble box-section to prevent distortion of the door opening in a three-quarters front collision, the most common of all. In consequence, the door, provided with an anti-burst safety lock, can be opened easily after an accident.

A sturdy box-section member (F), which, among other things, serves for the attachment of the hinges of the bonnet, which opens forward, distributes the force of the impact over as large an area of sheet metal as possible in the event of a head-on collision. The spare wheel (R), located towards the front of the engine compartment, also absorbs much of the force. This gives the occupants maximum safety, particularly if they use the safety belts DAF can supply and for which anchorage points are provided in the floor and pillars.

Finally, the short steering column (S) makes it possible for an obstacle to penetrate 70 cm into the car without the steering column coming dangerously close to the driver's chest.

Many SAE papers have been published over the years on the subject and every new development is documented in this manner. For many of these reports the reader is referred to reference 117 under the following subject index items.

A) Manufacturing and Production
B) Materials and Structures
C) Passenger Cars
D) Buses and Trucks

Designers often serve on inter-departmental committees with engineers from tooling and manufacturing departments. They are expected to coordinate their design effort, and if necessary, redesign to avoid difficult and expensive features, to assist tooling and manufacturing to meet production quantities and cost estimates.

BODY STRENGTH ELIMINATES THE FRAME

The all-steel body structure as we know it today provides its strength in the skin panels which are reinforced and stabilized by the inner panels to serve as fully enclosed, torsionally rigid, load carrying structural members. By careful design the body can serve as frame also. High bending loads as well as torsion loads are easily taken in such a structure which approximates the monocoque structure of an airplane fuselage, and like it, takes its loads in the skin of the shell.

Early frameless body designs suffered from additional road harshness and nose transmission from the suspension members which were directly attached to the body. They lacked the rubber pads between frame and body for isolation of shocks and noise from drive train and road surface.

In today's designs the engine, drive train, and suspension attachments are carried on a small sub-frame which is then isolated through its elastic pads at attachment points to the body structure.

SAFETY AND FEDERAL REGULATIONS

The all-steel body is a major step forward in safety for its occupants. Federal rules and regulations as indicated in chapter IX add test requirements for bumpers front and rear, roof crush resistance, and side impact resistance as well as steering wheel displacement and passive restraints of occupants. Other safety requirements are included but these deal primarily with the frame/body structure.

Bumpers, front and rear, have been redesigned in steel, aluminum, and plastic materials. References 118, 119, and 120 provide an insight in to how these design problems have been dealt with. Another interesting approach is used on the VIKING-4, described in the last chapter and in reference 121.

Many additional papers have been written on the safety problem and how designers dealt with it. Reference 117 is an excellent source for these.

LIMITED PRODUCTION FRAME/BODY STRUCTURES

The tremendous cost of tooling, especially body dies, and the massive facilities needed to produce stamped, spot-welded all-metal bodies cannot possibly be justified for limited production vehicles. As a result, designers must invent and be very ingenious in creating frames and bodies that can be produced at reasonable cost in limited quantities. In many ways this is a greater design challenge than the established designs for mass production.

A few approaches to this problem are discussed herein.

FIBERGLASS BODIES

Logically, the first attempts were in fiberglass and these bodies were mounted on existing chassis frames to create limited production automobiles of unique design. This is still a popular approach.

A complete male body buck is made of wood and plaster, full size. Then a female mold is laid up over the buck with separations along mold lines to facilitate assembly and disassembly during production. Then body lay-ups can be made in the mold and removed in one piece. This is a well established procedure that can be studied at any fiberglass body shop, boat shop, or in the available literature.

There are certain limitations that must be dealt with. First, the integrity of the body depends in large part on it remaining in one piece. When door openings and/or other cutouts are made, the body panels may warp or twist a bit due to internal stresses and the lack of stiffness. If an opening is to be made, provision for support or added stiffness in the form of flanges, added material thickness, or metal inserts, should be incorporated in the design and fabrication at the time of laying up the shell. Secondly, bulkheads, hinge points, hold-down attachment points, windshield and seat attachments, and other details of internal fastening must be incorporated in the body lay-up while still wet in the mold. Designing beautiful outside contours is not enough. It is strongly recommended that the designer either work with the material itself to learn these things directly, or coordinate his design in every detail with the fiberglass shop personnel and the chassis engineers.

STEEL TUBE FRAMES WITH ADDED FEATURES

Since outer skins, metal or plastic, require support in the form of an inner skin, or bulkheads for attachment, many designers prefer to have the major frame structure include extensions for this purpose.

Another problem occurs in that tooling dies are non-existent for special frames of limited production vehicles. Relatively simple jigs may be built to hold standard size tubes and standard tube bending equipment exists. The designer must plan his frame structure around these manufacturing methods along with arc and/or wire-fed welding methods.

The third problem involves safety. It is not enough to provide a torsionally rigid frame for the suspension and to serve as a platform for the body, occupants, engine, and other items. Roll-over structures are also needed for occupant protection. High loads from front and rear bumpers must be taken. Side impact loads are another federal requirement. The special frame must now provide the protection normally offered by the all-steel body in addition to its normal functions.

Space frames such as on the mid-engine experimental MUSTANG, the VIKING, and many race cars should be studied along with the federal rules and regulations.

ABS PANELS AND BODIES

ABS (Acrylonitrile-Butadiene-Styrene) is a very tough, yet hard and rigid thermoplastic material that can be obtained in standard sizes and thicknesses. It comes in sheet form as well as in other forms. It is quite popular for making panels and even complete body tubs for all-terrain vehicles, house trailers, truck cabs, and experimental bodies for motor vehicles of all kinds. We used it in sheet form on the DUO-DELTA three-wheeler. It is easy to form into complex shapes with heat and simple molds. It can be drilled, cut, sawn, trimmed, and machined quite readily. It may be obtained as a laminate with center plies or core for added strength and stiffness. Its lightness, along with its other features, makes it ideal for use in limited production vehicles as well as for mass-produced applications.

Chapter XVII

WEIGHT AND BALANCE

Many motor vehicles have been built without detailed weight and balance data – in the past. To undertake the design and development of a modern day motor vehicle without accurate and detailed weight and balance calculations followed up by actual weighing-in of each part for comparison and/or possible redesign may be a serious oversight.

WHO STARTED IT AND WHY WEIGHT CONTROL FOR MOTOR VEHICLES?

As you might have guessed, the aircraft and aerospace industries have developed this phase of engineering to the highest degree. Weight and balance are of the utmost importance to any vehicle expected to successfully lift itself off the ground and carry a payload at hundreds of miles per hour for a distance of over a thousand miles.

Next the trucking industry recognized the importance of weight control when gross vehicle weight (GVW) became limited by law many years ago in practically every state of this nation. Clearly, the less the vehicle itself weighs the more payload it can carry and still keep the GVW within the law.

Racing vehicles, including stock cars set up for racing, were next. Acceleration went up as the pounds came off. This is abundantly clear when we study performance calculations in chapter VI.

Now even mass produced passenger cars for every day use on city streets are being lightened up by the most detailed weight-watcher engineering techniques due to Corporate Average Fuel Economy (CAFE) regulations of the federal government. Lighter vehicles require less power, hence less fuel, for equal performance.

THE THREE WAYS WEIGHT CONTROL DATA IS USED

Major automobile companies have, for many years, purchased competitive vehicles and disassembled them carefully to study and weigh each part for comparison with the equivalent part on their own vehicles. So, actually, weight comparisons began in the automobile business more for competitive reasons than for fuel economy.

Today, the reasons have been expanded to include the following three primary reasons:

1) *A means for recognizing opportunities for overall weight reduction for better fuel economy.*

Weight reduction must be accomplished by every designer on every part, ounce by ounce, with no exceptions until each part represents a very efficient design at acceptable cost for its function.

Records of previous designs within the company as well as of the competition are most helpful. Alternative higher strength or lighter weight materials must always be considered. Redesign to combine functions in one part instead of two also helps. More detailed stress analysis to permit the elimination of excess material, often as much as a third is another approach to the problem.

2) *A means for determining CG location and polar moment of inertia.*

Center of gravity (CG) location and polar moment of inertia influence ride, handling and cornering, which are important to today's more sophisticated buyers as well as to racers.

This information is also needed to determine applied loads and for stress analysis.

3) *Detail weight estimates provide a "Target" figure for cost estimates of all parts*

Costing out a part is based, in part, on the amount of material required. The detailed weight estimates provide this information and eliminate the need for duplication of this effort by the cost estimator. Additionally, if the part is on the heavy side the designer may be able to reduce its cost by redesign as well as improve fuel economy and acceleration of the vehicle.

THE "TARGET" WEIGHT ESTIMATE AND ITS DETAIL BREAKDOWN

When a new vehicle is planned its weight is "targeted" for a start in estimating performance, fuel economy, and cost to manufacture. This target weight number is chosen based on experience and comparisons with other but similar vehicles. Improvements are taken into consideration and while the target is a figure to be seriously aimed for, it may be altered slightly up or down as the design develops, based on realistic reassessments and as actual parts are weighed.

The target weight for the overall vehicle is first broken down into a weight summary somewhat as indicated in Table XI of the FORD mid-engine MUSTANG described in chapter XXII. These weight allowances are then assigned to each design group responsible for the design of these portions of the vehicle along with a time schedule and budget for getting the job done.

TABLE XI

WEIGHT SUMMARY

GROUP NUMBER	NAME	TOTAL WEIGHT	FRONT LBS.	FRONT %	REAR LBS.	REAR %
1	BODY	466	198		268	
2	FRAME	132	63		69	
3	ENGINE	270	44		226	
4	FRONT SUSPENSION	54	54		0	
5	REAR SUSPENSION	78	75		3	
6	BRAKES	80	46		34	
7	TRANSAXLE AND CLUTCH	121	- 4		125	
8	WHEELS AND TIRES	90	57		33	
9	EXHAUST SYSTEM	16	- 1		17	
10	FUEL SYSTEM	28	34		- 6	
11	STEERING SYSTEM	18	13		5	
12	COOLING SYSTEM	19	4		15	
13	ELECTRICAL SYSTEM	59	42		17	
18	OILS	17	3		14	
	SHIPPING WEIGHT	1448	628	43.4	820	56.6
	GASOLINE	78	90		- 12	
	WATER	18	4		14	
	CURB WEIGHT	1544	722	46.8	822	53.2
	DRIVER ONLY					
	FULL TANK	1719	807	46.9	912	53.1
	EMPTY TANK	1641	717	43.6	924	56.4
	DRIVER AND PASSENGER	1894	892	47.0	1002	53.0
	DRIVER, PASSENGER ANC 50 LBS. LUGGAGE	1944	888	45.6	1056	54.4

Next, each portion or group number is further broken down in more detail for specific parts as we show in Table XII for the open sports car racer of figure 12 of chapter V. Each detail weight estimate is assigned to the designer responsible for the design of that part. When the design of the part is completed its weight according to the design will be calculated and compared with the estimate. If over the estimate every effort to reduce its weight must be considered. Occasionally supervision may decide the part as designed is efficient and satisfactory even though over the weight estimate. In such case the overage will best be removed from some other part or accepted as a legitimate increase in target weight. Such a decision is never made without very serious review by all concerned and final approval from department heads.

TABLE XII

SPORTS CAR RACER – FIGURE 12, CHAPTER V

1ST ESTIMATE – TARGET WEIGHTS

CHASSIS		
FRAME, ROLL BAR, FIREWALL, SEAT (Est.)		95.00
SUSPENSION		
Wheels (4) Mag @ 3.6 each		14.40
Tires (4) – 10. x 4		40.00
Shock & Spring 7 lb. ea. x 4		28.00
Upper Arms – 2 lb. ea. x 8 (est.)		16.00
Lower Arms – 2 lb. ea. x 8 (est.)		16.00
Hubs & Brgs. – 4 lb. x 4 (est.)		16.00
Brake Discs & Calipher – 3 lb. x 4		12.00
Spindle Assem. 2.7 lb. ea. x 4 (est.)		10.80
STEERING		
Rack & Pinion Unit	3.00	
Col. & Shaft & Keyways	2.80	
Wheel, Hub, and Center	3.50	
Tie Rods & Idlers	4.50 (est.)	
	13.80	13.80
PEDAL ASSEM. Complete with cables & Hose (est.)		12.00
SHIFTER MECHANISM to Gear Box (est.)		4.00
INSTRUMENTS & LINES (Tac, Speedo, Electrical) (Est.)		2.50
DRIVE TRAIN		
Chains, Sprockets, Timing Belt, Mounting (Est.)		18.00
Final Drive Center Section Unit		5.50
Half Shafts (2) and U-Joints (4) (Est.)		15.00
Reverse Motor, Belts, Sprockets, Jackshafts (Est.)		22.00
	TOTAL	341.00
POWER PLANT		
Engine & Transmission (Est.)	132.00	
Exhaust System (Est.)	30.00	
Fuel Tank (Plastic), Lines & Fittings (Est.)	3.00	
Ducting to and from engine (Est.)	6.00	
	171.00	171.00
BODY		
Complete with Windshield & Head Rest (Est.) (including mounting brackets & fittings)		100.00
BATTERY, RACK, CABLES & WIRING (Est.)		38.00
VEHICLE WEIGHT EMPTY		650.00
USEFUL LOAD		
Driver	165.00	
Fuel & Oil	25.00	
	190.00	190.00
GROSS VEHICLE WEIGHT		840.00

Table XIII – lists additional detail items required for the 2-passenger small sports racer GT shown in figure 11 of chapter V. Target weights have not yet been assigned to these items. This might be an interesting class assignment. Obviously, weight can vary considerably for each item, depending on how "lean" or how sophisticated the design is to be. Since nearly every item is not a structural item, weights will be determined with but little thought as to its stress analysis. In fact, many of these items are purchased complete as is from outside sources so actual weights cannot be influenced to any great direct degree.

TABLE XIII

ADDITIONAL DETAIL ITEMS FOR 2-PASSENGER G.T. (Street Legal)

WINDSHIELD
- **Windshield Wipers & Drive Motor**

ELECTRICAL
- **Headlights & Covers**
- **Stop, Turn, & Tail Lights**
- **Front Turn Signals**
- **Relays, Switches & Wiring**

MUFFLERS

BUMPERS
- **Front**
- **Rear**

SPARE TIRE & WHEEL

DOOR MECHANISM (2)

WINDOWS

ROOF & AFT ENCLOSURE

LICENSE PLATE & BRACKET

PASSENGER SEAT & TOE-BOARD

UPHOLSTERY AND INTERIOR TRIM

ONE OF THE LIGHTEST SMALL RACERS EVER BUILT – A COMPARISON

Discussions about how low the target weight should be are quite normal and comparisons are difficult and rarely completely fair. It may be argued that the author's target weight for the empty vehicle of 650 lb. is quite low. Actually, there is some evidence that the target allowance for the body and possibly the engine, along with a few other items, is a bit on the high side.

The Odengborg and Hamilton MERCURY SPECIAL described in reference 122 provides a delightful comparison. Its description follows:

Wheelbase = 64 inches – Front tread, 43 inches – Rear tread, 46 inches.

Space frame of 1020 tubing 3/4 and 5/8 od x 18 and 20 gauge.

Power train – Mid-engine mercury outboard 2-cycle of 747 cc. laid on its side – drives a BSA motorcycle type A gear box (1952). Carter model N carbs. Reworked cooling system – used a pair of automotive heater cores up front and joined to make an 8 x 20 x 4 inch cross flow radiator. A Jabco rubber impeller high speed marine water pump is pulley driven.

Weights – Dry 450 lb. – Race ready 490 lb.

Body consists of aluminum sheet belly pan, bulkhead, and seat with a fiberglass nose and aft body. Fiberglass portions weigh 26 lb. with windshield. Fiberglass is made of only 2 layers of 6 oz. cloth and resin. (This would be too light and flimsy for street use).

Race class – H modified, but not FIA legal as built. Today SCCA and other sanctioning bodies would require additional items which would increase the weight a bit, although it would still be remarkably light.

This light racer won nearly all of the races entered with its reported 125 mph top speed and outstanding acceleration. It also was stated to handle beautifully.

In summary comparison the MERCURY SPECIAL would have to weigh a bit more by today's standards and our little racer's target weight may be a bit on the high side for some items.

BALANCE CALCULATIONS CG LOCATION

To determine CG location we will use the principle of levers and moments. A drawing similar to figure 132 is needed with a reference line at the front as indicated. The center of gravity of each item is located on the drawing and measured to the reference line. This is the lever distance.

Next the data must be tabulated as we did in Table XIV with three columns (A) (B) & (C). Column A lists the weight of each item. Column B lists its distance from the reference line in inches. Column C provides its moment, obtained by multiplying–A x B = C.

At the bottom of the columns we have the totals. Divide column C (total moments) by column A (total weight) to determine the distance from the reference line for the center of gravity location.

TABLE XIV

BALANCE GASOLINE 3-WHEELER

	(A) (lb.) WEIGHT	(B) Distance From Ref. Line (in.)	(C) Moment (A)x(B)=C
Bumper	4	2.2"	8.8
Battery–(Front) & Frame	30	10.5	315.0
Pedal Assembly	12	18.8	226.0
Headlights	9	21.5	193.5
Fuel & Tank (5 gal.)	40	27.7	1,108.0
Steering Gear	6.5	32.8	213.0
Windshield Assem.	40	41.8	1,665.0
Steering Wheel	3.5	44.4	155.5
Body	175	52.5	9,400.0
Driver & Passenger	360	56.0	20,100.0
Frame Complete	196	58.0	11,380.0
Engine & Starter, Etc.	100	78.3	7,830.0
Transmission (70 lb.) & Exhaust System (30 lb.)	100	90.00	9,000.0
TOTALS	1,076.0		61,594.8

(C.G.) Center of Gravity = C divided by A = 57.2"
C.G. from Rear Axle = 52" (by scaling drawing)
Front Wheel Location – therefore W/B = 84.5"
Front Wheel Location to C.G. = 32.5" = 38.5% $\quad$ $32.5/84.5 = 38.5\%$
Rear Wheel Location to C.G. = 52.0" = 61.5% $\quad$ $52/84.5 = 61.5\%$
W/B = 84.5"

Notes:

(A) Calculations shown for horizontal position of C.G. only.

(B) Vertical position of C.G. determined in same manner by using ground line as reference and measuring up to each item to determine vertical moment, then divide by weight to determine C.G. distance above ground line.

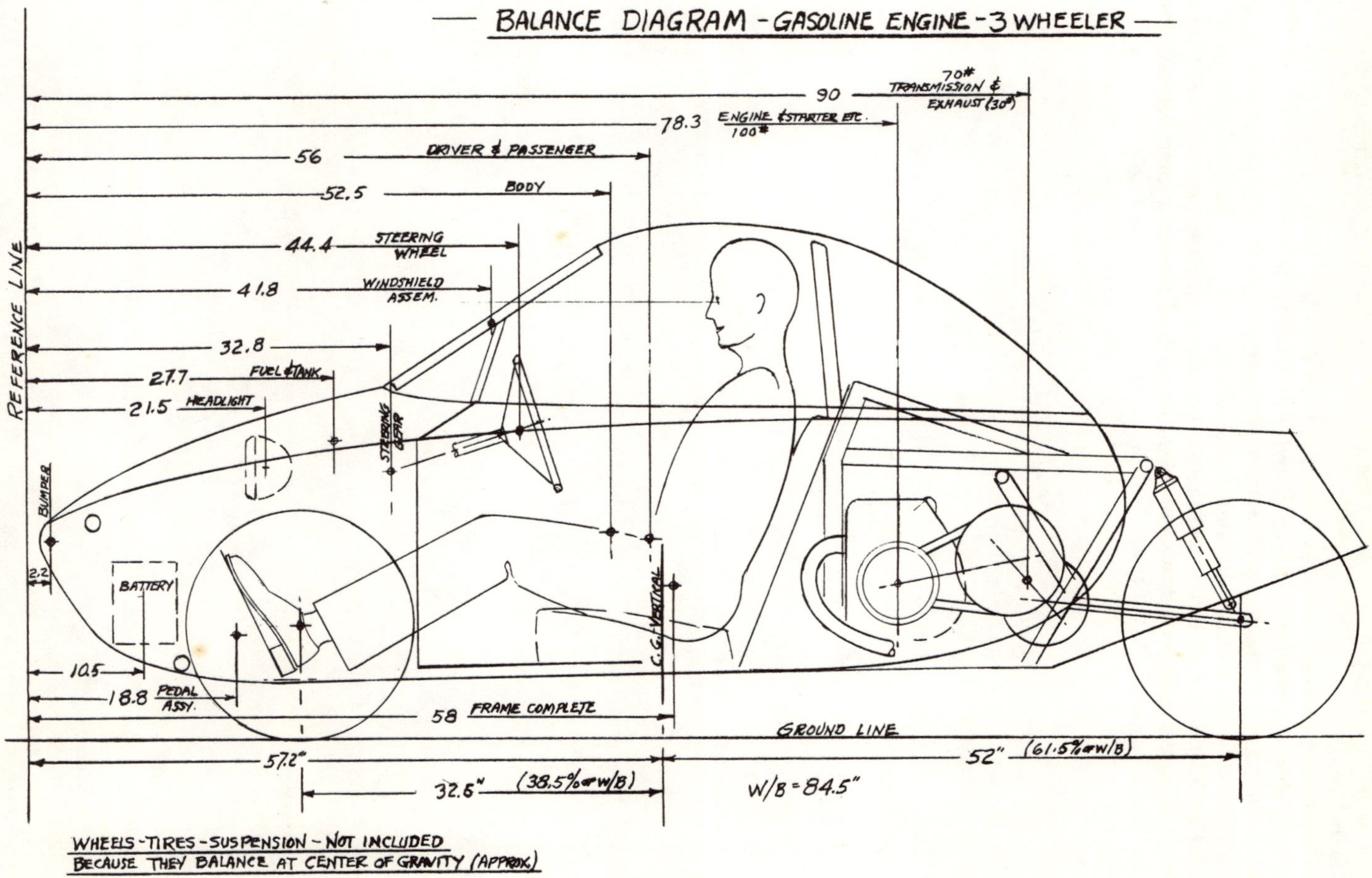

— BALANCE DIAGRAM - GASOLINE ENGINE - 3 WHEELER —
90
70# TRANSMISSION & EXHAUST (30#)
78.3
ENGINE & STARTER ETC. 100#
56
DRIVER & PASSENGER
52.5
BODY
44.4
STEERING WHEEL
41.8
WINDSHIELD ASSEM.
32.8
27.7
FUEL & TANK
21.5
HEADLIGHT
STEERING GEAR
REFERENCE LINE
BUMPER
2.2
BATTERY
10.5
18.8
PEDAL ASSY.
58
FRAME COMPLETE
C.G. VERTICAL
GROUND LINE
57.2"
52" (61.5% of W/B)
32.6" (38.5% of W/B)
W/B = 84.5"
WHEELS-TIRES-SUSPENSION - NOT INCLUDED
BECAUSE THEY BALANCE AT CENTER OF GRAVITY (APPROX.)

— FIGURE 132 —

This same method is used to determine the height of the CG from the ground line when it is used as the reference line. Simply measure vertically instead of horizontally. Each item has a vertical distance from the ground line. Use these distances to determine vertical moments.

The number of items in our example has been kept low for clarity and simplicity in making the drawing. For greater accuracy the work must be done in greater detail.

Although a three-wheeler is used for our example, exactly the same procedure is used for four-wheelers and/or other type vehicles.

The polar moment of inertia and radius of gyration use these same weights and their distances from the center of gravity and follow the laws of physics in their determination.

Chapter XVIII

STRESS – ANALYSIS AND STRENGTH OF MATERIALS

Rarely is the designer expected to prepare the stress analysis of a design. This engineering job is quite normally done by a specialist in structural analysis. The designer must be able to understand a few principles to be able to converse and absorb the information offered by structural engineers and that is about all that will be discussed in this chapter. With some basic knowledge in this field, hopefully the designer can do his job a bit better. The question seems to be – just how basic should this be and where is a logical place to start without getting too deep into all the details.

APPLIED LOADS

Let's start with a few ideas about applied loads. The gross weight of the vehicle is being supported by its wheels, front and rear, in proportion to how near the center of gravity is. Clearly this is a matter of leverage. While at rest the load is considered to be "static". The load becomes "dynamic" with vehicle motion such as acceleration, braking, and/or road irregularities which introduce shock, vibration, or a redistribution of the loads. The "static" load may be thought of as unity and the "dynamic" load is factored upward to insure that strength will be adequate. Measurements have been taken over the years to determine what multiple of "static" load is typical in every case. A review of applied mechanics will also indicate the forces and their directions. The type and size of the material should be more than adequate to insure a "positive factor of safety".

The first step is to tabulate the applied loads and by means of applied mechanics follow them upward from each tire, into the wheel, its bearings, spindles, suspension members, springs, into frame attachment points, etc. Concentrated loads such as the engine are worked into the frame at motor mounts. All this is quite involved and requires a great deal of work to determine the applied loads under every circumstance in every structural member of a motor vehicle. But it must be done in any stress analysis before the strength of the material being considered can be introduced into the study.

Elements of chassis design include gear tooth design, bearings, shafting, housings, fatigue problems, critical speeds, to name but a few related to rotating parts. Springs, frame deflections, bumper loads front and rear, federal regulations, body attachments; all have their own unique stress analysis problems.

RIGIDITY IS A MAJOR DESIGN CRITERION FOR A BODY

Body design often is more related to thin sheet analysis. With many of the bodies now expected to serve as the frame also, loads into body structures become more complicated. Often the major design criterion becomes rigidity or stiffness. This is true for two reasons. It is a principle of dynamics that the stiffer a beam or structure, the *higher the frequency* of vibration and the *lower the amplitude* of vibration. The amplitude of vibration is greatly affected by how close the natural frequency of the structure corresponds to the frequency of excitation (spring and tire rates, unsprung parts, etc.). Vehicle weight, wheelbase, suspension types also influence the stiffness.

The other reason for rigidity or stiffness is distortion at openings such as at doors, hood, and deck lid. Also the windshield frame and backlight frame. Deflections must be limited.

Squeaks and rattles are usually related to body rigidity too.

Distortion from applying a jack to a bumper or some part of the body should also be investigated.

The side framing of the body, top and bottom, essentially take bending and torsional loads so must be adequately tied together.

Since the body represents a major portion of vehicle weight the designer must proceed most carefully in weight reduction programs to retain minimum deflections, bending and torsional strength, and ample rigidity. Close coordination with structural engineers and their detailed stress analysis, as well as with the detailed weight and cost analysis, is needed to achieve a balanced design.

It may appear that once the load is known and the material is chosen, the designer has but limited freedom to influence the design, reduce weight, or conserve cost. We have an example in aircraft design which is worth repeating now that the automobile industry has become so weight conscious, especially in competition vehicles such as racers. Let's assume we have a pin ended strut five feet long, made from a 2" diameter rod of 2024 ST6 aluminum alloy. As a column it will carry a 22,000 pound load. If we redistribute this same amount of material into a tube of ideal proportions (6.41" diameter x .160 wall) it will then carry 120,000 pounds as a column. The cost for material as a tube may be a little more than double the cost as a rod, but it takes more than five times as much load. If, instead of a tube, the material is used in a stiffened cylinder 20" diameter x .040 wall with stiffeners at .725 spacing, the weight remains the same but it will then, as a column support 152,300 pounds and probably cost about four times as much as the tube and about nine times as much as the rod.

The designer does have choices, and efficiency at reasonable cost is still the name of the game. Lots of rods have been used in automobiles, many of which should now be tubes at higher cost. Rarely would stiffened cylinders pay off in a car considering material and labor costs.

It is strongly recommended that the serious student of design acquaint himself with a glossary of terms used in stress analysis, study applied mechanics, and the latest in structural design to begin to realize his full potential as a designer. A start can be made with references 123 through 126. Additionally, most libraries have excellent books on the subject and articles and SAE papers are available and new ones are coming out every few months.

STRENGTH OF MATERIALS (SOURCES)

The strength of materials is generally available in tables on every conceivable material and updated regularily as new alloys and additional materials are developed so will not be repeated here. Supplier catalogs regularily list such data along with all the properties and characteristics of their product. The metallurgist with a large company has this data at his finger tips. The references for the next chapter, under materials and processes, will include sources for such information. If you like the information in one document, there are numerous engineering handbooks available with tables under every well known material.

Chapter XIX

PRODUCTION ENGINEERING

Ideas remain only ideas until designed in detail to see how practical they *might* become. Then detail designs cannot become manufactured products until *all aspects of production* are dealt with in a completely satisfactory manner. This is true if only a few products are to be produced or if mass-production is to occur. The Production Engineering Department is the bridge the design must go over to reach the manufacturing stage. It cannot be avoided nor treated lightly. An inexperienced designer may object to the modifications and compromises he must make for the sake of production. If he does he is only exposing his inexperience and ignorance.

In some companies this function is shown under the Master Mechanic's office, while in others it is a separate department as we show it on our functional chart. In either case, the Production Engineering Department is a staff function that serves all designers and engineers on all projects. It consists of the following groups: STANDARDS, MATERIALS AND PROCESSES, MANUFACTURING METHODS, AND VALUE ENGINEERING. These group titles may vary a bit with companies and small companies may combine them to some degree, but the functions remain even if handled by one person on a part-time basis. These are the functions that help the designer create in a manner that is within the state-of-the-art (or science) of the manufacturing branch. The production engineering people also indicate to manufacturing people the long-term trends of design so they can plan their research to prepare in advance to meet the coming challenges.

Early in the designer's experience he will also be told "Don't re-invent the wheel". It is nearly unbelievable what has gone before; is refined and reduced to practice, and fully documented. It's all there just for the looking. The place to look comes next.

STANDARD PARTS, COMPONENTS, AND DESIGN STANDARDS

All designers and engineers in the motor vehicle industry soon learn to become well acquainted with the Society of Automotive Engineers Handbook (Reference 128). It now comes in two volumes, part 1 and part 2. It contains SAE Standards, Recommended Practices, and Information Reports. It is revised annually. These handbooks and much more are available in the Standards Group in the Production Engineering Department.

This group also has current catalogs from suppliers of standard parts hardware such as bolts, nuts, rivets, pins, fasteners, and fittings of all sorts. The data includes the strength of materials mentioned in chapter XVIII for the items listed. The Standards Group contacts the suppliers and provides all coordination between designers and suppliers of these items. The Group also provides membership on committees, industrywide, to initiate new standards and revise and update existing standards as required. All engineering service on these items required in Manufacturing, Procurement, and Quality Control is also provided by this group.

Components such as electrical items, electronics, hydraulic brakes, wheels, and engine accessories are also handled by this Group. Again, they provide the coordination with suppliers, Procurement, Quality Control, and Manufacturing as well as industry membership on committees to initiate new and revise existing components.

A "Design Standard" is a drawing that provides a "standardized design" of something such as a bead, a lightening hole, a flange, a corrugation, or whatever is produced in the shop with standard tools. This also includes machining operations such as splines and thread cutting. By adhering to this "Design Standard" existing manufacturing tooling can be used.

MATERIALS AND PROCESSES

All raw materials and semi-finished materials such as sheet metal are the domain of the Materials and Processes Group. These will include ferrous and non-ferrous metals, plastics, elastomers, and other engineering materials of all kinds. Strength of these materials as well as any other characteristics can be learned in this group. Again, the SAE Handbook provides a great deal of information. Reference 129 is also recommended.

The processing of materials in manufacturing is also coordinated by the engineers in this group to assure that engineering requirements are met. This will include heat treatment, storage, care in handling, and any other processing that may be used by the manufacturing division or by a supplier.

As with the Standards Group, Materials and Processes people contact their suppliers and others related to these items and provide membership to update and revise information.

MANUFACTURING METHODS

This group has specialists who can help a designer with his problems in the design of a casting or forging, the limits of formability with any alloy in any metal, unique welding and machining problems, and with the latest manufacturing methods in plastics.

If a design has questionable features that should be researched or a development program is needed to insure that production will run smoothly, these specialists can alert the designer and others and initiate the program. These people are in constant contact with suppliers in their specialty and are fully aware of the latest state-of-the-art. Often a supplier who specializes in a certain manufacturing method and has superior facilities can produce what is needed in quantity and at a lower price than if the part is kept "in house".

It is important that the designer become knowledgeable in all phases of production engineering and use these specialists to avoid difficulties with his designs when they reach the shop.

VALUE ANALYSIS

Although we show this group in Engineering it may be located in Product Planning or even Finance in some companies. The important thing is that this function serves the designers in Engineering to indicate how their designs compare with the competition in Value. This function has been well established in the automotive industry for many decades.

During and after World War II the aircraft industry studied value analysis as practiced by the automobile industry and developed their own version. The military services encouraged this with all contractors and it is now an integral part of practically all government bids in some form.

Reference 130 provides an excellent description of how Value Analysis works in the automobile industry. Essentially "competitive analysis techniques" are used. Each company obtains new competitor vehicles, dismantles each completely, and in about ten days all the parts are cleaned, weighed, and mounted on display boards along with all relevant data. These displays are then compared with the company's own similar parts. Weights, number of pieces in an assembly, complexity of design, difficulty to manufacture, type of material used, and direct cost to manufacture are compared. Of course, the part function should be identical. If not, allowance is made for differences. Thus, a value analysis is made. If the competitor's part is better, costs less, or weighs less, for the same function, it is considered to have greater value. In such case the designer is advised and redesign is considered.

The above technique is not practical in the aircraft industry because of the immense cost of the competitor's aircraft and the inordinate expense of dismantling, to say nothing of the time involved. So we approached the problem a bit differently. This way is also of interest to small manufacturers with limited rates of production, it is believed.

We have Value Engineers, specialists in all aspects of Production Engineering, who "cruise the drawing boards". They are immediately available to designers for consultation while the design is in progress. Studies include alternative methods of manufacture such as: 1) a casting or forging, 2) would it be better or lighter or less costly as a built-up part – possibly welded in places? The Value Engineer makes his recommendations based on the quantity of parts needed. If the quantity changes, possibly due to an additional contract, then a review may indicate a superior method is preferable and would save money. This greater value enhances production, improves profits, and often lowers selling price to become more competitive. This is a continuous process and Value Engineers not only assist the designer to create a design that can be manufactured with the least difficulty but also saves many times their wages in the recommendations they make. They also serve the designer by being the coordinator for information from all other specialists in the Production Engineering Department. If a small company lacks these people the designer must then assume the responsibility for learning about manufacturing too.

Chapter XX

COST ANALYSIS

Determining the cost of a product is accomplished in several different ways depending on the type of information desired. Costs are analyzed in different divisions of a company for their own specific purposes and, at times, bear no relation to data in other divisions. The prediction of costs at some future time and based on increased or decreased production rates is also practical and accomplished in its own unique way.

The designer needs to understand the principles of each approach to cost and how his efforts have a direct bearing on the cost of the product.

THE ENTREPRENEUR'S APPROACH TO COSTS

Let's first think of costs from the standpoint of the entrepreneur who is starting a new company. He needs to know:

1 – ONE TIME COSTS – COMPANY WIDE. Things such as costs of land, buildings, machinery, and other facilities.

2 – CASH FLOW PROBLEMS. Start-up costs before income (sales) begin to balance outgo (expenses).

3 – HOW MUCH MONEY IS REQUIRED. How much will be raised by sale of stock or by borrowing.

4 – WHAT THIS WILL COST IN DILUTION OF HIS SHARE, or in interest on borrowed money.

5 – PRODUCT COST IN BROAD TERMS. With reasonable accuracy and at some point in the future when production quantities are reached.

6 – HOW VARIOUS ALTERNATIVE COSTS CAN BE DEFERRED OR "TRADED-OFF" TO REDUCE RISK. This is needed to optimize the overall cost picture and is also tied into time schedules.

The head of Engineering and his key designers will be expected to provide some inputs to management to assist in answering questions. Other division heads such as in Manufacturing, Procurement, and Marketing must also provide their inputs.

THE FINANCIAL BRANCH'S APPROACH TO COSTS

After the company is in operation the Financial Branch must check actual costs against predictions, constantly monitor the data from all other Branches and advise management of changes, up or down, not only as they occur but insofar as possible, anticipate trends. This Branch has its own unique way of figuring costs.

The Cost Estimating Department in the Financial Branch also estimates costs and profits for new models and new contracts. This then goes to top management for approval and/or revision based on other factors such as competition and expected changes in economic conditions.

To somewhat understand the many elements of overall costs looked at by the Financial Branch, the pie chart shown in figure 133 was prepared. Incidentally, the dealer discount may be as high as 25% for luxury cars, 24% for top medium priced cars, 22% for so-called standard priced cars, 19% for intermediates, and as low as 15% for compacts and sub-compacts. The highest figure, 30%, could only occur for special, limited production vehicles.

Traditionally, labor costs on mass-produced cars are roughly one-third of total production costs. Even this figure may vary a bit depending on quantities and just where the product is on the "learning curve" as we shall learn later. By now it should be clear that costs do not remain fixed nor always turn out as predicted.

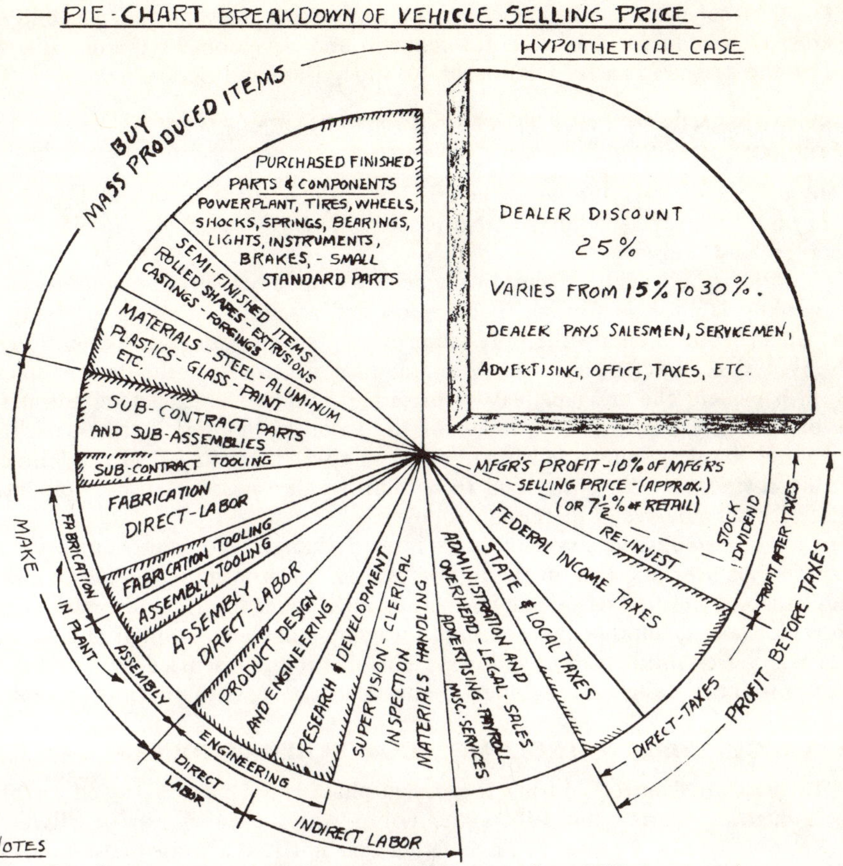

NOTES

1- SUB-CONTRACT ITEMS MAY BE MADE IN OTHER DIVISIONS OF PARENT COMPANY OR BY OUTSIDE JOB SHOPS ON CONTRACT BASIS.

2- PERCENTAGES HAVE NOT BEEN ALLOCATED TO ANY ITEMS OF COST TO MANUFACTURE BECAUSE THEY VARY BY PRODUCT AND COMPANY

3- PERCENTAGES AND ACTUAL COSTS ARE INFLUENCED BY DESIGN UNDER DESIGNER/ENGINEER AS WELL AS BY MANAGEMENT

– FIGURE 133 –

So much for the big picture. Now let's see how the Engineering Branch of the company looks at costs.

THE ENGINEERING BRANCH'S APPROACH TO COST

Those elements of cost that can be influenced by the designer are separated out and subjected to detailed analysis in the Engineering Branch. These elements will now be discussed in greater detail. How they fit in the overall picture can always be reviewed by referring to the pie chart, figure 133.

The items that are not influenced directly by the designer are on the right side of the chart. Dealer discount, profit, taxes, administration and overhead are influenced primarily by management and others outside the company. These items often account for half or more of the retail price of a product such as a motor vehicle.

Items in the upper left hand corner of the pie chart under "buy" are influenced by the designer in an indirect and sometimes minor way through his choice of items in the design. He should be well acquainted with the price and value of these items so that those chosen are in keeping with the market segment of his product and its competition.

The designer has some influence on the engineering costs through his ability to meet budgets and schedules. The need for research and development is only slightly influenced by the designer as is his relationship to supervision and clerical help.

THOSE ELEMENTS OF COST THAT CAN BE INFLUENCED BY THE DESIGNER

His designs directly influence the cost of manufacturing parts and assemblies as indicated under the segment "make". Notice this includes items made "in-plant" and those sub-contracted for elsewhere.

The primary elements of cost influenced by the designer are materials, fabrication and assembly labor, and tooling. Fabrication and assembly labor are normally referred to as "direct labor" and are the only labor elements we consider as being influenced directly by design. Material, labor, and tooling are considerable in the motor vehicle industry so it is evident the designer plays a major part in influencing cost and ultimate selling price of a vehicle. Additionally, many of the so-called "indirect" and/or slightly influenced items are determined and usually limited to a "percentage" of the "direct manufacturing costs", so final costs tend to be built up like an inverted pyramid from what the designer has a degree of direct control over.

In a well established company with limited changes and improvements from year to year, as has been the case in the motor vehicle industry in the past, the value analysis method of cost control described in chapter XIX under Production Engineering works quite well. There is another method used for completely new products and for new companies not yet established in an industry. This method is discussed next and may be of interest for those designers working on advanced concepts in the motor vehicle field.

THE "TARGET COST CONTROL" METHOD FOR NEW PRODUCTS

"Target Cost Control" during the design phase is not new. It is well established in many industries. During the 1950's the writer was the supervisor of PRODUCTION DESIGN ANALYSIS, a group in Engineering that initiated a "target cost control" method in the aircraft industry. In the aircraft industry, it is necessary to arrive at the selling price of a new airplane before it is even designed in detail. While this procedure is the reverse of most industries, it is necessary because the initial investment is so great that few, if any, companies can take the risk without tangible protection in the form of orders. With the stepped-up developments in the motor vehicle industry, as well as in others, a somewhat similar situation, just short of advanced orders is approaching. Reference 131 describes in considerable detail how we went about it. The following brief description will convey the ideas here.

To avoid any misunderstanding it should be stated that, compared to producers of other transportation equipment, the aircraft industry has to incorporate a high degree of operational reliability and structural efficiency in its products. To attain this, the industry has given and will always give its designers considerable freedom from the restrictions of designing within specific cost limits. This is particularily noticeable when "dollars per pound" is used as a basis for comparison between the airframe industry and the manufacturers of other transportation equipment of all kinds.

It is a matter or record that aircraft *weight and cost increase as the design progresses from its inception.* No doubt this is also true in other industries. In our case, if weight increases were not recorded and a detailed effort not made to control it, performance would soon suffer so much that the airplane could not be sold for half its contract price. The automobile industry now has a strict weight control problem due to the CAFE requirements.

Since the Target Cost function somewhat parallels the operation of a weight and balance function it is convenient and logical to use the same tabulated components and parts breakdown as they use.

We provided a TARGET **COST** ALLOWANCE for each group of parts just as the weight people provided a TARGET **WEIGHT** ALLOWANCE. As the design developed both the cost and the weight allowances were refined, breaking them down into more and more detail.

As a part or component was completely designed its final weight was carefully calculated and listed for comparison with the TARGET WEIGHT ALLOWANCE. Its cost, material, labor, and tooling, was also calculated, listed, and compared with the TARGET COST ALLOWANCE.

In both cases, weight and cost were known as soon as an individual item was designed and quick enough to initiate corrective action in the form of redesign if over weight or over cost. We never had to wait until the vehicle was completely designed to know if we had a problem with either weight or cost.

In a very few cases the over weight and/or cost was justified and differences (overage) could be removed from surrounding or other items to remain within our overall targets.

Historical data on previous models provides much needed information on both weight and cost. The difficulty comes when we start using more exotic materials which are not only initially more expensive but usually also require more processing and are more difficult to fabricate. The Financial Branch is of great assistance with historical data and original cost estimates from which we worked down to our targets. The Manufacturing Branch together with certain research data provides a basis for estimating previously untried materials and advanced designs. Material size can be readily determined from the Bill of Material on the drawing. Labor hours required come from standard estimating tables for various operations, together with a "planning sheet" that lists in sequence the plan of manufacture starting with all the usual cut and shear, through forming or machining for fabrication, to be followed by the assembly operations.

Tool design and make is usually rather straight-forward and historical estimating data is available.

It is also necessary to agree on a "design quantity". This is the expected number of vehicles to be manufactured. Once established, the designer, production engineers, and the cost analyst can determine the preferred design for the given quantity.

THE LEARNING CURVE AND HOW IT AFFECTS LABOR COSTS

Job shops and sub-contractors quite normally use standard time study figures for every operation performed in their shop when estimating a job for someone else. They know from experience that there is much confusion when a new job is started, often delays, scrapped material, and inexperienced new employees cannot perform an operation as quickly or accurately as an "old hand". By the end of the production run everything is coming along fine. If they can just get a follow-on contract before they lose their momentum the company will make a higher profit the next time, even at the same bid price.

This is the way it is in every manufacturing industry. It is predictable within reasonable limits for different kinds of work.

The first man to recognize that this learning process follows a pattern and could be predicted was Mr. T.P. Wright in the aircraft industry. He constructed his first "learning curves", tables, and the mathematical methods in 1937. Since then the name has been changed to "Progress Curves" in many companies and some excellent books on the subject can be found in the library. References 132 and 133 are examples.

With everything now computerized, forecasting direct labor hours at any point in production can be done much quicker. Additionally, the learning curve concept is used throughout the financial branch of most large companies for many other forms of forecasting and financial planning.

THE THEORY

Now for the theory. A worker learns as he works; and the more often he repeats an operation, the more efficient he becomes, with the result that the direct labor input per unit declines. This is true whether the industry is aircraft, automotive, metalworking, textile, or something unrelated like cutting the grass or candy making. Every athelete, musician, and painter, knows he improves as he practices, just as does a typist. Fitting the pieces of a jig saw puzzle together becomes easier and faster every time it is repeated. *The key is that the rate of improvement is predictable.*

The rate of improvement for the direct labor operations in certain manufacturing industries are such that, once production gets under way, the 4th unit requires about 80% as much direct labor as the 2nd; the 10th, 80% as much as the 5th; the 200th unit, 80% as much as the 100th; and so on. In each case, when quantities are doubled the direct labor drops 20%. This 80% figure holds true within close limits for a company in which fabrication labor is one-quarter and assembly labor is three-quarters of the total direct labor.

The fewer opportunities there are for learning, the flatter the curve, or higher the percentage number. In the case of operations made up of about three-quarters machine time (fabrication of parts) and only one-quarter assembly time, the rate of learning may be found to be about 90% rather than 80%. In other words, direct labor hours only drop about 10% with each doubling of units.

WHAT THE CURVE MIGHT LOOK LIKE

A hypothetical learning curve is shown in figure 134 as it might normally be drawn on log-log graph paper. Log-log graph paper is used so that the lines are essentially straight for ease in plotting. We have shown a 20% reduction in labor between all doubled quantities. This is referred to as an 80% learning (or progress) curve. The lower line (unit curve) represents the direct labor hours for a particular unit. In our example the 100th unit is projected up to the unit line, then horizontally to the direct labor ordinate showing the expected amount of direct labor for that particular unit.

The dash line projected to the cumulative average curve above and then horizontally across to direct labor hours shows what the *expected average cost in unit hours is to that point,* the 100th unit.

Since the mind does not visualize in terms of log-log paper, figure 135 shows what the same 80% learning curve might look like on ordinary graph paper. It dramatically depicts the rapid learning that occurs in the early stages of production and nearly levels off as production quantities increase. Although the curve appears nearly flat at higher quantities, the same lowering of hours with doubling of quantities occur. Note, too, that a hand-built prototype, although similar to a production article, is not included on a learning curve. Its direct labor hours bear no relationship to what happens in production.

CONTENTS OF THE LEARNING CURVE

The operator of the machine and the assembler are not the only people who "learn". The designers and engineers who design the product introduce changes to improve production. So also do the planners and the tool designers. The supervisor is also constantly seeking ways to improve efficiency in his department. All this can be referred to as "rising productivity" and is not due entirely to the workman on the line. The various elements of learning might be shown as we have in figure 136. Note that "pure learning" of workers is shown as 95%, a rather flat curve. This is the rate of improvement due to the *average worker without benefit of tools or other elements of mass production.* It also assumes he will be permitted to work without interruptions and without changes or modifications (good or bad) in his way of doing his job except as he initiates. Again this is "average". Some workers are more ambitious and/or intelligent than others, while others are only

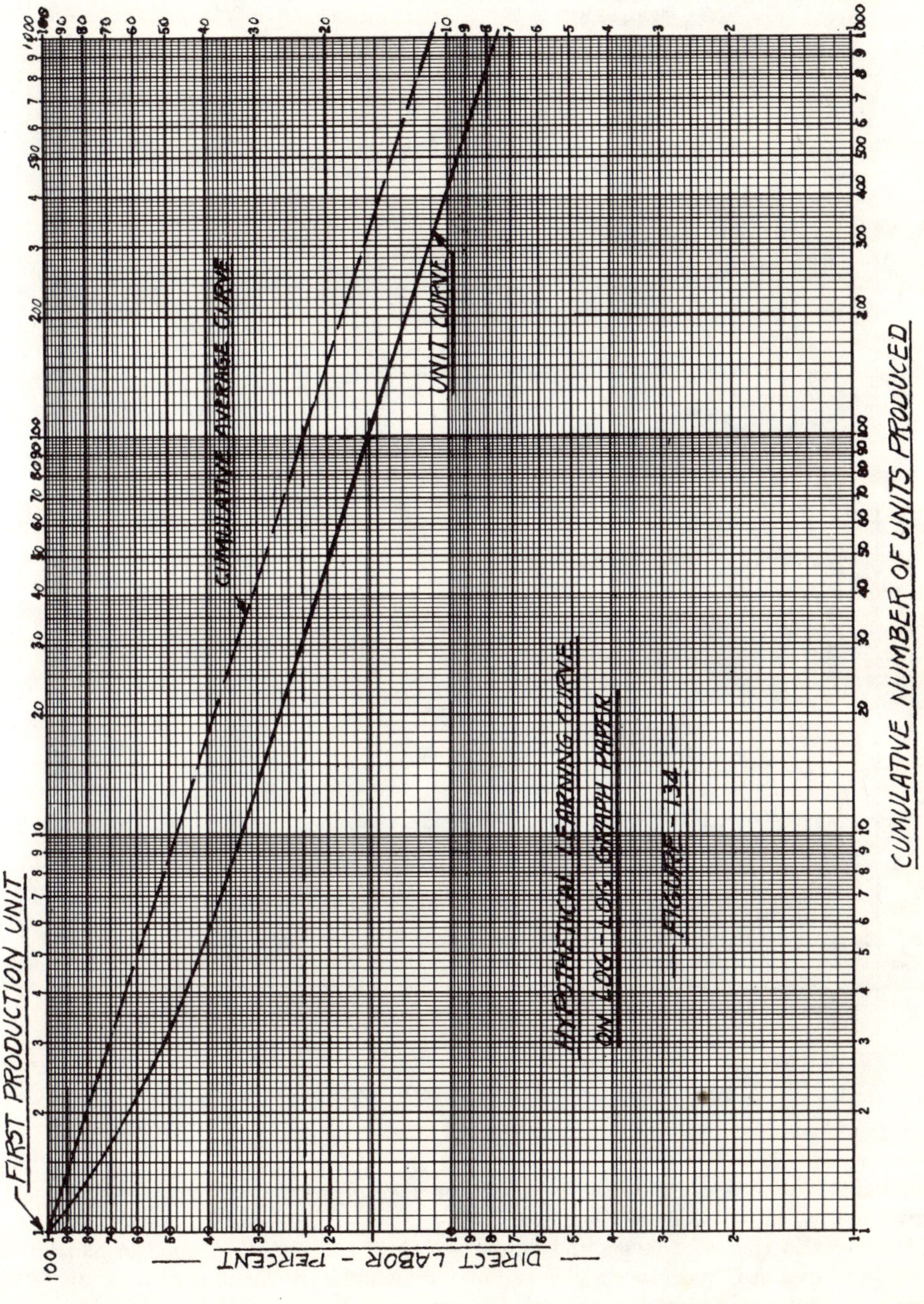

HYPOTHETICAL LEARNING CURVE ON LOG-LOG GRAPH PAPER

— FIGURE - 134 —

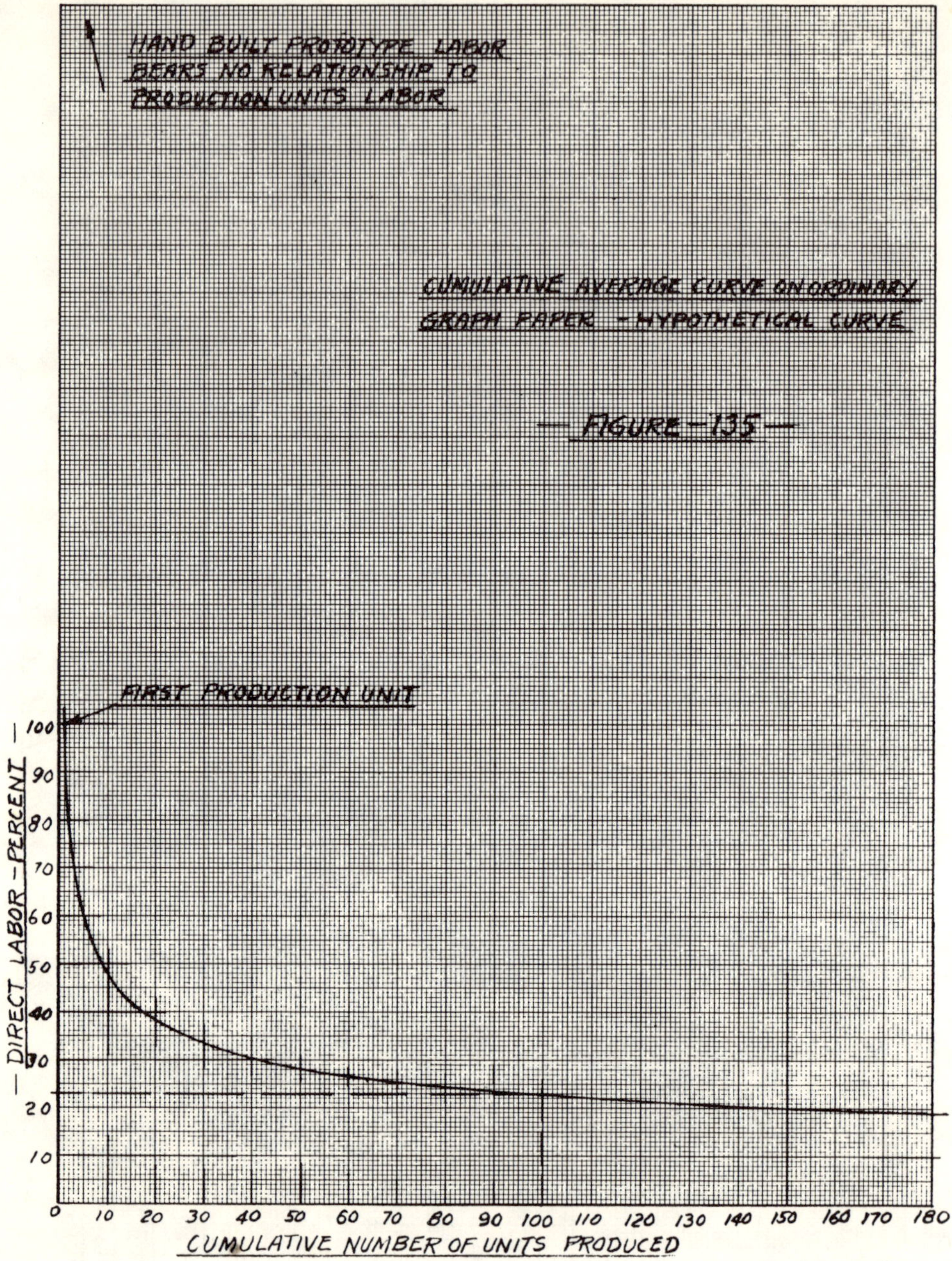

marginal at best. When the labor market is "tight" and companies are hiring, the "average" goes down, because of the higher percentage of "marginal" employees. Some are never motivated to rise above "marginal".

Note also that all changes and "improvements" may be either good or bad for the learning curve. An "engineering change" may be mandatory for reasons of safety of the product. This has nothing to do with lowering cost – in fact, it usually adds cost. It also has a "domino effect" in that a planning change then becomes necessary, which in turn requires a tool redesign and tool remake. Methods and material often are affected, with shortages then sometimes occurring.

Supervisors must also attend meetings to learn how to deal with the new design and its tools, etc. Thus must then be passed on to the worker. Clearly "learning" or progress is delayed to say the least.

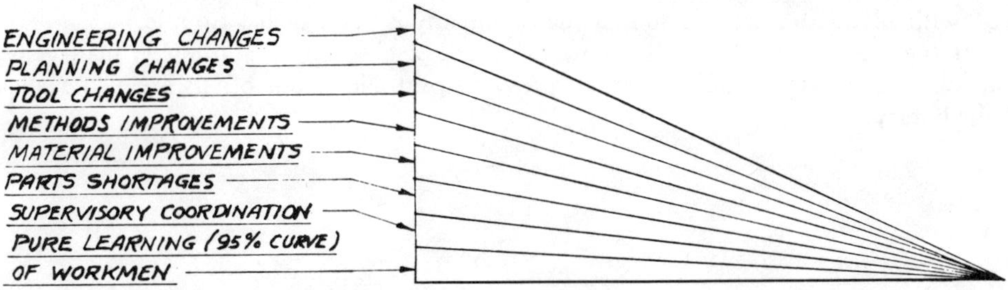

CONTENTS OF THE LEARNING CURVE
– FIGURE 136 –

ACTUAL PRODUCTION IS INFLUENCED TO MOVE EITHER ABOVE OR BELOW THE PREDICTED LEARNING CURVE

Experienced managements soon learn that what is predicted does not happen automatically. Conditions can occur that may make it possible to do better than "averages" or conversely everyone is in such a state of uproar that labor hours never seem to go down according to any planned pattern. A few of the reasons will now be discussed.

Even if a change is "helpful" in that it will improve efficiency or eliminate a bottleneck to eventually reduce labor hours its immediate effect is to disrupt "learning". There will be a rise or bump in the curve before it can start back down at (hopefully) a steeper rate. If changes, even "helpful" changes, continue to occur, they will continue to disrupt "learning" and the curve can never go down. Direct labor hours will remain level or even go up until all this "new learning" can be absorbed.

If the rate of production is to be increased rapidly new employees must be hired *and trained* before being expected to become productive on the job. This too is a disruption to others and the learning curve, even if nothing else changes. Fifteen percent (15%) increase in the number of direct labor employees per month is considered a rapid rate of increase.

If sales fall then the *rate* of production must also be reduced. This means layoffs. Often, in an effort ot keep the best employees, they will be transferred to another department to replace a marginal employee who is to be laid off. The better employee must then "learn" the new job, and that means a temporary increase in direct labor hours until he is up to rate.

The only thing worse than an inexperienced, untrained, poorly supervised work force is an uncoordinated, inexperienced management team that keeps changing signals while the ball (production) is in motion. "Learning" becomes impossible, morale hits a new low, and production plummets – no matter how many new "trainees" are added.

Learning or progress requires stability with confident, trustworthy, direction toward *achievable* goals.

Nearly all changes must be worked out on the "pilot production line" if not before. That is the purpose of a pilot line.

If the rate of production is to increase faster than normal, "learning" with a stable work force, it must be planned very carefully. Only a few new trainees can be absorbed by a producing group and its supervisor at any time. Preferably not more than 15% more per month. Do not overload either the group or its supervisor with "green help". If so, the added confusion will cause delays, rejected parts, and near stoppage of production until the new employees become productive.

If changes *and* new employees are introduced simultaneously into a production line pure chaos can only result. Goals cannot be reached until the changes and the new employees are settled in.

Experienced managements who recognize these situations therefore set *achievable goals* with adequate financing so that their company is never in any trouble.

If the reader would like to know more about accounting and higher management control, references 134 and 135 may prove helpful. No doubt others are also available in the library.

Chapter XXI

INVENTIONS, PROPRIETARY RIGHTS, AND PATENTS

Engineers in general and designers in particular are just naturally inventors. It is their job to create new ideas and to think up new ways of accomplishing something. Companies have clauses in their employment agreements that set forth their "rights" to inventions and ideas while an engineer is on and off a project, often with a means for additional payment in certain situations. However, surprisingly, engineers and designers seldom fully understand what they should do relative to every idea they come up with and the day-to-day steps required to protect these ideas for the company and themselves.

Of course, every idea is not necessarily an invention, and not all inventions are patentable. But this is not for the engineer and designer to decide for the company. Supervisors normally review the idea first, and if believed to be valuable to the company, either on an existing project or possibly at some future date, asks the inventor to then fill out a "disclosure form" and contact the company patent attorney, all of which we will discuss a bit later.

PROPRIETARY RIGHTS

"Know-how" and detailed ways of doing things are often unique to a company, quite valuable, and should be protected, especially from the competition. An engineering drawing depicts, often in great detail, salient features of a design in which the company has invested considerable time and money and which can be recognized as of a "proprietary" nature. It is quite normal to indicate this with a proprietary note somewhere on the drawing, usually in the upper left hand corner. To assure uniformity it is often stamped on. Such a note might read as in figure 137. The author has used this for a number of years. Variations can be found in nearly every company, but in every case the intent is essentially the same. Designers working alone might use their own name and reword to more nearly serve their own purposes.

> THIS IS A PROPRIETARY DESIGN AND MAY NOT BE COPIED, MFGR'D, OR SOLD WITHOUT WRITTEN PERMISSION OF__________ __________ __________

INSERT COMPANY NAME

PROPRIETARY NOTICE
– FIGURE 137 –

PATENTS – THE FIVE STEPS TO TAKE BEFORE FILING

If you have an idea for an invention, you or your company, (when you assign the idea to the company) may apply for a patent. If granted, the patent provides an exclusive statutory right to the inventor (or his company) by the U.S. Patent Office to exclude others from making, using, or selling, the invention during the seventeen-year term for which the patent is granted. The steps between coming up with an idea and obtaining a patent are involved, somewhat costly, and certainly time consuming.

The first thing that should be done is to start keeping excellent, detailed records of progress. The idea should be fully described with detailed sketches to illustrate the idea, all prepared in ink, dated, and signed by the inventor and two witnesses who understand the idea. Companies normally have "disclosure forms" for this purpose and the company patent attorney and employee's supervisor usually serve as witnesses. It is important to do this promptly because this serves to establish the date of conception by you. If someone else has the same idea and also applies for a patent, interference occurs, and the one who can prove earliest conception obviously then has the best chance of receiving the patent.

The second step is to avoid "abandonment" (expressed or implied). This is the failure to "continuously" act to bring an invention from an idea state, to a reduction to practice, to filing of a patent application. Keep working on it. Show progress in your records with sketches, improvements, and dates.

Inexperienced inventors often become anxious to get the patent application off to Washington, D.C. at the earliest possible moment. This is exactly the wrong thing to do. If the disclosure is properly prepared, adequate temporary protection is provided and three additional steps should be taken before filing. In fact, most companies insist on these steps.

The first idea may be the best – but not always. The third step for the inventor is to think about his idea some more. What additional ways of accomplishing the same thing can he now come up with? The patent application should provide broad coverage. Every conceivable way of doing the same thing should be included.

The fourth step is to initiate a search of the prior art. This is mainly information of issued and expired patents. Clearly you will not be granted a patent when one has already been issued on essentially the same idea. This may even help your own thinking a bit.

The fifth and last step before applying for a patent is to be sure the idea is workable. Often when an idea is incorporated into a machine or a vehicle, the inventor begins to see a better way to do the same job. Possibly lighter, stronger, fewer parts, less costly, easier to manufacture, or even more functional than the original idea. In any event, the company wants to use what is found to be best for the purpose. As long as it can be shown that the idea has not been abandoned, filing can be held up to be sure the idea is workable and practical.

SOURCES OF INFORMATION ABOUT PATENTS

One short chapter on the subject of patents could not possibly do the subject justice. Our purpose is to indicate importance to the designer and outline the five initial steps before filing, along with the note of precaution. As always, a good place to start to learn more about the subject is the library; either company, school, or public library. Reference 136 has been helpful to the writer.

The Patent Office published a booklet called "GENERAL INFORMATION CONCERNING PATENTS". A single copy is free and may be obtained from:

Commissioner of Patents
Washington, D.C. 20231

If you would like to have your invention evaluated before trying to patent it, and providing it is energy-related, you may use the services (free) of the OERI. Write for general information and evaluation request form (NBS–1019) to:

Office of Energy Related Inventions
National Bureau of Standards
Washington, D.C. 20234

While you may make your own patent drawings, write the description, specification, and claims, and do your own filing, it is seldom recommended this be done without the assistance of a patent attorney. If a local patent attorney is hired, he will almost certainly have an associate in Washington, D.C. who is used to contact the Patent Office directly, investigate, and plead the case as required. For this reason some inventors prefer to be in contact with the Washington attorney directly, rather than go through a local man. As you read up on the subject this problem and many others can be dealt with.

Chapter XXII

NOTEWORTHY EFFICIENT DESIGNS A VARIETY OF INTERESTING SPECIAL PURPOSE EXPERIMENTAL VEHICLES

A variety of interesting motor vehicles are included in this, the last chapter, to show the reader how others have solved unique design problems. No production cars are included because such data is readily available elsewhere. All these vehicles are of an experimental nature aimed at specific purposes.

The first two examples were designed and built with limited resources and clearly show that innovation in the motor vehicle field is not restricted to the large manufacturers of this or any other country. If you have the desire, ambition, and stick-to-it-iveness a way can be found.

(A) – WORLD LAND SPEED RECORD GOLDENROD

On November 12, 1965, Mr. Robert S. Summers drove the GOLDENROD to a two-way average speed of 409.277 mph for a new world land speed record for wheel driven cars. This brought the record back to the United States for the first time since 1928. It remains the world record to this date, 1980.

This is a dual success story. First: the Summers brothers, Bill and Bob, launched a very successful business shortly after their 1965 record runs as mentioned briefly in Chapter I. This is an inspiration to all young people of limited means who intend to make their career in the motor vehicle industry. Many "special situations" can always be recognized just as did these brothers. Second: the vehicle itself was and remains quite a success story. This development will now be discussed.

Once the GOLDENROD program officially got under way responsibility was assigned as follows:

Bob Summers — **Technical Management and Testing**
Responsible for technical decisions relative to choice of components, qualifications, installation, and testing. Also responsible for chassis design in conjunction with W. Korff.

Bill Summers — **Administration, Coordination, and Materials**
Responsible for procurement of components, raw materials, and parts through all required tasks, sub-contracting, transportation, services, costs, scheduling and inspection.

Walt Korff — **Engineering and Aerodynamics**
Responsible for preliminary design of chassis, development of body shell shape and wind tunnel program.

GENERAL ARRANGEMENT AND SPECIFICATION

Bob preferred the four engines in a row with the driver aft in the "sling shot" position. This arrangement was ideal for my aerodynamic approach to the problem in that I wanted minimum frontal area and to design the shape so that at least half of the air would pass around the sides, considerable under the vehicle; and to achieve some down load by basic body shape alone when set at a slight negative angle in side view. With the cockpit aft the parachute tubes would blend into the shape nicely and hopefully, no additional fin area would be needed.

Bob also designed a transfer case with a pair of engines back-to-back so that their torque would counteract each other. This was duplicated to use the four engines and to avoid engine torque weight transfer problems. The first drawing was in planview to show

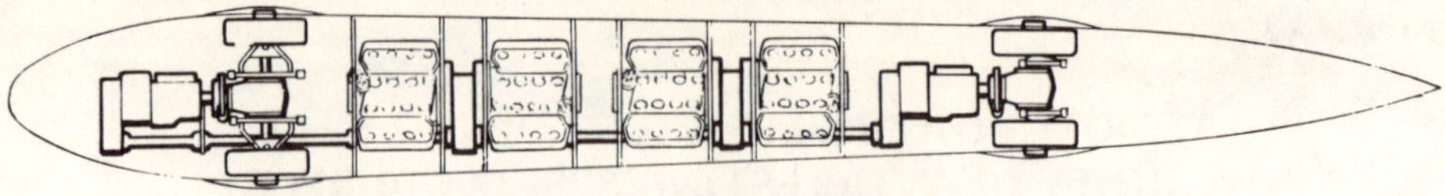

GOLDENROD ENGINE AND DRIVE TRAIN LAYOUT

— FIGURE 138 —

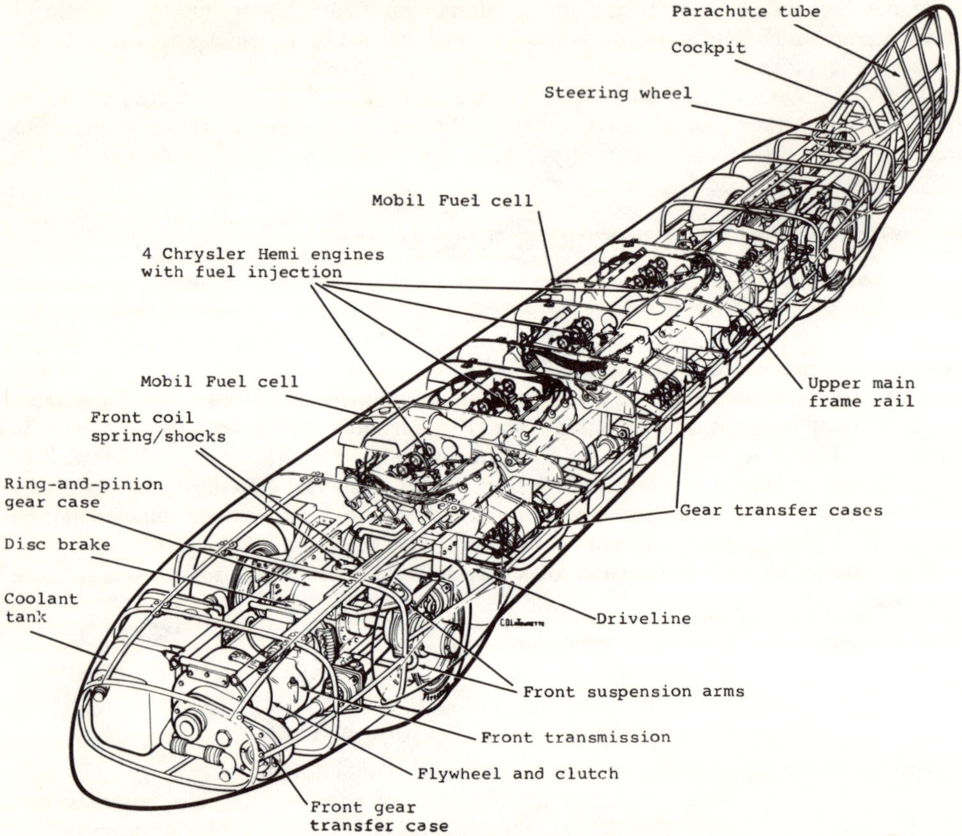

— FIGURE 139 —

the drive train layout as indicated in figure 138. Front tread was 36" and rear tread 24" to give me the taper I wanted.

Figure 139 shows the detail arrangement of all parts and how well everything was packed into the minimum of space. Table XV, the specifications, provides all the mechanical information.

AERODYNAMIC DESIGN AND WIND TUNNEL TEST PROGRAM

The aerodynamic design and the wind tunnel test program will now be described. Figure 140 is a scale drawing of half of the cross section together with the areas of both sides combined for the model and the full size vehicle. The model was one-fifth scale.

The wind tunnel program was thoroughly coordinated with Mr. William Bettes, Director of the 10 foot wind tunnel at the California Institute of Technology, Pasadena. The complete procedure is outlined in reference 137 and portions will be included here to indicate what was done. Table XVI lists the nineteen runs and configuration changes

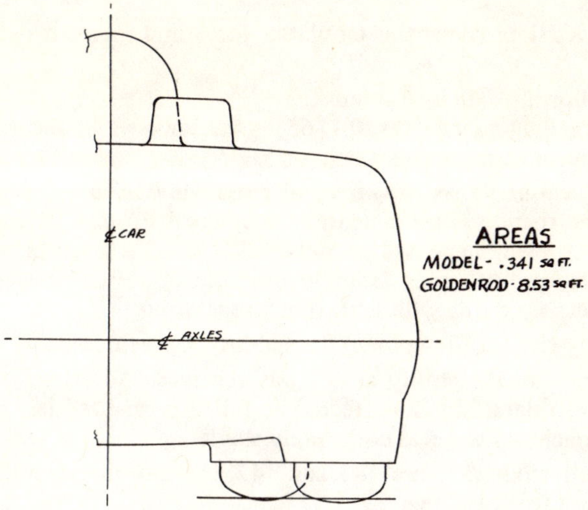

— FIGURE 140 —

Specifications of Goldenrod

TABLE XV

Chassis and Body

Overall Length	32 ft (384 in.)
Wheelbase	207 in.
Tread (Front and Rear)	36 and 24 in.
Overall Height	To top of tailfin 42 in.; to top of engine hood 28 in.
Overall Width	48 in.
Ground Clearance	5 in.
Frame	Mild steel. Lower rails: 2 in. diam. × 1/8 in. wall. Upper rails: 2 × 6 in. rectangular tubes, 3/16 in. wall, tubes mounted vertically for maxmum rigidity.
Body Material	Harvey aluminum: alloy 3003, 0.064 in. thick.
Wheels	Aluminum. Hurst designed, forged by Harvey Aluminum; 16 in. diam × 6-1/2 in. rim width. Demountable rim.
Tires	6.50 × 16 Firestone tubeless nylons; special low-profile treadless design. Width of tread contact area 4 in. Static diam 23 in.; inflation 250 psi; design speed 550 mph.
Braking Systems	Airheart triple spots mounted on front and rear pinion gear coupler flanges for use at 100 mph and below. Deist parachutes for high speeds; pilot chute 8 ft diam first stage for maximum speeds; 24 ft final chute for 250 mph and below, plus emergency 8 ft chute. Automatic systems with manual override.

Engines and Drive Train

Engines	Four Chrysler Hemi V-8s, mounted in-line and coupled in pairs, back to back, front pair driving front wheels, rear pair driving rear wheels. Engine rpm's synchronized between pairs by mechanical coupling. Except for dry sump, lubrication, and fuel injection (required for low overall height), engines are to stock specifications throughout.
Bore and Stroke	4.25 × 3.75 in.
Displacement	426 cu in. each engine
Estimated Horsepower	650 at 6600-6800 rpm, each engine
Fuel	Racing Methanol, 105 Research octane, supplied by Mobil Oil Co.
Lubricants	Mobil Oil co. products
Fuel Supply	5 gallons per engine in four tanks
Transmissions	Two 5-speed Spicer units, utilizing top four gears. Ratios: 2nd -- 2.6:1, 3rd -- 1.5:1, 4th -- 1.19:1, and 5th -- 1:1. Simultaneous synchronized shifting via special Hurst shifter
Clutches	Two Schiefer double-disc, hydraulically actuated
Ring and Pinion Ratios	1:1, locked rear ends
Final Drive Ratios	May be varied. Anticipated: 0.95:1 to 1.05:1
Steering	Chrysler, hydraulically actuated, 10 deg limit
Suspension	Fully independent, all four wheels via upper and lower A-arms. Upper arms pivot on frame, act on four special Monroe coil spring/shock units mounted inboard.

and Table XVII provides the tabulated force and moment data in coefficient form for each run.

The following data is of interest.

a) The drag coefficient, 0.1165, is the lowest ever recorded for a four-wheel land vehicle.

b) Run four shows the effect of cross winds. From 0 degrees through 4 degrees of yaw there was no increase in drag coef. From 4 degrees to 15 degrees the drag coefficient increased to only 0.1210. This small increase indicates the air was effectively flowing laterally over the upper and lower rounded corners to the downwind side with little if any separation.

c) Negative lift (download) for stability was achieved.

d) Directional stability (yaw) was achieved. Additional fin area was added in run 18 to determine its effect. The full size car used less than shown to enclose the parachutes and was completely stable.

e) Tuft studies, figure 141 and 142, indicate the smooth flow that was obtained, both front and rear.

TABLE XVI Wind Tunnel Tests

Run No.	Configuration	Dynamic Pressure q	Remarks
1	Droop nose { Model shakedown and	30	Front and rear wheels 1/4 in. from ground plane
2	Droop nose { Reynolds No. checks	40	Front and rear wheels 1/4 in. from ground plane
3	Droop nose	50	Front and rear wheels 1/4 in. from ground plane
4	Droop nose	50	Directional stability runs; rotate model through 1-15 deg increments
5	Droop nose	50	Raise model; front and rear wheels 3/4 in. from ground plane
6	Droop nose	50	Nose up pitch; front wheels 3/4 in. and rear wheels 1/4 in. from ground plane
7	Droop nose	50	Nose down pitch; front wheels 1/4 in. and rear wheels 3/4 in. from ground plane
8	Droop nose with transition strip 1-5/8 in. back	50	Wheels 3/4 in. from ground plane; add transition strip 0.007 in. thick
9	Droop nose with transition strip 0.333 in. back	50	Wheels 3/4in. from ground plane; add transition strip 0.007 in. thick
10	Droop nose with two transition strips 0.333 in. and 19 in. back	50	Wheels 3/4 in. from ground plane; add transition strip 0.007 in. thick
11	Symmetrical nose	50	Wheels 3/4 in. from ground plane
12	Symmetrical nose	50	Nose down pitch; front wheels 3/4 in. and rear wheels 1-1/4 in. from ground plane
13	Symmetrical nose	50	Nose up pitch; front wheels 3/4 in. and rear wheels 1/4 in. from ground plane
14	Symmetrical nose	50	Neutral position; both wheels 1/4 in. from ground plane
15	Droop nose with tufts	50	Neutral position; both wheels 1/4 in. from ground plane
16	Droop nose with probe only	50	Visual-low speed; same position as above
17	Droop nose with spoiler 0.17 in. high and 0.65 in. back	50	Neutral position; both wheels 1/4 in. from ground plane
18	Droop nose with tail fin upper and lower	50	Neutral position; both wheels 1/4 in. from ground plane
19	Droop nose with tail fin plus spoiler No. 2 (0.40 in. high and 1 in. forward of main body)	50	Neutral position; both wheels 1/4 in. from ground plane

Notes: Engine air inlets were faired with clay for smooth airflow over inlet upper surface.
All model gaps, including strut entry points, were sealed to eliminate model internal airflow.

— FIGURE 141 —

— FIGURE 142 —

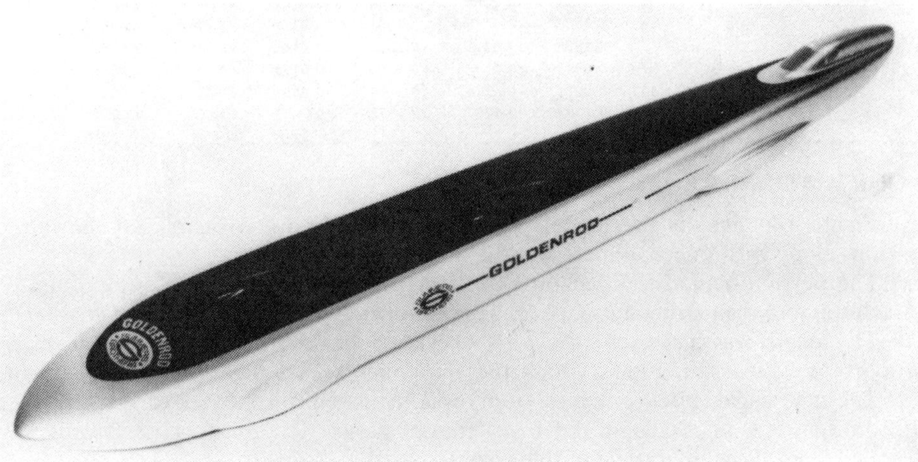

— FIGURE 143 —

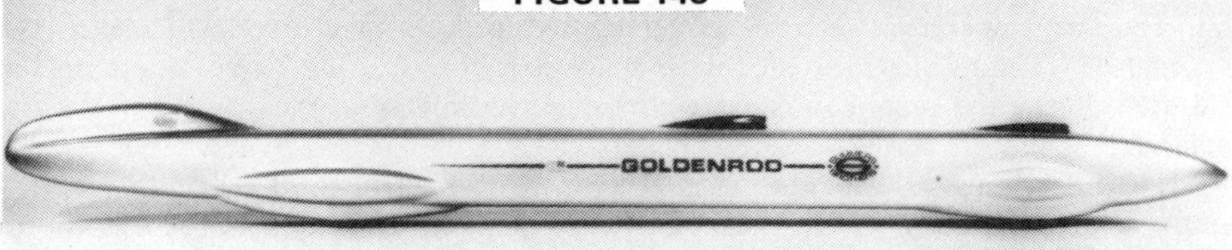

TABLE XVII

Tabulated Force and Moment Data in Coefficient Form for Car Runs

				Stability Axes Coefficients					
				Forces			Moments		
		Deg.		Lift	Drag	Cross Wind	Pitch	Roll	Yaw
Run No.	q, psf	a_g	ψ_g	$C_{L_s} = C_L$	C_{D_s}	C_{C_s}	$C_{m_s\,c.g.}$	$C_{i_s\,c.g}$	$C_{n_s\,c.g}$
1	30	0	0	-0.2216	+0.1160	-0.0419	+0.0251	-0.0003	+0.0034
2	40	0	0	-0.2181	+0.1170	-0.0635	+0.0248	-0.0005	-0.0014
3	50	0	0	-0.2164	+0.1165	-0.0343	+0.0229	-0.0003	-0.0042
4	50	0	-1	-0.2247	+0.1165	+0.0693	+0.0235	+0.0006	+0.0120
			0	-0.2164	+0.1165	+0.0093	+0.0229	+0.0001	+0.0063
			+1	-0.2335	+0.1165	-0.0157	+0.0223	+0.0007	+0.0101
			2	-0.2164	+0.1165	+0.0018	+0.0229	+0.0010	+0.0140
			3	-0.2429	+0.1165	+0.0198	+0.0175	+0.0023	+0.0231
			4	-0.2003	+0.1165	+0.0809	+0.0182	+0.0031	+0.0375
			5	-0.1665	+0.1192	+0.1163	+0.0177	+0.0033	+0.0488
			7	-0.1249	+0.1192	+0.1959	+0.0131	+0.0045	+0.0759
			10	-0.0039	+0.1210	+0.2865	+0.0245	+0.0071	+0.1232
			15	+0.0764	+0.1210	+0.5394	+0.1076	+0.0123	+0.1803
5	50	0	0	-0.1806	+0.1165	-0.0260	+0.0233	-0.0002	-0.0067
6	50	+.73	0	-0.2098	+0.1103	-0.0039	+0.0635	-0.0001	-0.0067
7	50	-.73	0	-0.1991	+0.1146	-0.0529	-0.0265	+0.0006	-0.0045
8	50	0	0	-0.1731	+0.1201	-0.0227	+0.0185	-0.0002	-0.0087
9	50	0	0	-0.1682	+0.1165	-0.0428	+0.0192	-0.0003	-0.0053
10	50	0	0	-0.1696	+0.1210	-0.0260	+0.0190	-0.0002	-0.0067
11	50	0	0	-0.1715	+0.1156	-0.0429	+0.0180	-0.0003	-0.0071
12	50	-.73	0	-0.1652	+0.1186	-0.0512	-0.0270	-0.0003	-0.0046
13	50	+.73	0	-0.1666	+0.1144	-0.0241	+0.0677	-0.0019	-0.0050
14	50	0	0	-0.2069	+0.1192	-0.0343	+0.0194	-0.0003	-0.0042
17	50	0	0	-0.2050	+0.1237	0	+0.0204	0	0
18	50	0	-1	-0.1975	+0.1200	+0.0042	+0.0238	-0.0024	+0.0067
			0	-0.1975	+0.1165	-0.0258	+0.0237	-0.0013	-0.0032
			+1	-0.2094	+0.1165	+0.0172	+0.0238	+0.0016	+0.0021
			2	-0.2164	+0.1165	+0.0597	+0.0229	+0.0031	+0.0022
			3	-0.2221	+0.1165	+0.1154	+0.0197	+0.0054	-0.0013
			4	-0.1936	+0.1165	+0.1861	+0.0178	+0.0079	-0.0081
			5	-0.1535	+0.1183	+0.2516	+0.0143	+0.0101	-0.0173
			-7	-0.0998	+0.1219	+0.3705	+0.0114	+0.0152	-0.0371
			10	+0.0277	+0.1174	+0.5214	+0.0298	+0.0215	-0.0600
19	50	0	0	-0.1949	+0.1736	-0.0312	+0.0290	-0.0019	-0.0087

MODIFICATIONS REQUIRED

Figure 143 depicts the model as originally conceived and figure 144 shows the car as built, along with Bob Summers and the writer.

The portion behind the cockpit was increased slightly to provide more room for the parachutes and their firing mechanism. This is shown in figure 145, the rear view.

The length was increased 10% with wheelbase and overhang, front and rear, kept in proportion. This was needed because the components were slightly larger than expected.

The rake angle (pitch down in front) was increased 0.3 degrees to assist, along with the droop nose, in enclosing the front transmission. The droop nose is only slightly drooped from the symmetrical in side view.

The air inlets were moved further aft for the engines. This became a concern on the salt. The small openings were the concern. Because of a malfunction of one or two of the engines (running rich) the air inlets' small size were suspect. Larger inlets were made and used during the record runs. After the new record was set the original air inlets were replaced and another run was made – faster. The speed increased to 425 mph.

It is interesting to also note that the record was set in third gear, not fourth. The course was exceedingly wet at both ends which limited the length for acceleration to 5 miles or less on both sides of the measured mile. The car was still accelerating as it went through the timing traps and it never got into fourth gear. A drizzling rain began to make matters worse – but the new record was officially set.

– FIGURE 144 –

– FIGURE 145 –

(B) – VIKING SERIES RESEARCH VEHICLES

Dr. M.R. Seal, Director, Vehicle Research Institute, Western Washington University provided us with the data and photos of these noteworthy cars. The students, under his direction, designed and built these vehicles. They had the opportunity to see theory and design transformed into hardware as well as personally having the hands on experience during construction and later, the pleasure of driving and testing their creations.

VIKING II

This two-passenger, mid-engine, rear drive sports type vehicle is powered by a four-cylinder 1969 Subaru engine converted to propane. It was entered in the 1975 Student Engineered Economy Design (S.E.E.D.) rally from Bellingham, Washington to Los Angeles, California. S.E.E.D. was sponsored cooperatively by Western Washington State College (WWSC) and the University of California at Los Angeles (UCLA). The vehicle attained 58.18 mpg (24.5 Km/m) on the 1400 mile rally and later, with a newly fabricated carburetor, fuel economy at a steady 48 mph increased to 77.5 miles per gallon of gasoline equivalent.

Because of the success of this vehicle the Subaru factory has given the project two examples of their latest stratified charge (SEEC-T) engines and a five-speed transaxle with suitable final drive for these components. The fuel system is also to be gasoline.

Side and planview drawings of VIKING II appear in figure 146 along with dimensions and features. The car weighs 1,180 pounds, has an overall length of 12 feet, width of 5 feet, and is 42 inches high.

The body skin is made of fiberglass reinforced plastics. Front and rear bumpers are self-skinning polyurethane foam, to resist 5 mph damage. The chassis is of semi-monocoque construction – framing members and body panels are integrated to form a rigid, lightweight structure.

The wind tunnel tested shape has a drag coefficient of 0.33. It requires but 7 HP at 50 mph. Figure 147 is a tuft study showing the excellent flow around all portions of the body. Figure 148 shows what happens to the tufts when the headlights are raised. This also cuts fuel mileage by 10 miles per gallon. Figure 149 provides an excellent rear quarter view of the vehicle and its internal arrangement.

ADDITIONAL PROJECTS

A stratified charge Wankel rotary engine is to be developed and installed in an all-aluminum monocoque version of VIKING II. It will be VIKING IV. Another version of the rotary engine has been adapted to a VW transaxle which has the crown wheel inverted to allow the mid-engine placement.

An 1184 twin-cylinder IZUSU diesel engine is to be mounted in a fiberglass monocoque vehicle with the same lines as the VIKING IV, the all-aluminum car mentioned before. This is to be VIKING V as shown in figure 150 and 151. Note the increased tail or aft body length in figure 150. A record attempt at Bonneville is planned for the larger tail version as it appears in figure 152. Note the excellent airflow as indicated by the manner in which the tufts lay back. Figure 153 shows how the canopy lifts.

Another project, VIKING VI, will be a safety car (or cars) for NHSTA with government funding. The body shape is to be revised with the front wheels enclosed and wheel fairings added to the underbody.

Additional information can be found on these interesting cars in references 138 and 139.

As this chapter goes to the publisher we learned that VIKING IV, now with a VW Rabbet diesel engine, was rated the most economical vehicle in the SCORE competition with 88 mpg. SCORE stands for Student Competition On Relevant Engineering. VIKING IV is reported to have a top speed of 120 mph with a potential up to 200 mph.

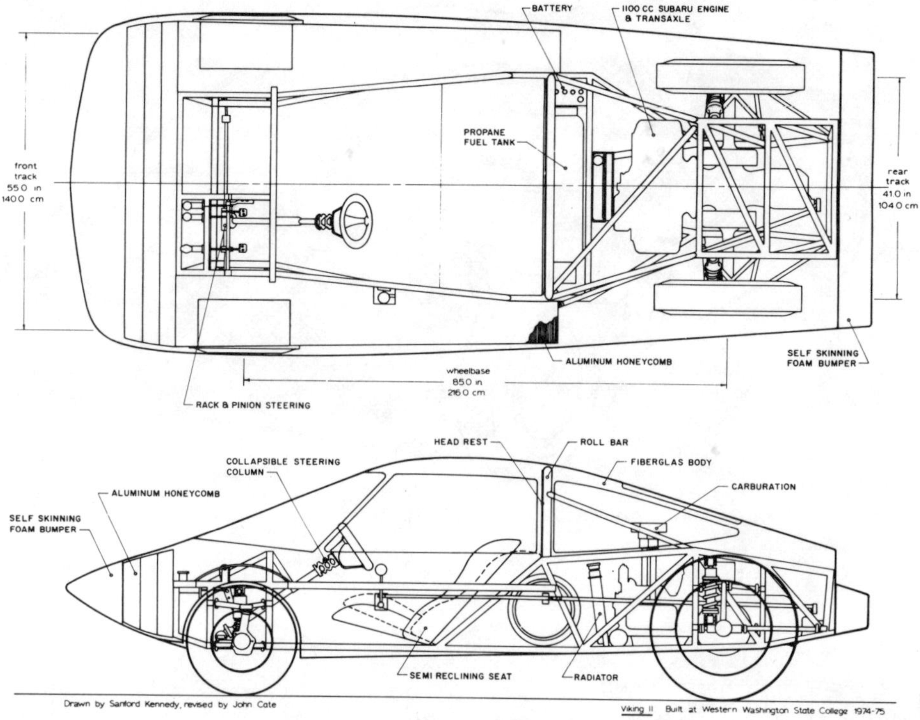

– FIGURE 146 –

– FIGURE 147 –

– FIGURE 148 –

— FIGURE 149 —

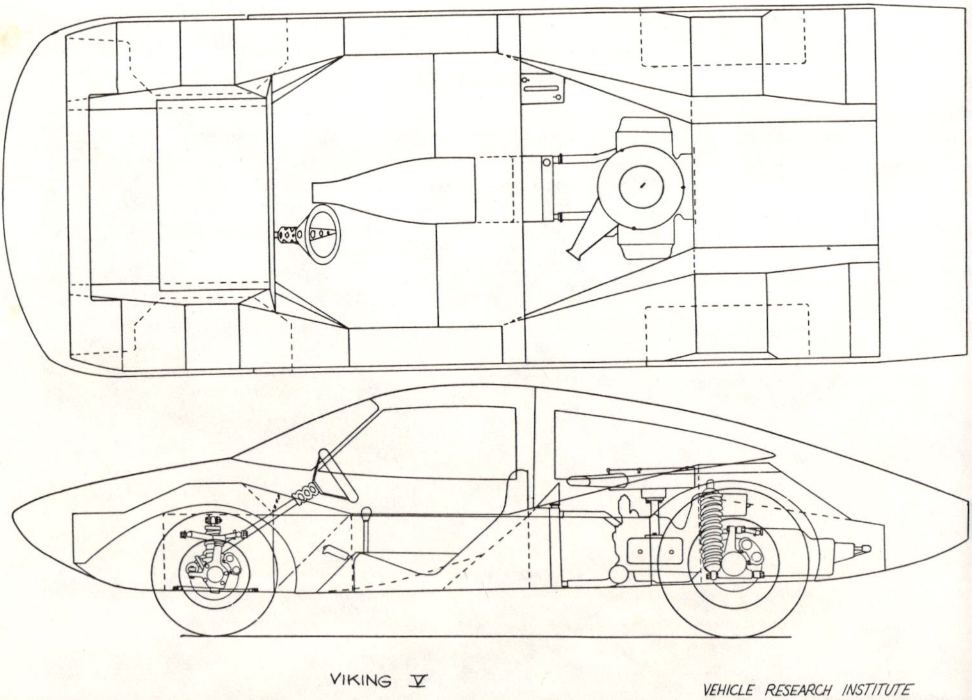

— FIGURE 150 —

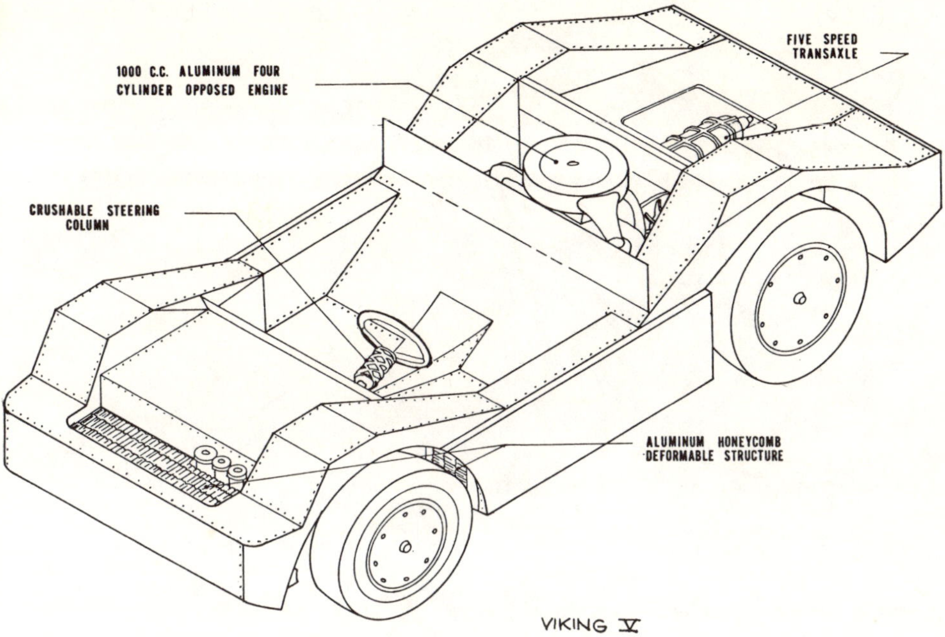

— FIGURE 151 —

— FIGURE 152 —

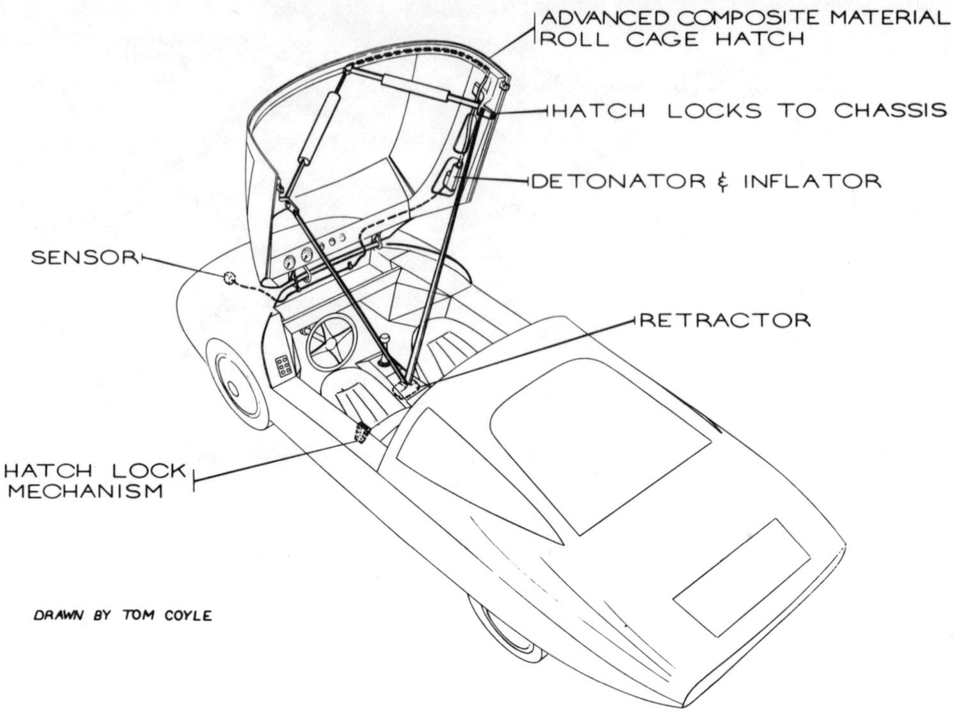

– FIGURE 153 –

(C) – THE MID-ENGINE MUSTANG – FORD'S EXPERIMENTAL SPORTS CAR. A CONCEPT THAT MAY BE REPEATED IN THE 1980's

Back in 1962 FORD built an experimental mid-engine sports car and called it the MUSTANG. It had nothing in common with the production MUSTANGS that followed other than its name. The car is noteworthy insofar as this book is concerned, in that it used a production front-wheel drive V-4 engine and transaxle and transplanted it to the mid-engine position to create a successful sports car that looked great, handled and performed very nicely, and was well received on both sides of the Atlantic. This design concept may be somewhat copied now, two decades later, with all the new front-wheel drive power units becoming available and the love of sports cars not fully satisfied by domestic manufacturers. This seems reason enough to include it. Possibly it will start the creative juices flowing for this generation of designers.

CONCEPT AND DESIGN – FRONT WHEEL DRIVE UNIT RELOCATED

Mr. R.C. Lunn of the Ford Motor Company was responsible for this project and prepared reference 140 in which the car was described in complete detail. To keep the cost down it was decided to use an engine and transmission from an existing high volume production line. The English Ford 4-cylinder with conventional 4-speed transmission was compared with the newer (of that time) German Ford Taunus 12M front-wheel drive unit. Early on it was decided the FWD concept, while suitable for sedans, restricted the development of a low-drag front end silhouette in a sports car. The studies were therefore confined to a conventional drive-line and a mid-engine rear drive. The comparisons are shown in figure 154. While the wheelbase is 9.5 inches greater for the mid-engine arrangement, the overall length is only 4.7 inches greater and the weight distribution was a little better together with favoring roadability due to a lower polar moment of inertia.

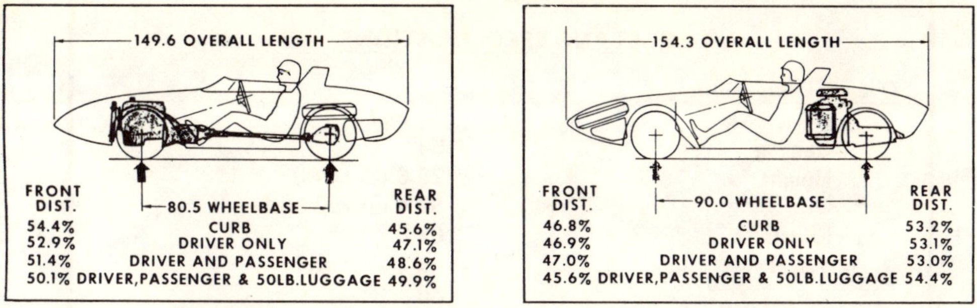

Conventional

Midship

– FIGURE 154 –

The drive-line was eliminated and the exhaust also was simplified with engine mid-ship. For these and other factors it was decided to pursue the mid-ship approach.

The V-4 engine and transaxle as used in the Taunus 12M appeared in chapter X, figure 39. The heavy transverse leaf spring was not used, only the engine and transaxle.

The vehicle package in greater detail appears in figure 155. The space frame by itself is shown in figure 156. The frame was made of 1" diameter .064 wall 4130 steel tubing welded as shown. The body skin was hand-formed from .060 aluminum sheet. The seats are part of the body skin, all aluminum welded together for additional structural strength.

The weight summary appears in chapter XVII, table XI.

The specifications for the car, including the engine, are listed in table XVIII.

The complete car is shown in figure 157. Since the seats are fixed the pedals and steering wheel are adjustable.

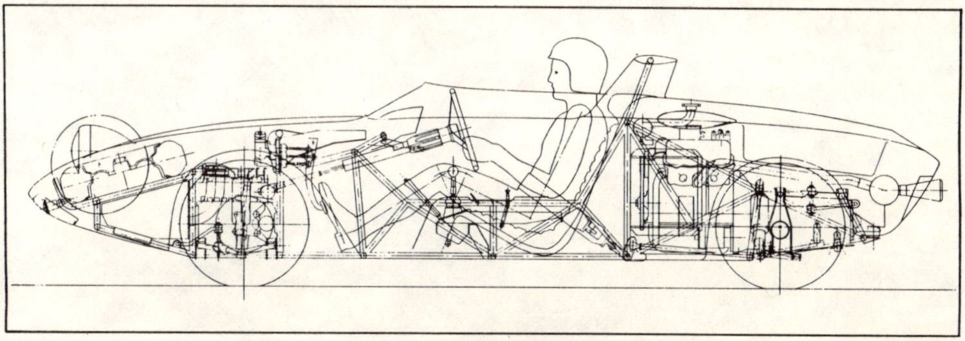

Vehicle Package

– FIGURE 155 –

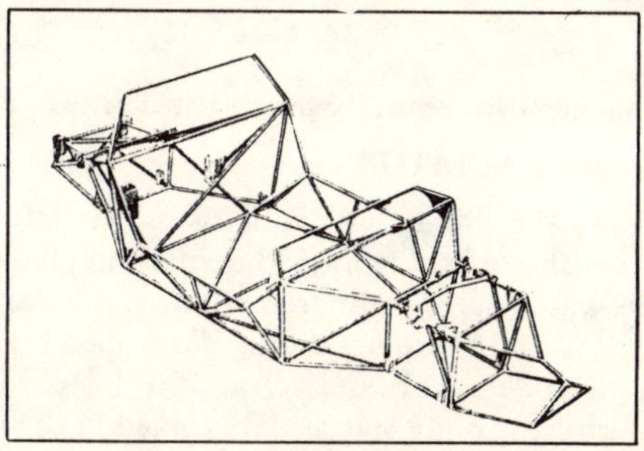

Frame

– FIGURE 156 –

TABLE XVIII

MUSTANG SPECIFICATIONS

Overall Dimensions		
	Wheelbase	90 inches
	Length	154
	Height	28.8 (at cowl)
		39.4 (at roll bar)
	Width	61.0
	Tread	48 (Front)
		49 (Rear)
	Curb Weight	1500
Engine		
	Position	Mid-ship
	Type	60° V-4
	Displacement	91.4 cu. inches
	Bore	3.54
	Stroke	2.32
	Horsepower	89 @ 6600 rpm (Road Version)
		109 @ 6400 rpm (Competition)
	Torque	89 @ 3600 rpm (Road Version)
		99 @ 5200 rpm (Competition)
	Compression Ratio	11.0 to 1
	Axle Ratio	3.30 to 1

– FIGURE 157 – THE MUSTANG—FORD'S EXPERIMENTAL SPORTS CAR

UNIQUE TWIN SIDE MOUNT RADIATORS

Figure 157 clearly shows the side mount radiator on the left side. Actually, there is one on each side just behind the seats. Figure 158 depicts, in planview, how the radiators are diagonally placed, the water pump and circulation, and separately, a heater adaptation. The electrically operated and thermostatically controlled fans are only on when the water temperature reaches 200 degrees F and off at 190 degrees F. The radiator scoops are located in pressure areas of the body side as determined in the wind tunnel. The mean velocity of the air through the scoops is about 68% of vehicle velocity at 60 mph.

ADJUSTABLE PRECISE SUSPENSIONS – FRONT AND REAR

In order to achieve a high standard of precise roadability the suspension members, front and rear, use metal-to-metal ball-joints as shown in figures 159 through 162. Note also that they are adjustable to alter roll center height and camber angles.

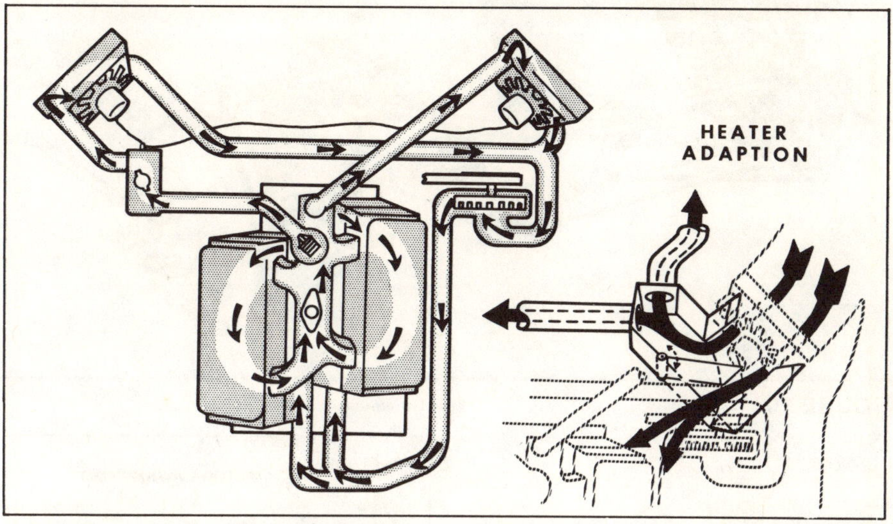

Cooling System

– FIGURE 158 –

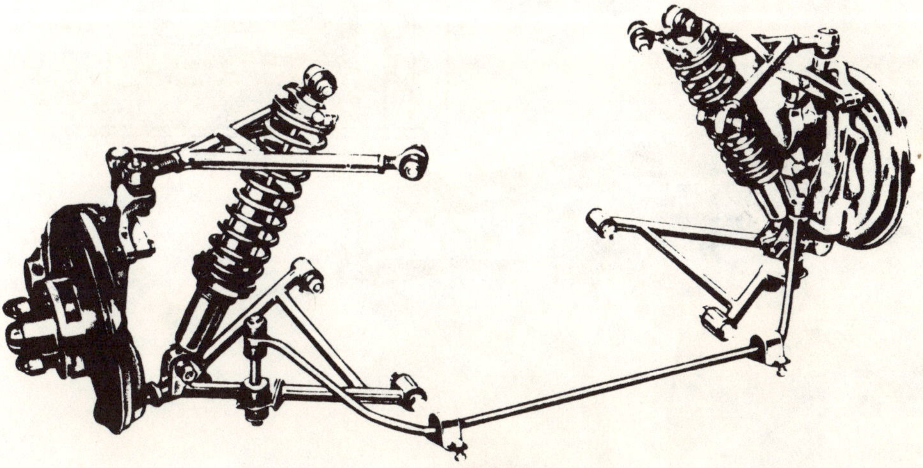

Front Suspension

– FIGURE 159 –

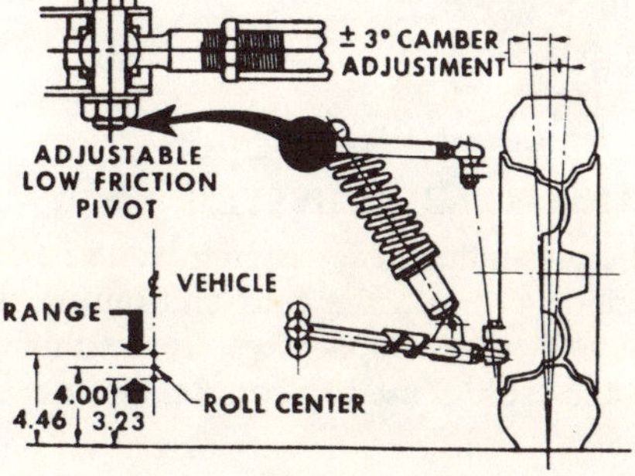

Front Suspension Adjustments

– FIGURE 160 –

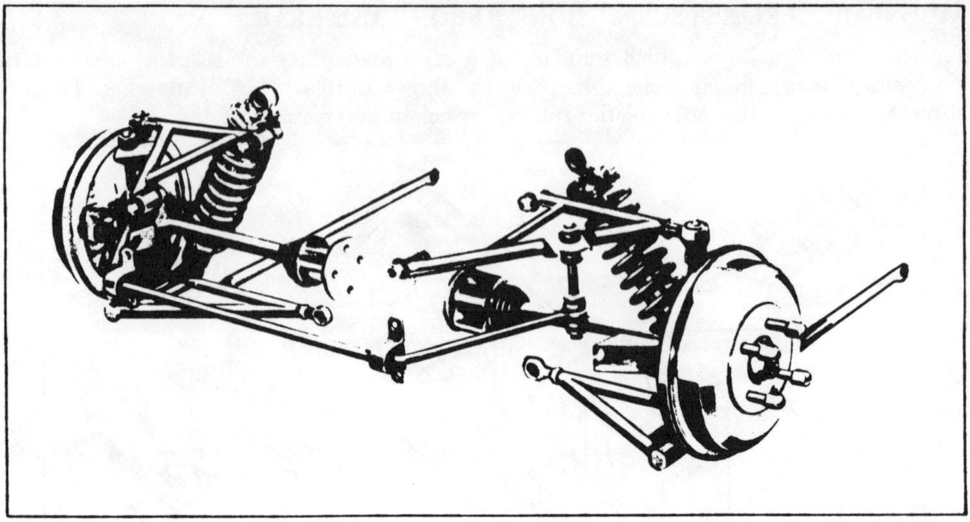

— FIGURE 161 —

Rear Suspension

Rear Suspension Adjustment

— FIGURE 162 —

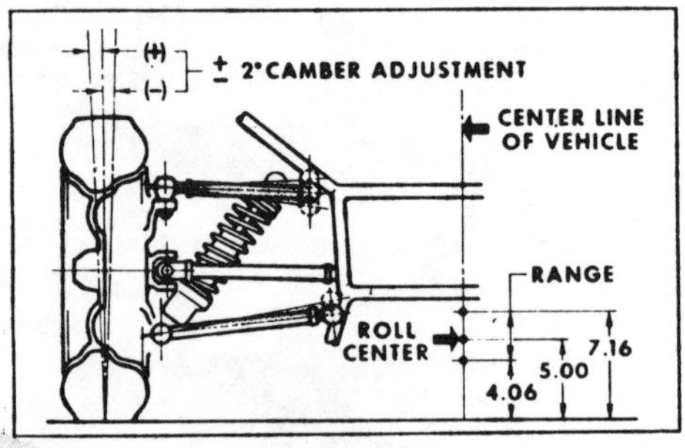

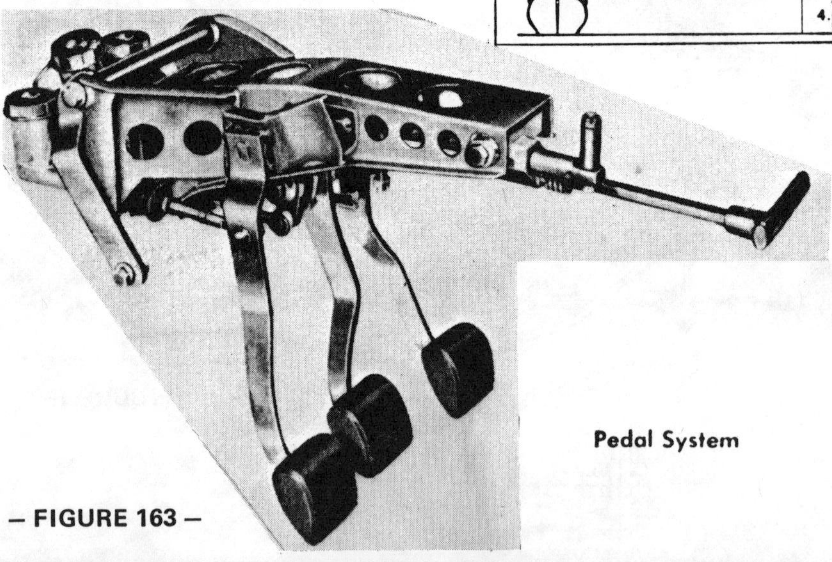

Pedal System

— FIGURE 163 —

ADJUSTABLE PEDALS AND STEERING SYSTEM

Figure 163 depicts the suspended pedal assembly and its dash mounted lever for fore and aft adjustment. Figure 164 shows the rack and pinion steering which provides 2.9 turns lock-to-lock. Each wheel turns 35 degrees from straight-ahead with no ackerman (Jeantaud) effect. A flexible cable is used to connect the steering column through a 23.5 degree bend to the pinion shaft.

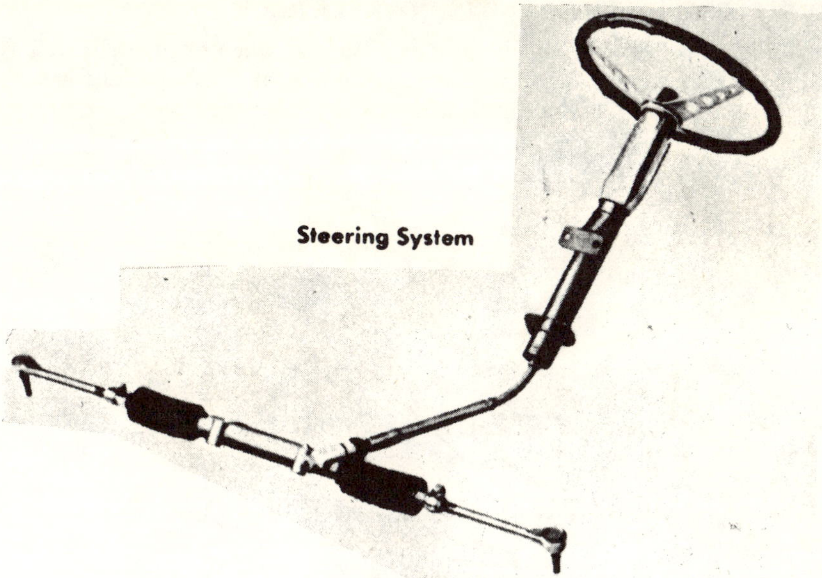

Steering System

— FIGURE 164 —

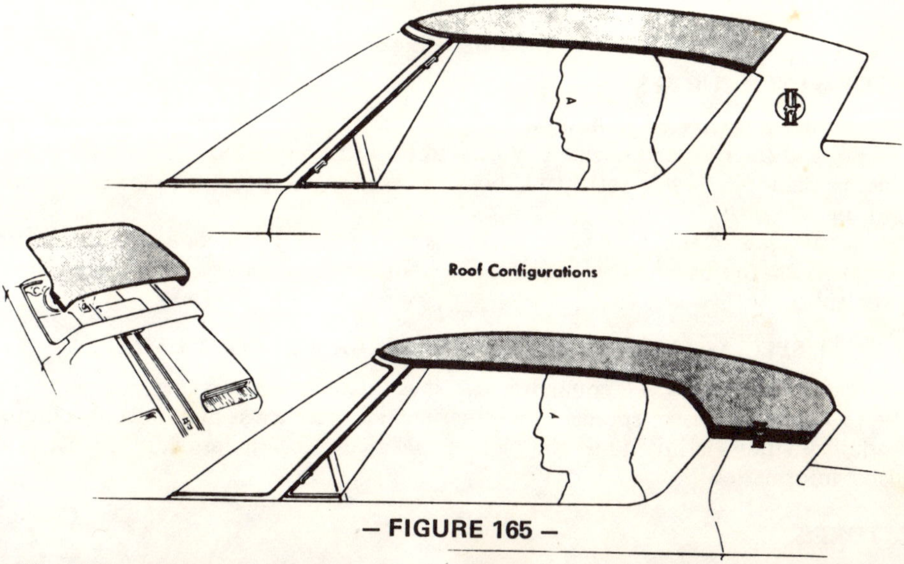

Roof Configurations

— FIGURE 165 —

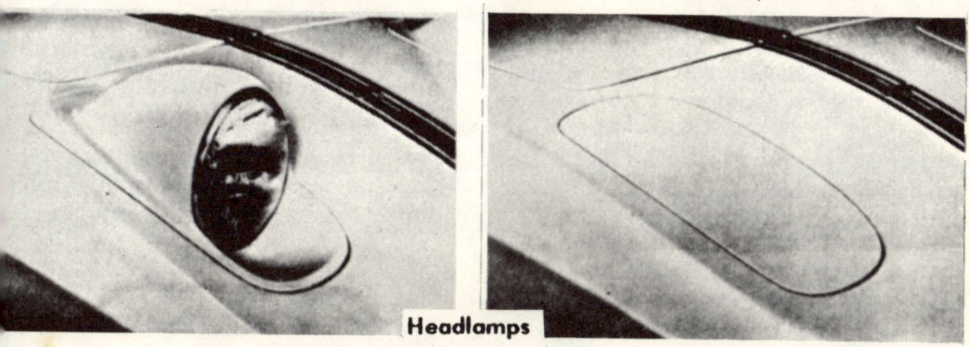

Headlamps

— FIGURE 166 —

RETRACTABLE HEADLIGHTS AND REMOVABLE ROOF

Figure 166 shows how nicely the headlights appear when exposed, as well as in the smooth retracted position. Two different removable roof configurations are shown in figure 165 along with a full windshield. Although designed these were not, to my knowledge, ever built.

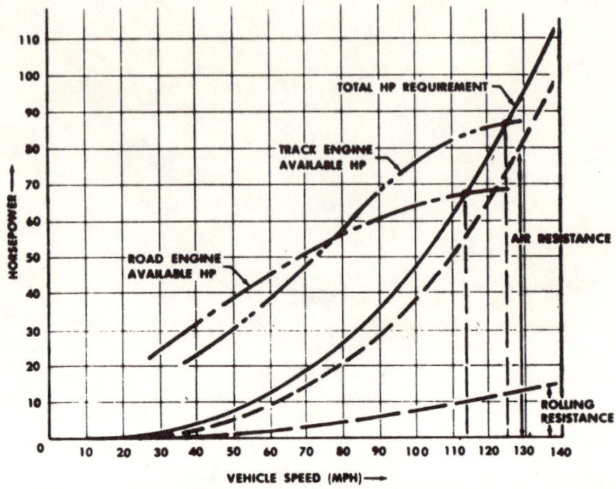

Horsepower Vs. Vehicle Speed

— FIGURE 167 —

PERFORMANCE CURVES

Everyone is interested in how well a sports car performs. Acceleration is not known but top speed for the road engine is shown to be 112.5 mph and 125 mph for the track engine, in figure 167. It is anticipated both values may improve due to lower drag with a closed top.

Since interest in this type vehicle remains high, possibly someone will develop a new small sports car in this country around the newer production power units that are becoming available.

(D) – G.M. SPECIAL PURPOSE THREE WHEEL CAR FOR COMMUTING

Referred to as the 511 commuter car, it is intended for carrying two people from home to work at freeway speeds. This experimental vehicle was built in connection with a feasibility study, is not quite complete as we shall see, but provides a great deal of valuable information.

FEATURES

It is noteworthy for its many unusual features such as three wheels, a unique backbone frame, a single canopy, handlebar steering, an unsprung frame and powertrain, semi-reclining seats, a low center of gravity, and a fanless cooling system.

It is without doubt one of the most beautiful three-wheelers ever built, as can be seen in figures 168, 169, and 170. The forward swinging canopy provides excellent access as shown in figure 171. The handlebar steering and adjustable pedals and other interior details appear in figure 172. Mr. Edward A. Rishavy led the project, appears in these pictures, and prepared reference 141 which provides a thoroughly detailed report on this vehicle.

Figure 173 shows in phantom view the interior arrangement of frame and powertrain as well as the suspension system designed to eliminate lean while cornering.

Figure 174 depicts how far forward the complete canopy swings, shows the semi-reclining seats, and the overall low center of gravity, which is only 13.5 inches above the ground for exceptional stability and maneuverability. This low center of gravity was achieved by tilting the engine at 45 degrees and putting the passengers practically on the floor in a semi-reclining position. Ground clearance is but 4.5 inches and height 40 inches.

– FIGURE 168 –

– FIGURE 169 –

– FIGURE 170 –

– FIGURE 171 –

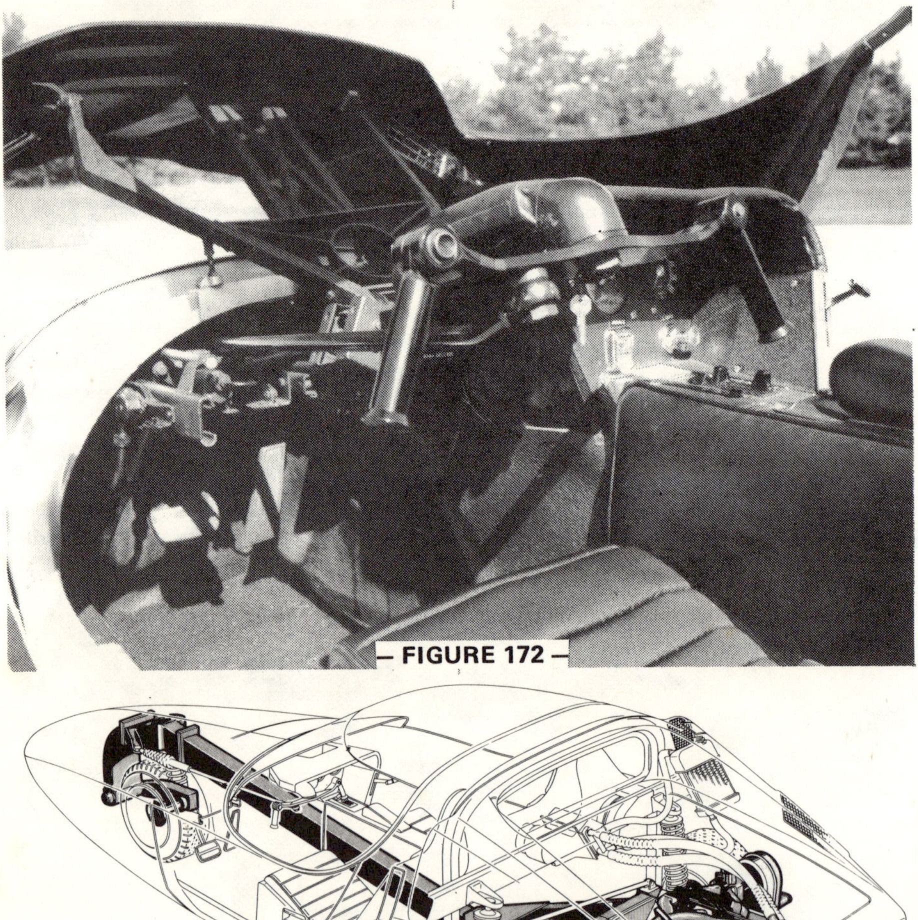

– FIGURE 172 –

– FIGURE 173 –

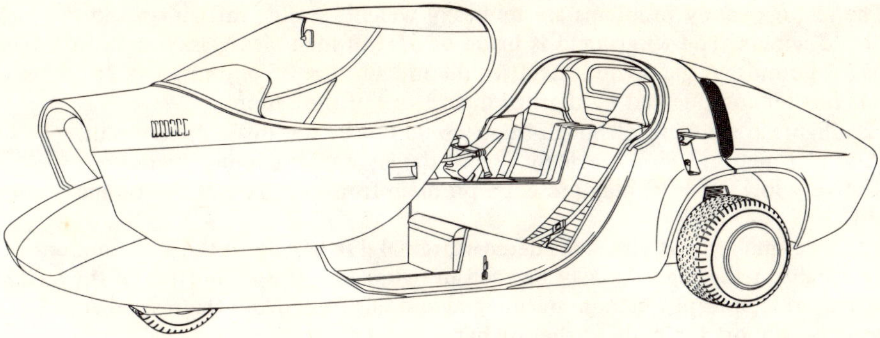

– FIGURE 174 –

GENERAL SPECIFICATIONS

Table XIX provides its performance and general specifications. Originally the target weight was planned at 1,000 pounds. As built, the curb weight came to 1,300 pounds, which in turn, reduced acceleration. The 0 to 60 mph figure of 16.5 seconds would have dropped to about 13 seconds had the target weight been achieved.

TABLE XIX

GENERAL SPECIFICATIONS

PERFORMANCE	0-60 MPH	16.5 sec
	Top Speed	80 mph
	City Economy	30 mpg
DIMENSIONS	Wheelbase	86 in.
	Rear Track	54 in.
	Length	149 in.
	Width	63 in.
	Height	40 in.
	Grd. Clear.	4.5 in.
	CG Height	13.5 in.
	Frontal Area	14.7 ft^2
WEIGHT	Curb-Full Tank ...	1300 lb
	% Front-Rear	21-79%
ENGINE	4 Cyl., Water Cooled, 45° Slant	
	Displ.	66 cu. in.
	C.R.	9.5:1
	Max. Gross hp	67 hp @ 6000 rpm
TRANSMISSION	3 Speed Automatic	
	9 In. Dia. Converter	
REAR AXLE	Axle Ratio	3.55
	N/V Ratio	64
TIRES	Rear 5.00 x 10 ..	22 psi
	Front 6.50 x 8 ..	22 psi
BRAKES	7 In. Dia. Drums in All Wheels	
RIDE RATES	Front	80 lb./in.
	Rear	40 lb./in.
STEER RATIO	Handlebar	2.1:1
	Steering Wheel	9:1
CAPACITIES	Fuel Tank	7.5 gal.
	Water	6 qt.
	Engine Oil	3 qt.

The major canopy problems are its heavy weight, sealing difficulty, and the lack of windshield wipers. The windshield is made of 3/16th inch plexiglass formed as shown. It weighs 95 pounds. A glass windshield would add another 25 pounds. The 25-foot sealing line was further complicated by thermal distortions of the canopy.

Headlights are lacking, although these would not be too difficult to install.

Poor rear quarter vision is evident due to the large inlet cooling ducts.

The very low eye-level may present a problem from the glare of on-coming headlights at night.

The ride quality was also considered marginal due largely to the great amount of unsprung weight at the rear. As may be seen in figure 173, the aft portion of the backbone frame, rear axle, and powertrain are integral and unsprung from the rear wheels. The coil springs only support the body at the roll bar.

(E) – G.M. MODEL 512 SPECIAL PURPOSE URBAN CARS

These three small two-passenger urban cars were designed and built in 1969 and indicate how long ago General Motors was thinking about the specialized urban car market. Since one is an electric and another a hybrid, they are quite timely for today's thinking also. In fact, many urban vehicle specialists consider these vehicles benchmarks to measure later designs by. For a complete analysis of these designs, the reader is referred to reference 62.

CONCEPT AND BASIC DESIGN

One basic body was designed to incorporate either the electric, gasoline, or hybrid powerplant. The short overall length of but 86.3 inches permits the car to be driven straight in, 90 degrees to the curb, with the front wheels against the curb. The short car *length* is no greater than a standard size passenger car's *width* from the curb. Three of these urban cars can be parked in the space of a conventional car. Entry and exit is from a single front door and raised canopy, as may be seen in figure 175. These cars are somewhat reminiscent of the old BMW ISETTA of 1956, which also featured front entry and exit from a single door.

Width is 56 inches, height 51.9 inches, front tread of 44.7 inches, rear tread 48.6 inches, and the frontal area is 15.5 square feet.

Special 6.55 x 9 tires are used and the vehicle can turn in an 18-foot circle.

A steel underbody and roll bar form the basic body structure with external body panels of fiberglass.

– FIGURE 175 –

THE ELECTRIC VEHICLE

The electric is shown in three-quarter front view in figure 176. The general arrangement of components is shown in figure 177.

The D.C. motor produces 8½ horsepower at 4000 rpm and weighs 54 pounds. Maximum short time output is 13 HP at 2100 rpm.

– FIGURE 176 –

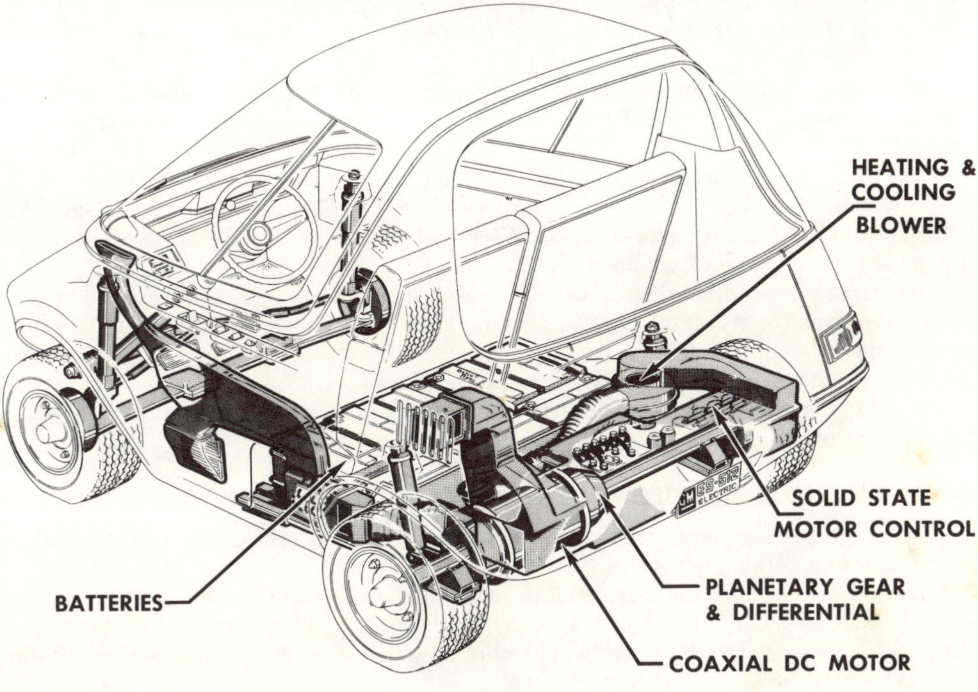

– FIGURE 177 –

Two different control systems were tested. The solid-state SCR design was combined in one unit with the smoothing inductor and dual charger for a combined weight of 63 pounds and located across the rear of the vehicle.

An alternate step control system was also evaluated in this vehicle. It was more efficient, less complex, and cost less. It is less smooth because of the discrete voltage steps. Seven steps are used.

The 84 volt battery pack, used with the SCR controller, had 7 batteries rated at 85 ampere hours (20 hr. rate) and each weighed 47 pounds. An auxiliary 12-volt battery was used for accessories.

It is understood this interesting little car is still used for testing the newer, more powerful batteries now available.

GASOLINE POWERED VEHICLE

In figure 175 the gasoline powered car is in the foreground with its canopy removed to convert it to a roadster.

A two-cylinder four-cycle water-cooled horizontal-opposed aluminum engine of 19.6 cu. in. displacement is used. Maximum horsepower of 13.8 at 4400 rpm dropped to an as-installed power of 11.8 HP at 4400 rpm with 15.8 lb/ft torque at 3200 rpm. The difference was due to losses from the intake silencer, exhaust system, alternator, cooling blower, and air injection pump for emissions control.

The variable ratio belt-type transmission was chosen since it was considered to be simple, inexpensive, and compatible with a low power output engine. It was modified to include a vacuum assist during full throttle operation and during braking. It appears this type of transmission may have a place in small urban cars of tomorrow.

THE HYBRID GASOLINE-ELECTRIC

This vehicle is also quite easy to enter or leave as may be evident in figure 178 and 179. The side view, figure 180 again shows the remarkably short length with a wheel at each corner. Figure 181 shows how all the components fit in the rear. It is also the most complicated with two powerplants.

The engine is two-cylinder, horizontally opposed, of 11.7 cu. in. displacement and 5 HP at 5000 rpm. It is considered a bit marginal, but used because it was commercially available.

The electric motor is essentially the same as used on the electric with a weaker field (series wound) to match the 72-volt battery pack. Six 12-volt automotive-type 32-ampere hour (20 hr. rate) form the power supply. Each battery weighs 22 pounds. These also supply the accessories through a converter.

Recharging is accomplished in one of two ways. By using a 90-volt, gasoline engine driven alternator or by drawing power from a 115-volt AC 15-ampere outlet. The first would generally apply during a cruising period of the vehicle, although the engine can operate with the car stopped to recharge from the alternator.

The car may be driven as an electric alone, or as a gasoline powered unit, or together for acceleration.

SUMMARY AND COMPARISON

In the proper environment these small-size cars would be a real asset and would minimize traffic and parking problems.

The electric is emission-free, smooth and quiet, and of simple design. However, its range is limited and it most likely would have heating problems (lack of heat) in cold climates. Today's better batteries would improve range and a separate gasoline heater (now available) would provide heat.

The gasoline engine provides unlimited range through rapid refueling but it requires

– FIGURE 178 –

– FIGURE 179 –

emissions controls, a transmission, is noisier than an electric and produces some vibration.

The Hybrid design combines some of the advantages of the gasoline engine and the electric drive such as unlimited range, passenger heat, and reduced emissions by optimizing the gasoline engine operation. It is more costly, however.

— FIGURE 180 —

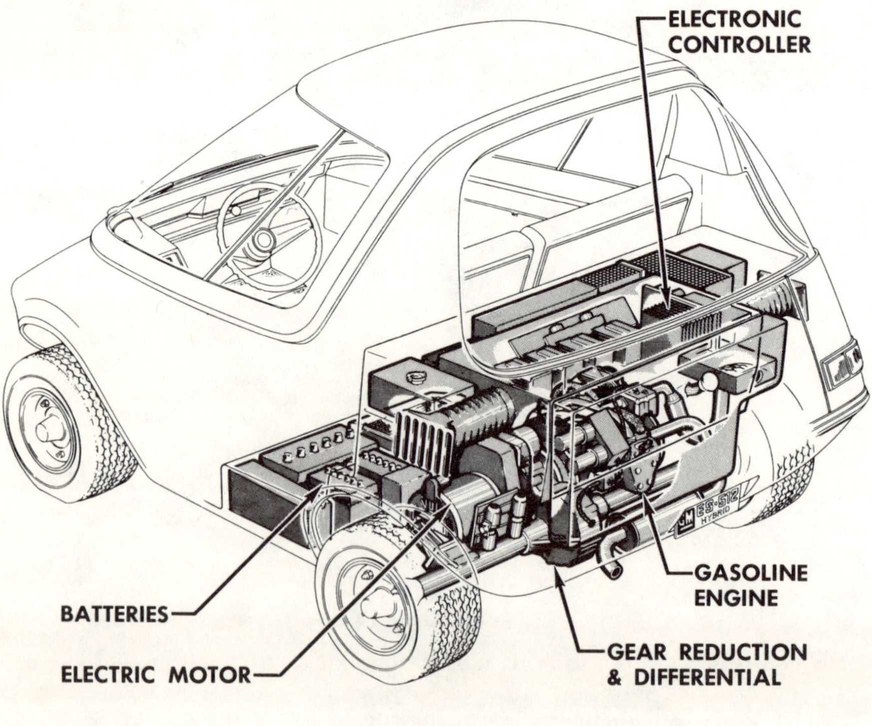

— FIGURE 181 —

Table XX provides a comparison of the performance characteristics of the three 512 cars. The operating cost was, of course, based on 1969 prices for fuel and battery replacement prices.

TABLE XX

	GASOLINE	ELECTRIC	HYBRID
CAR WEIGHT-LBS.	950	1250	1250
POWERTRAIN WEIGHT-LBS.	325	575	585
TOP SPEED-MPH	45	45	40
PERFORMAMCE (0-30 MPH)-SEC.	19	12	16
RANGE-MILES			
30 MPH-CONST. SPEED	290[1]	47	160[1]
CITY SCHEDULE	188	31	15[2]
OPERATING COST[3]-CENTS/MILE			
30 MPH-CONST. SPEED	.31	2.2[4]	.52
CITY SCHEDULE	.49	3.5[4]	—

1 DEPENDENT ON SIZE OF FUEL TANK
2 DEPENDENT ON SYSTEM RECHARGE CAPABILITY
3 COST EXCLUDES ALL TAXES
4 INCLUDES BATTERY REPLACEMENT AT O.E.M. PRICES

(F) – THE G.E. CENTENNIAL ELECTRIC – WITH TODAY'S COMPONENTS

In September 1978 the General Electric Research and Development Center introduced their "CENTENNIAL ELECTRIC". This experimental car is noteworthy in that it represents the state-of-the-art in electrics built with off-the-shelf components and battery systems now commercially available (fall of 1978). Drive motors, solid-state controls, plastic side and rear windows and other plastic parts, headlamps, and numerous other components were supplied by more than a dozen G.E. product departments.

Batteries were supplied by Globe-Union Inc.

– FIGURE 182 –

– FIGURE 183 –

The front-wheel drive test car was designed "from the ground up" and built to G.E. specifications by Triad Services Inc. of Dearborn, Michigan.

Figure 182 shows the four-passenger car in three-quarter front view. Curb weight, including battery, is 3,250 pounds and gross weight comes to 3850 pounds. Wheelbase is 92 inches, width 66.1 inches, height 53.6 inches, and length is 160 inches.

Frontal area is 19.9 square feet and the drag coefficient at 0 degree yaw is 0.337. This, when weighted for yaw (cross winds), goes up to 0.367.

Figure 183 reveals the "engine compartment". The 24 hp DC series traction motor, solid-state controller, power conditioner, and blower for cooling the motor and electronic components are shown. In front of the motor is the auxiliary 12-volt battery for headlights, fans, and accessories.

Figure 184 shows how the 18 six-volt lead-acid batteries are installed along the cedterline of the car. Six to eight hours from 220 volts is required to recharge them.

Performance is as follows:

Range – Urban stop and go – 45 miles
40 mph constant speed – 75 miles
55 mph constant speed – 50 miles

Acceleration – 0 to 30 mph – 9 seconds
0 to 40 mph – 14 seconds
0 to 50 mph – 30 seconds

Passing speed – 60 mph.

Maximum grade – 30%

– FIGURE 184 –

(G) – AN ELECTRIC CAR FOR TOMORROW – THE ETV-1

This is the first car to be built under the U.S. Department of Energy's Near Term Electric Vehicle Program. ETV-1 stands for ELECTRIC TEST VEHICLE-1. General Electric was the prime contractor and Chrysler Corp. was the major sub-contractor.

This vehicle seems appropriate for finalizing this book on designing tomorrow's cars. It is more aerodynamic, uses the latest chassis components, and is smaller and lighter than today's standard size cars. It is an advanced urban electric with improved performance that is amenable to mass-production by the mid-1980's. In quantities of 100,000 units the consumer price goal is $6,400 in 1979 dollars. The estimated life-cycle cost of the vehicle for its projected 10 years life is less than 18 cents per mile. This includes 2 cents per mile for scheduled maintenance.

Figures 185 and 186 provide three-quarter front and rear views of the sporty two-door hatchback for four people.

Figure 187 shows the key chassis features. These include a separately excited direct current motor which drives the front wheels, and the 18 high density lead-acid batteries located in a tunnel down the center of the vehicle. The Power Conditioning Unit (PCU) weighs 87 lbs. and provides the electronic controls for the armature and separate field control. The power transistor module in the chopper circuit weighs less than 1 lb. and would reportedly cost less than $50 if produced in quantity. The microprocessor controls propulsion sequencing, "fuel" gauge computation and display, and programmed battery charging. A regenerative braking system recovers braking energy to recharge the propulsion batteries.

— FIGURE 185 —

— FIGURE 186 —

— FIGURE 187 —

U.S. DEPT. OF ENERGY
Near-Term Electric Vehicle Program
GE/Chrysler Vehicle

GLOBE-UNION INC.

EV2-13 LEAD-ACID
ELECTRIC VEHICLE BATTERY

- 6 Volt
- 27.2 kg (60 lb)
- Unconventional, Computer-Designed Cell Geometry
- Left-Hand and Right-Hand Models

KEY

1. Thin, Lightweight, Durable Polypropylene Container and Cover Thermally Welded for a Leak-Free Assembly
2. Single-Point Watering System with Safety Venting
3. Low-Resistance, Through-the-Partition Intercell Welds
4. High-Efficiency, Computer-Designed Radial Grids
5. Optimized Active Materials
6. Submicro Polyethylene Envelope Separators with Glass Mat

– FIGURE 188 –

The transmission is a double reduction chain drive transaxle.

Figure 188 depicts the improved lead-acid battery. Note that the plates are oriented parallel to the long dimension of the case, rather tha perpendicular. This battery has the same length and width as a golf-cart battery. Energy density of 17 Wh/Lb has been achieved and a cycle life of 500 discharge cycles is anticipated (70% discharge). Each battery weighs 60 lbs. The battery system provides higher energy density than commercially available lead-acid batteries – an increase of about 30%.

GENERAL SPECIFICATIONS

Curb weight – 3320 lbs. – Gross weight 3920 lbs.
Wheelbase 98 inches - Tread 55.5 inches front and rear.
Width 65.7 inches – Height 51.6 inches – Length 169.4 inches.
Ground clearance – 6 inches – Tire size P175/75 R 13.
Frontal Area – 19 sq. ft. – Cd (weighted for yaw) 0.31
Motor – Separately excited DC 20 hp continuous – 41 peak.
Voltage – 98 volts – Current rating 175 amps.
Speed range 0 to 5000 rpm – weight 214 lbs. Efficiency 89%.
Size – 12 inch diameter by 20 inches long.

Power Conditioning Unit (PCU) – Transistorized chopper
Armature control electronics – 0 to 2500 rpm
Current rating – 200 amps. continuous – 400 amps transient
Frequency – 100 Hz to 2 Hz variable.

Field control/Battery charger electronics – 2500 to 5000 rpm
On-board charging capacity – 110 volt/15 to 30 amps.
Off-board charging capacity – 220 volts/60 amps.
Recharge time – 10 hours at 30 amps.

PREDICTED PERFORMANCE

Range – 35 mph constant – 122 miles
45 mph constant – 102 miles
SAE J227 driving cycle "D" – stop/start – 45 mph max. – 75 miles
Acceleration – 0 to 30 mph – 8.9 seconds – 25 to 55 mph – 17.6 seconds.
Speed – Cruising 55 mph - Passing 60 mph.
Grade – Maximum – 17%

Reference 142 was the source of information about ETV-1 as well as the CENTENNIAL ELECTRIC.

A FINAL FEW WORDS TO THE DESIGNERS OF TOMORROW'S CARS

This chapter on experimental vehicles provides some insight into the thinking of others in their quest for solutions to the transportation needs of tomorrow. Clearly, cars must continue to become smaller to meet CAFE requirements. Rapid increases in fuel costs make the buying public even more sensitive to the need for greater efficiency as well as smaller vehicles. Just smaller and lighter isn't enough. Unique concepts of revolutionary as well as evolutionary designs are now thinkable and will be welcomed by the avant garde.

True urban cars must appear. Three-wheelers, properly designed, have a place and offer advantages. Much smaller engines will provide stiff competition for Electrics and Hybrids, both in performance and initial and operating costs. Better batteries and electronics (both at reduced costs) will make the electric car practical for mass-markets.

Aerodynamics can no longer be treated as a sales gimmick – the big steps must be taken now for all cars, trucks, vans, motorcycles, and large tractor-trailer rigs.

Bold designers of vision, working from a solid technical base, have unlimited opportunities for creating the cars of tomorrow.

– REFERENCES –

1 "The development of the Porsche", R. von Frankenberg, AUTOCOURSE Magazine, November 1953 issue, London, England.

2 SEMA (Specialty Equipment Manufacturers Association) 11001 E. Valley Mall, Suite 200, El Monte, CA 91734.

3 "Energy in America's Future", Resources for the Future, reported by Associated Press, in the Los Angeles Times, August 26, 1979 under title "Ideal Energy Plan Seen Possible"

4 EV FOCUS Magazine, May 1, 1978

5 "Polaroid – a leap from concept to product" by Mark Stevens 5-9-78. Reprinted by permission of Dr. Land and The Christian Science Monitor, copyright 1978 The Christian Science Publishing Society, all rights reserved.

6 CHROME DREAMS – Automotive Styling since 1893, Paul C. Wilson, Chilton Book Co., Radnor, Penna.

7 AERODYNAMIC DRAG, Hoerner, Published by S.F. Hoerner Dr.–Ing., 148 Busteed Dr., Midland Park, New Jersey

8 THE AUTOMOTIVE CHASSIS, P. M. Heldt, Nyack, N.Y. (Chapter II)

9 "Electronic Wizardry", Bob Irwin, March 31, 1979, AUTOWEEK Magazine.

10 SAE HS-J782b – Seating Manual – Motor Vehicles

11 Recommended Practices, SAE HANDBOOK, Part II, section 34, Passenger Car Components & Systems (a) SAE J826b, Devices for use in defining and measuring vehicle seating accommodation; (b) SAE J963 Anthropomorphic Test Device for use in dynamic testing of motor vehicles.

12 SAE #750356 – "Describing the driver's work space: Eye, Head, Knee, and Seat Positions"

13 "Are we choking on the politics of smog", Wm. S. Everett, Ventura County Hearing Board, Los Angeles Times, October 4, 1979

14 "Air Pollution and Human Health", Lester H. Lave and E. P. Seskin, John Hopkins University Press, reported by J. M. Callahan in Automotive Industries, April 1978 under "Car Smog Role is Debunked"

15 "What about Natural Air Pollution?", Automotive Industries, November 1, 1975

16 "Smog Pattern Shifting but Quality of Air Improves", Sandra Blakeslee, Los Angeles Times, June 25, 1978. Copyright 1978, Los Angeles Times, reprinted by permission.

17 "It's Time to Slow Down", Automotive Industries, January 1, 1976

18 "No light at the end of tunnel in smog war", Los Angeles Herald Examiner, September 16, 1979

19 "What the Pollution Audit Found", J. M. Callahan, Automotive Industries, January 1979

20 "A Brave Experiment in Pollution Control", Fortune, February 12, 1979

21 National Safety Council, Chicago, Illinois 60611

22 ACCIDENT FACTS, 1976 edition, National Safety Council, Chicago, Illinois

23 Near Term Hybrid Vehicle Program Pre-solicitation Industry Briefing, Jet Propulsion Laboratory (JPL), Pasadena, CA, November 18, 1977

24 Petroleum Industry Research Foundation Study

25 "Supplying Gasoline for Tomorrow's Cars", SAE Automotive Engineering, September 1977

26 "After 10 years U.S. - - - - - ", Edward S. Lechtzin, Los Angeles Times, December 1, 1977. Reprinted by permission of United Press International.

27 "Carter's Automotive Overkill", Automotive Industries, April, 1978.

28 "NHTSA Reveals Five Year Plan", Automotive Industries, April, 1978

29 "Performance Standards for Electric and Hybrid Vehicles" (DoE), May 30, 1978, NEWS by Booz, Allen, and Hamilton, Bethseda, Maryland 20014

30 Standards Development and Supplies Branch, Environmental Protection Agency, 2565 Plymouth Road, Ann Arbor, Michigan 48105

31 Air Resources Board, 9528 Telstar Ave., El Monte, CA 91731

32 Noise Control, 514, Department of Health, State of California, 2151 Berkeley Way, Berkeley, CA 94704

33 "Want to Build a Kit Car? – Here are a few legal pitfalls you should know about", Road and Track Magazine, June 1979

34 Automotive Industries, Chilton Co., Chilton Way, Radnor, PA 19089

35 Two-stroke Tuner's Handbook, Gordon Jennings, HP Book #41-ISBN 0-912656-41-7, HP Books, Box 5367, Tucson, Arizona 85703

36 The Two-stroke Engine – Its Design and Tuning, K.G. Draper, G.T. Foulis & Co., Ltd. 1 - 5 Portpool Lane, London, England E.C. 1

37 Small Wankel Engines, H. Keller, Fichtel and Sacks AG, SAE paper #680572, September 1968

38 Suzuki Production Rotary Engine, model RE-5 for powering motorcycles, Kamiza and Shirasagi, SAE paper #770190, February 1977

39 An Update of Direct Injected Stratified Charge Rotary Combustion Engine Developments at Curtiss-Wright, Jones, Lamping, Myers and Lloyd, SAE paper #770044, February 1977

40 Scientific Design of Exhaust and Intake System, P.H. Smith, Robt. Bentley Inc. 993 Massachusetts Ave., Cambridge 38, Mass.

41 Development of Toyota Lean-burn Engine, Noguchi, Sanda, Nakamura, SAE paper #760757, 1976

42 Powertrain Control: A convergence of LSI Technologies, R.N. Noyce, SAE paper #780831

43 Single Chip Microcomputers in Automotive Systems, Knowles, Hill, Silvestri, and Rubinstein – SAE paper #780120

44 A Low Vibration Engine with Unique Counter-balance Shafts, Nakamura, SAE paper #760111, 1976

45 Spicer Axles – Spicer Clutch Div., Dana Corp. P.O. Box 191, Auburn, IN 46706

46 The Technique of Front Drive in Europe, Sezne and Lagrange SAE paper #750013 1975

47 Development of a Front Drive Car, Subaru SAE #770803, 1977

48 Efficient Vehicle Packaging with Front Wheel Drive, Renault, SAE paper #78031, 1978

49 AE-7 Universal Joint and Driveshaft Design Manual SAE

50 Synchronous Belts for High Power Drives, Machine Design, November 9, 1978

51 Salsbury Corporation, 1010 E. 62nd Street, Los Angeles, CA 90001

52 Electromatic Drive Corporation, 5521 Old Bendbrook Rd., Fort Worth, TX 76126

53 Machine Design, Mechanical Drives, Annual Reference issue, Penton/IPC, Penton Plaza, Cleveland, Ohio 44114

54 "State of the Art Assessment of Electric and Hybrid Vehicles", January 1978, NASA. Available from National Technical Information Service (NTIS), U.S. Department of Commerce, 5285 Port Royal Road, Springfield, Virginia 22161 – Printed copy $13.75

55 Basic Electricity by Van Valkenburgh, Nooger and Neville, Inc., John F. Rider, Publisher Inc. New York

56 Audel's Electric Motor Guide, E. P. Anderson, Theo. Audel & Co., Div. of Howard W. Sams & Co., Inc., 4300 W. 62nd St., Indianapolis, IN 46206

57 Electric Vehicle News, Porter Corporation, P.O. Box 533, Westport, CT 06880

58 Electric Auto Association, 1674 Merill Drive, No. 12, San Jose, CA 95124

59 "Electronic Commutation Promises New Breed of DC Motors", E. Bahm, page 97, Machine Design, February 10, 1977

60 "Low Loss DC Synchronous Multi-Pole Traction Motor", page 45, SAE Automotive Engineering, Volume 85, Number 1, January 1977

61 Automotive Electronics Books, SAE, 400 Commonwealth Dr., Warrendale, PA 15096
SP-417 Recent Advances in Automotive Electronics
SP-418 The Application of Automotive Sensors
P-68 The 1976 International Conference on Automotive Electronics and Electric Vehicles
SP-404 Sensors for Electronic Systems – I

SP-388 Automotive Electronics I
SP-393 Automotive Electronics II
P-57 The International Colloquim on Automotive Electronic Technology – Convergence 74

62 Special Purpose Urban Cars, Gumbleton, Frank, Genslak, and Lucas, General Motors Corp., SAE paper #690461, May, 1969

63 Sundancer: A test bed electric vehicle, McKee, Borisoff, Lawn, and Norberg, SAE paper #720188, January 1972

64 A Quiet, Efficient Controller For Electric Vehicles, Armstrong and Pocobello; SAE paper #750470

65 The Efficiency Improvement of Electric Vehicles by Regenerative Braking, Kasama, Naito, Katada, and Shibata of Hitachi. SAE paper #780291, February 1978

66 Electric Flywheel Vehicle for Postal Service Applications, T. A. Norman, SAE paper #780297, February 1978

67 "Energy Sources", F. J. Port, Electric Vehicle News, November 1976

68 Tire Testing for Rolling Resistance and Fuel Economy, Glemming and Bowers – SAE paper #750457, February 1975

69 Motor Vehicle Engineering – The Chassis By Ethelbert Favary, McGraw-Hill Book Co., Inc. (1922)

70 The Automotive Chassis – (Without powerplant) by P. M. Heldt, P. M. Heldt, Nyack, N.Y. (1948)

71 Manual on Design and Application of Leaf Springs, SAE J-788a (HS-J788a)

72 Leaf Springs for Motor Vehicle Suspension – SAE – J-510C

73 Metric Bar Sizes for Leaf Springs – SAE – J1123a

74 Torsion Bars, F. E. Burdette, Northrup Corp. Machine Design, page 124, October 1, 1959

75 Helical Compression and Extension Spring Terminology – SAE J-1121

76 Helical Springs – Specification Check Lists – SAE J-1122

77 Rated Spring Capacity – SAE J-274

78 Pneumatic Spring Terminology – SAE J-511a

79 Ride and Vibration Data – A set of reference charts – SP-6 Special publications department, SAE (may no longer be available)

80 HS-J670d Vehicle Dynamics Terminology – SAE Recommended practice SAE

81 HS-J795 Manual on Design and Application of Helical and Special Springs SAE

82 The New Packard Torsion Level Suspension, F. R. McFarland report #537, June 1955, SAE (may no longer be available)

83 Packard Torsion-Level Suspension – Automotive Engineering, SAE, July 1975, Volume 83, Number 7

84 The 1963 Chevrolet Corvette, Arkus-Duntov, Hansen, and Jakust, GMC, SAE report 611b, January 1963

85 Laboratory Testing Machine and Procedures for Measuring the Steady State Force and Moment Properties of Passenger Car Tires, SAE J-1106, J-1107 (HS 210)

86 Passenger Car Tire Performance Requirements SAE J-918c

87 Tire Peak and Locked Wheel Traction SAE J-345a

88 "Around and Around We Go to Find Out About Cornering Power", Ron Wakefield, Road and Track Magazine, April 1969

89 "Pipeline" – The Great Sidecar Controversy – Cycle Magazine, April 1979

90 Morgan – Car and Driver Magazine, July 1965

91 The World of Automobiles, Vol. 12, Columbia House, 51 West 52nd St., New York, N.Y. 10019

92 Automobile Quarterly, Spring 1963, Automobile Quarterly, 245 West Main Street, Kutztown, Penna.

93 Airplane Design – Aerodynamics, E. P. Warner, McGraw-Hill Book Co., Inc. New York, N.Y.

94 The Body Engineer's Role in Automotive Aerodynamics, W. Korff, SAE paper #649b, 1963

95 Aerodynamic Drag – Hoerner, DR. S.F. Hoerner, 148 Busteed Dr., Midland Park, N.J.

96 Road Vehicle Aerodynamics, Scibor-Rylski, Halsted Press, Div. of John Wiley and Sons, New York, N.Y.

97 Some Considerations of Automotive Lift and Drag, J. J. Cornish III, SAE paper #948b, January 1965

98 The Body Shape of Minimum Drag, Morelli, Fioravanti, and Cogotti, SAE paper #760186, February 1976

99 Aerodynamics for Body Engineers, Kelly and Hocombe, SAE paper #649a, January 1963

100 Automotive Aerodynamics – A Progress Report, SP-180 SAE 1960.

101 Auto Aerodynamics, C. J. Heltemes, Jr. – CAR LIFE – September 1961

102 Fairmont/Zephyr – Engineered for Lightweight and Improved Fuel Economy, Gutherie, SAE paper #780134, February 1978

103 The Optimization of Body Details – A Method For Reducing The Aerodynamic Drag of Road Vehicles, Hucho, Janssen, Emmelman, VW, SAE paper #760185, February 1976

104 An Effective Aerodynamic Program in the Design of a New Car, L. Nedley, SAE paper #790724 (1979)

105 The Porsche 924 Body – Main Development Objectives, Burst and Srock – SAE paper #77031, February 1977

106 The Copper Electric Town Car, Pocobello and Armstrong, SAE paper #760071, February 1976

107 The Aerodynamics of Sports and Competition Automobiles, Vol. 7, Symposium, American Institute of Aeronautics and Astronautics (AIAA) 1968, Published by Western Periodicals Co., 1300 Raymer St., North Hollywood, CA 91605

108 Second Symposium – Aerodynamics and Racing Cars (AIAA) same publisher as ref. 107 above

109 Aerodynamic Characteristics of Sixteen Electric, Hybrid, and Sub-compact Vehicles, D. W. Kurtz, JPL publication 79-59, June 30, 1979. Prepared for U.S. Department of Energy through agreement with National Aeronautics and Space Administration by Jet Propulsion Laboratory, California Institute of Technology, Pasadena, CA

110 Truck Aerodynamics, Flynn and Kyropoulos, SAE paper #284a, January 1961

111 Development of Devices to reduce the Aerodynamic Resistance of Trucks, Lissaman, SAE paper #750702, August 1975

112 Aerodynamic Drag Reduction Tests on a Full Scale Tractor-Trailer Combination and a Representative Box-Shaped Ground Vehicle, Steers, Montoya, Saltzman, SAE paper #750703, August 1975

113 Comparisons of Effectiveness of Commercially Available Devices For The Reduction of Aerodynamic Drag on Tractor-Trailers, Buckley and Sekscienski, SAE paper #750704, August 1975

114 Wind Tunnel Development of Dragfoiler–A System for Reducing Tractor-Trailer Aerodynamic Drag, W. T. Mason, Jr., SAE paper #750705, August 1975

115 Aerodynamic Devices Can Significantly Reduce The Fuel Consumption of Trucks: Experience with CECA Designs, Servais and Bauer, SAE paper #750707, August, 1975

116 U.S. Dept. of Energy Press Conference, Washington, D.C., June 22, 1979

117 Publications Catalog (Annually) , SAE, 400 Commonwealth Drive, Warrendale, PA 15096

118 The 1973 Ford Impact Absorbing Bumper System, SAE paper #730032

119 Urethane Energy Absorbers For Automobile Bumpers, SAE paper #730025

120 "Bumpology", ALCOA Aluminum Company

121 Technical Features of the Viking 4 Research Vehicle, SAE paper #770689

122 "H-Bomb From McMinnville–Mercury Special" – Sports Car Graphic, Petersen Publishing Co., 8490 Sunset Blvd., Los Angeles, CA 90069

123 Racing and Sports Car Chassis Design, Costin and Phipps, Robt. Bentley, Inc., 993 Massachusetts Ave., Cambridge 38, Mass.

124 An Introduction To Structural Analysis For Management, SP-387, SAE

125 Vehicle Analytical Design For Structural Criteria – An Overview, SP-392, SAE

126 Structural Analysis of The Vehicle Design Process, P-71, SAE

127 Vehicle Structural Mechanics, P-83, SAE

128 SAE Handbook, SAE, 400 Commonwealth Drive, Warrendale, PA 15096

129 Machine Design – Materials issue (March annually), Penton/IPC Inc., Penton Plaza, Cleveland, OH 44114

130 Value Analysis From The Standpoint of Product Planning, C. F. Horner, SAE paper #614c, January 1963

131 "Control of Manufacturing Cost During Design Development", Korff and Papen, an SAE paper presented at the National Aeronautics Meeting of the SAE in Los Angeles, CA in 1957

132 Production Engineering in the Aircraft Industry, A. B. Berghell, McGraw-Hill Book Co., Inc. (1944)

133 Production Efficiency Curves, Hughes and Golem, San Diego, CA. Arts and Crafts Press (1944)

134 Basic Accounting and Cost Accounting, E. L. Grant, McGraw-Hill Book Co., 330 W. 42nd St., New York, N.Y.

135 Higher Management Control, Rose and Farr, McGraw-Hill Book Co., N.Y.

136 Inventor's Handbook, Fenner and Everett, Chemical Publishing Co. Inc., New York, N.Y.

137 The Aerodynamic Design Of The Goldenrod – To Increase Stability, Traction, and Speed, W. Korff, SAE paper #660390, June 1966

138 Viking II – Experimental Research Vehicle, J. L. Burwell, SAE paper #760060, 1976

139 Technical Features Of The Viking 4 Research Vehicle, Paul Tiffany, SAE paper #770689, 1977

140 The Mustang – Ford's Experimental Sports Car, R. C. Lunn, SAE paper #611f, January 1963

141 Special Purpose Car For Commuting, E. A. Rishavy, SAE paper #690462, May 1969

142 General Electric Research and Development Center, P.O. Box 8, Schenectady, New York, 12301

– INDEX –